THERMALLY STABLE AND
FLAME RETARDANT POLYMER NANOCOMPOSITES

Polymer nanocomposites have revolutionized material performance, most notably in the plastics, automotive, and aerospace industries. However, to be commercially viable, many of these materials must withstand high temperatures. In this book, leaders in the field outline the mechanisms behind the generation of suitable polymer systems, pulling together recent research to provide a unified and up-to-date assessment of recent technological advances. The text is divided into two clear sections, introducing the reader to the two most important requirements for this type of material: thermal stability and flame retardancy. Special attention is paid to practical examples that walk the reader through the numerous commercial applications of thermally stable and flame retardant nanocomposites. With a strong focus on placing theory within a commercial context, this unique volume will appeal to practitioners as well as researchers.

VIKAS MITTAL is a polymer engineer at The Petroleum Institute in Abu Dhabi, UAE. Dr. Mittal is well known within the academic and industrial sectors for his work on polymer nanocomposites. He has authored several papers and book chapters on the subject, and his research interests also include novel filler surface modifications, thermal stability enhancements, and polymer latexes with functionalized surfaces.

THERMALLY STABLE AND FLAME RETARDANT POLYMER NANOCOMPOSITES

Edited by

VIKAS MITTAL

The Petroleum Institute, UAE

CAMBRIDGE UNIVERSITY PRESS
Cambridge, New York, Melbourne, Madrid, Cape Town,
Singapore, São Paulo, Delhi, Tokyo, Mexico City

Cambridge University Press
The Edinburgh Building, Cambridge CB2 8RU, UK

Published in the United States of America by Cambridge University Press, New York

www.cambridge.org
Information on this title: www.cambridge.org/9780521190756

First published 2011

Printed in the United Kingdom at the University Press, Cambridge

A catalogue record for this publication is available from the British Library

Library of Congress Cataloguing in Publication data
Thermally stable and flame retardant polymer nanocomposites / edited by Vikas Mittal.
p.   cm.
Includes bibliographical references and index.
ISBN 978-0-521-19075-6 (hardback)
1. Fire resistant polymers.   2. Nanocomposites (Materials)   I. Mittal, Vikas.   II. Title.
TH1074.5.T535 2011
677′.689 – dc22
2010048227

ISBN 978-0-521-19075-6 Hardback

Additional resources for this publication at www.cambridge.org/mittal

Contents

Contributors

Günter Beyer
Kabelwerk Eupen AG, Malmedyer Strasse 9
B – 4700 Eupen, Belgium

S. Bourbigot
ISP-UMET
UMR-CNRS 8207
ENSCL, BP 90108
59650 Villeneuve d'Ascq, France

G. Camino
Politecnico di Torino
Sede di Alessandria – Viale Teresa Michel 5
15121 Alessandria, Italy

Mauro Comes-Franchini
Dipartimento di Chimica Organica
"A. Mangini," Facoltà di Chimica Industriale
Viale Risorgimento 4
40136 Bologna, Italy

Marius C. Costache
Department of Chemistry and Chemical Biology
Rutgers University, 101 Life Sciences Center
Piscataway, NJ 08854
USA

José-Marie Lopez Cuesta
CMGD
Ecole des Mines d'Alès

6 avenue de Clavières
30319 Alès, France

M. A. Delichatsios
School of Built Environment
University of Ulster
BT37 0QB, UK

Zhengping Fang
Laboratory of Polymer Materials and Engineering
Ningbo Institute of Technology
Zhejiang University
Ningbo 315100, P. R. China
and
Institute of Polymer Composites
Zhejiang University
Hangzhou 310027, P. R. China

A. Fina
Politecnico di Torino,
Sede di Alessandria – Viale Teresa Michel 5
15121 Alessandria, Italy

Jin Uk Ha
Otto H. York Department of Chemical, Biological, and Pharmaceutical
Engineering
New Jersey Institute of Technology
Newark, NJ 07102-1982
USA

Matthew J. Heidecker
Emerson Climate Technologies
1675 W. Campbell Road
Sidney, OH 45365
USA

Yuan Hu
State Key Laboratory of Fire Science
University of Science and Technology of China
96 Jinzhai Road, Hefei
Anhui 230026, P. R. China.

Musa R. Kamal
Department of Chemical Engineering
McGill University
Montreal, Quebec, Canada H3A 3R1

Abdelghani Laachachi
AMS
Centre de Recherche Public Henri Tudor
66 Rue de Luxembourg, BP 144
L-4002 Esch-sur-Alzette, Luxembourg

Tie Lan
Nanocor, Inc.
2870 Forbs Ave.
Hoffman Estates, IL 60192
USA

A. Leszczyńska
Department of Chemistry and Technology of Polymers
Cracow University of Technology
ul. Warszawska 24
31–155 Kraków, Poland

Evangelos Manias
Polymer Nanostructures Lab–Center for the Study of Polymer Systems (CSPS)
and Department of Materials Science and Engineering
The Pennsylvania State University
325D Steidle Bldg.
University Park, PA 16802
USA

Massimo Messori
University of Modena and Reggio Emilia Department of Materials and
Environmental Engineering (DIMA)
Faculty of Engineering
Via Vignolese 905
41100 Modena, Bologna

Vikas Mittal
The Petroleum Institute
Chemical Engineering Department
Abu Dhabi, UAE

Hiroyoshi Nakajima
Sumitomo Chemical Co., Ltd.
Petrochemicals Research Laboratory
2–1 Kitasode, Sodegaura
Chiba 299–0295, JAPAN

S. Nazaré
Centre for Materials Research and Innovation
University of Bolton
Bolton BL3 5AB, UK

J. Njuguna
Department of Sustainable Systems
Cranfield University
Bedfordshire MK43 0AL, UK

Guido Ori
University of Modena and Reggio Emilia Department of Materials and
Environmental Engineering (DIMA)
Faculty of Engineering
Via Vignolese 905
41100 Modena, Bologna
Italy

Joshua U. Otaigbe
School of Polymers and High Performance Materials
The University of Southern Mississippi
118 College Drive #10076
Hattiesburg, MS 39406
USA

Seongchan Pack
Samsung Cheil Industries, Inc.
332–2, Gocheon-Dong
Uiwang-Si
Gyeonggi-Do 437–711, Korea

K. Pielichowski
Department of Chemistry and Technology of Polymers
Cracow University of Technology
ul. Warszawska 24
31–155 Kraków, Poland

Longzhen Qiu
Key Laboratory of Special Display Technology
Ministry of Education
Academe of Opto-Electronic Technology
Hefei University of Technology
Hefei, Anhui Province, P. R. China

Baojun Qu
Department of Polymer Science and Engineering
University of Science and Technology of China
Hefei, Anhui Province, P. R. China

Miriam H. Rafailovich
Department of Materials Science and Engineering
State University of New York at Stony Brook
Stony Brook, NY 11794–2275
USA

F. Samyn
ISP-UMET
UMR-CNRS 8207
ENSCL, BP 90108
59650 Villeneuve d'Ascq, France

Cristina Siligardi
University of Modena and Reggio Emilia Department of Materials and
Environmental Engineering (DIMA)
Faculty of Engineering
Via Vignolese 905
41100 Modena, Bologna
Italy

Lei Song
State Key Laboratory of Fire Science
University of Science and Technology of China
96 Jinzhai Road, Hefei
Anhui 230026, P. R. China

Pingan Song
Institute of Polymer Composites
Zhejiang University

Hangzhou, 310027, P. R. China.
and
College of Engineering
Zhejiang Forestry University
Lin'an, 311300, P. R. China

Qilong Tai
State Key Laboratory of Fire Science
University of Science and Technology of China
96 Jinzhai Road, Hefei
Anhui 230026, P. R. China

Jorge Uribe-Calderon
Department of Chemical Engineering
McGill University
Montreal, Quebec, Canada H3A 3R1

Charles A. Wilkie
Department of Chemistry and Fire Retardant Research Facility
Marquette University
535 N. 14th Street
Milwaukee, WI 53213
USA

Marino Xanthos
Otto H. York Department of Chemical, Biological, and Pharmaceutical
Engineering
New Jersey Institute of Technology
Newark, NJ 07102–1982
USA

Vladimir E. Yudin
Institute of Macromolecular Compounds
Russian Academy of Sciences
199004 Saint-Petersburg
Bolshoy pr. 31, Russia

J. Zhang
School of Built Environment
University of Ulster
BT37 0QB, UK

Preface

The aim of this book is to provide comprehensive information about the two most important facets of polymer nanocomposites technology, thermal stability and flame retardancy. These two effects ensure a large number of potential applications of polymer nanocomposites. This book provides information regarding their mechanisms of action, as well as practical examples of recent advances in the generation of polymer nanocomposites that are thermally stable and flame retardant.

Polymer nanocomposites revolutionized research in the composites area through the achievement of nanoscale dispersion of the inorganic filler (clay platelets) in the polymer matrices after suitable surface modification of the filler phase. A large number of polymer matrices were tried, and nanocomposites with varying degrees of success were achieved with these polymer systems. In many instances, the generation of nanocomposites requires the use of high compounding temperatures for uniform mixing of the organic and inorganic phases. Conventional filler surface modifications are generally not stable enough for such high compounding temperatures, which initiate degradation reactions in the modification as well as in the polymer matrix. Apart from this, for the successful use of polymer nanocomposites in a number of applications, the materials should withstand high temperatures, making thermal stability a prime requirement for these materials. Similarly, flame retardancy is also of immense importance. A number of advances in thermally stable and flame retardant nanocomposite systems have been achieved in recent years. More thermally stable surface modifications for the fillers have been reported, and corresponding composites with superior thermal resistance have been obtained. Similarly, by incorporation of inorganic clay platelets, nanotubes, oxide nanoparticles, and other materials, the flame retardancy of polymer nanocomposites has been improved.

The first section of the book deals with the thermal stability of layered silicates and polymer nanocomposites. Chapter 1 provides an overview of layered silicates as fillers, organic surface modification of such layered silicates, and thermal stability considerations in relation to the surface modification molecules ionically exchanged on the filler surface. Chapter 2 provides in-depth information on the mechanisms of thermal degradation of layered silicates modified with ammonium and other thermally stable filler surface modifications. Chapter 3 provides the example of generating thermally stable polystyrene nanocomposites using thermally stable layered silicates as fillers. Chapter 4 focuses on the generation of thermally stable PET nanocomposites. Thermally stable polyimide nanocomposites are described in Chapter 5. Use of clays modified with thermally stable ionic liquids for the generation of polyolefin and polylactic acid–based nanocomposites is presented in Chapter 6. In the second section, dealing with flame retardancy considerations, Chapter 7 provides an introduction to the flame retardancy of polymer–clay nanocomposites. Chapter 8 describes flame retardant nanocomposites using polymer blends. Flame retardancy of polyamide–clay nanocomposites is presented in Chapter 9. Chapter 10 reports self-extinguishing polymer–clay nanocomposites. Chapter 11 describes the use of fullerenes as fillers for the generation of flame retardant polymer nanocomposites. Chapter 12 describes flame retardant polymer nanocomposites with alumina as filler. Layered double hydroxides are shown as fillers for the generation of flame retardant polymer nanocomposites in Chapter 13. Flame retardant SBS–clay nanocomposites are described in Chapter 14.

It gives me immense pleasure to thank Cambridge University Press for their kind acceptance of this book. I dedicate this book to my mother for being a constant source of inspiration. I express heartfelt thanks to my wife Preeti for her continuous help in editing the book, as well as for her ideas on how to improve the manuscript.

Part I

Thermal stability

1

Polymer nanocomposites

Layered silicates, surface modifications, and thermal stability

VIKAS MITTAL

The Petroleum Institute

1.1 Introduction

Inorganic fillers have conventionally been added to polymer matrices to enhance their mechanical strength and other properties, as well as to reduce the cost of the overall composites. Layered aluminosilicates, also popularly described as clays, are one such type of filler, which are responsible for a revolutionary change in polymer composite synthesis as well as for transforming polymer composites into polymer nanocomposites. Aluminosilicate particles consist of stacks of 1 nm–thick aluminosilicate layers (or platelets) in which a central octahedral aluminum sheet is fused between two tetrahedral silicon sheets [1, 2]. Owing to isomorphic substitutions, there is a net negative charge on the surface of the platelets that is compensated for by the adsorption of alkali or alkaline earth metal cations. Because of the presence of alkali or alkaline earth metal cations on their surfaces, the platelets are electrostatically bound to each other, causing an interlayer to form in between. The majority of the cations are present in the interlayers bound to the surfaces of the platelets, but a small number of cations are bound to the edges of the platelets. Though the use of layered aluminosilicates has been documented in some older studies [3, 4], indicating their potential for substantially improving polymer properties, reports from Toyota researchers in the early nineties attracted serious attention [5, 6]. In these studies, polyamide nanocomposites were synthesized by in situ polymerization in the presence of clay with organic modifiers.

Polymer nanocomposites are materials in which the filler phase is dispersed in the polymer matrix at nanoscale and at least one dimension of the filler is less than 100 nm. The nanoscale dispersion of the filler leads to tremendous interfacial contact between the organic and inorganic phases, completely changing the morphology of the composite from that of conventional microcomposites where the polymer and inorganic phases are only mixed at macroscale. As a result, the polymer performance is enhanced at much lower filler volume fractions, allowing the polymer matrix to retain its transparency as well as its low density, traits that are completely lost in conventional microcomposites. Apart from this, as the aluminosilicate particles are platelike in nature, with two finite dimensions and thickness roughly 1 nm, these crystalline platelets are more effective in improving gas barrier, and thermal barrier, and other properties, and are also effective in stress transfer

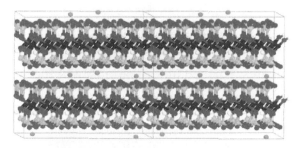

Figure 1.1 Molecular representation of montmorillonite. Reproduced from [16] with permission from American Chemical Society.

when dispersed at nanoscale. Initial studies of polyamide were quickly followed by a large number of studies of other polymers, and significant enhancements in mechanical performance, thermal stability, gas permeation resistance, flame retardancy, and so forth were reported [7–15]. Although different types of aluminosilicates such as mica, vermiculite, and montmorillonite have been incorporated into polymer matrices, the majority of studies have been of montmorillonite. Montmorillonites has a general unit cell formula of $M_x(Al_{4-x}Mg_x)Si_8O_{20}(OH)_4$ [1, 2] and a mean layer charge density of 0.25–0.5 eq.mol^{-1}. The layer charge density is a function of the number of substitutions in the silicate crystals and also indicates the strength of the electrostatic forces holding the platelets together. Figures 1.1 and 1.2 show molecular representations of montmorillonite and mica, respectively [16, 17]. Because of the relatively low mean charge density of montmorillonites, their platelets are held loosely together and can be delaminated in water. This exposes the alkali and alkaline earth metal cations in the aqueous phase and provides the opportunity to achieve exchange of these cations with other organic cations. This type of cation exchange on the surface of the platelets is required because in the pristine form, the surface of the platelets is very polar and possesses high surface energy, hindering their compatibility with the polymers, which are generally hydrophobic and have low surface energies. The exchange of organic cations on the surface of the platelets thus helps in two ways: First, it organophilizes the surface of the platelets, thus lowering the surface energy, and enables more uniform mixture of the fillers with the polymer. Second, the exchange of the long-chain organic ions also increases the interlayer spacing between the platelets by pushing them apart and causing weakening of the electrostatic forces between them. This again helps to intercalate the polymers in the interlayers, allowing the delamination of the filler [18–22]. Conventionally, long-chain alkylammonium ions have been exchanged on the surface of the platelets to organophilize them [18–22]. Alkylammonium ions such as octadecyltrimethylammonium, dioctadecyldimethylammonium, trioctadecylmethylammonium, and benzyldodecyldimethylammonium have been commonly used for the organic modification of silicates. Figure 1.3 shows a representation of the surface modification process [23].

(a)

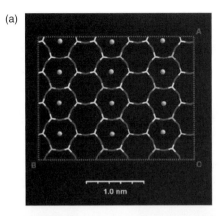

(b)

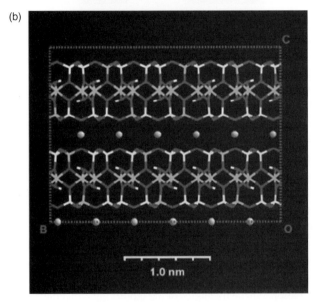

Figure 1.2 (a) Top view of a cleaved mica sheet along the *c* direction and (b) side view along the *a* direction. Reproduced from [17] with permission from American Chemical Society.

As has been mentioned, the montmorillonite substrate has a relatively low charge density, which allows easy delamination in water and subsequent cation exchange, but owing to the presence of a smaller number of ions on the surface of the platelets, less organic matter can be exchanged on the surface, which leads to a lesser expansion of the interlayer spacing. It is possible to employ higher–charge density (1 eq.mol^{-1}) minerals such as mica, which have much smaller area available per cation owing to the presence of a large number of cations on the surface. Exchange on the surfaces of such minerals would lead to much greater basal plane spacing. They also suffer from another limitation: The presence of a large number of cations generates very strong electrostatic forces in the interlayers, which

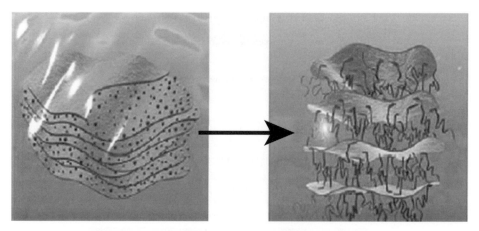

Figure 1.3 Schematic picture of an ion-exchange reaction. Reproduced from [23] with permission from Elsevier.

hinder the complete swelling of these minerals in water and do not allow optimal cation exchange. Aluminosilicates such as vermiculite with medium charge densities of 0.5–0.8 eq.mol^{-1} offer a better alternative, owing to their partial swelling in water and cation exchange causing much higher basal plane spacing in the modified mineral. The chemical constitution of the unit cell of vermiculite is (Mg, Al, Fe)$_3$(Al, Si)$_4$O$_{10}$(OH)$_2$Mg$_x$(H2O)$_n$ [24, 25] and the negative charges on the vermiculite layers are compensated for mainly by hydrated Mg^{2+} as interlayer cations. The benefit of the greater interlayer spacing in the modified mineral is that the forces of interaction are further reduced when the spacing between the platelets is increased, providing better conditions for the filler to delaminate in the polymer matrix.

The structural positioning of organic molecules on the surface of the platelets also affects the properties of the organically modified clay. The chemical architecture of the surface modification also significantly affects structure formation on the surface of the platelets. The structure formation and the tilt angles of the modifying molecules estimated from X-ray diffraction and thermal studies as a function of their increasing chain density were reported in a recent study [26]. Figure 1.4 demonstrates these structures for octadecyltrimethylammonium, dioctadecyldimethylammonium, trioctadecylmethylammonium, and tetraoctadecylammonium modifications on the montmorillonite surface. Molecular dynamics studies have also been used to predict the mechanism of monolayer formation on the surfaces of filler platelets. Figure 1.5 shows the formation of surface modification layers for two montmorillonites with different layer charge values as a function of the length of the alkyl chains present in the surface molecules [27].

Polymer nanocomposites are synthesized by a variety of methods, which include in situ polymerization, solution polymerization, and melt intercalation. The route for nanocomposite synthesis suggested by Toyota researchers was also based on in situ monomer polymerization in the presence of filler. Subsequently, Giannelis and co-workers [28, 29]

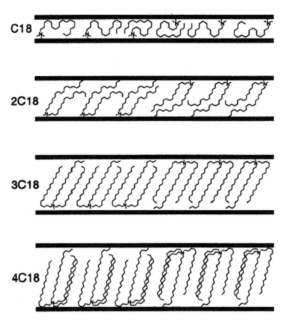

Figure 1.4 Schematic representation of possible structures of the organic modifiers as a function of increasing chain density in the chemical structure. Reproduced from [26] with permission from American Chemical Society.

reported the route of melt intercalation for the synthesis of polymer nanocomposites. In this method the direct use of high–molecular weight polymer can be achieved with mixers and compounders used for the generation of conventional composites. The polymer is melted at a higher temperature and the filler is then added to the polymer melt under shear and kneaded well to efficiently mix it with the polymer. Owing to its simplicity, this method has gained tremendous interest for the synthesis of polypropylene, polyethylene, and polystyrene nanocomposites, to name a few. Figure 1.6 shows a representation of the melt intercalation process. The use of compatibilizer to enhance compatibility between the organic and inorganic phases in the case of nonpolar polymers is also very common, and the compatibilizer is also added along with the polymer for nanocomposite synthesis. However, the melt-compounding approach generally uses high temperatures to achieve optimal mixing of the various components. The use of high operational temperatures can be of concern for the thermal stability of ammonium ion–based organic modifications, as these modifications have an onset of degradation near 200 °C, which is also the temperature commonly used for the melt-compounding of polymers such as polypropylene. It has been observed that the degradation of even a small amount of this modification can have a serious impact on the composite microstructure development and hence properties. The radicals generated during the degradation can also react with the polymer chains, thus lowering the molecular weight and hence impairing the performance. The interfacial interactions

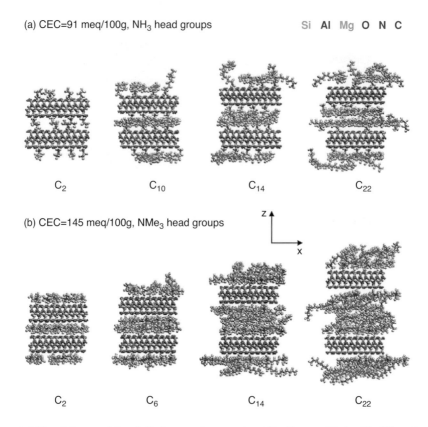

Figure 1.5 Simulation models of alkylammonium montmorillonites modified with different–chain length organic cations, (a) CEC = 91 meq/100 g and (b) CEC = 145 meq/100 g. Reproduced from [27] with permission from American Chemical Society.

of the modified filler with the polymer can also completely change owing to degradation, thus changing the interfacial dynamics that must be achieved to cause filler delamination. Thus, it is important to choose a compounding temperature that is high enough to mix the components by attaining optimal viscosity, but not so high that it severely degrades the organic modification. The compounding time should also be similarly controlled so that it is enough to achieve mixing but not so high as to degrade the modification and polymer. But in many cases, the use of high temperatures and longer compounding times cannot be avoided; it is thus important in these cases to use more thermally stable organic modifications. Also, the commonly used compatibilizers often have very low molecular weights; therefore, consideration should also be given to these molecules similar to that of the organic modifications. It is also important to analyze the thermal behavior of the nanocomposite in order to quantify the effect of compounding on the composite material.

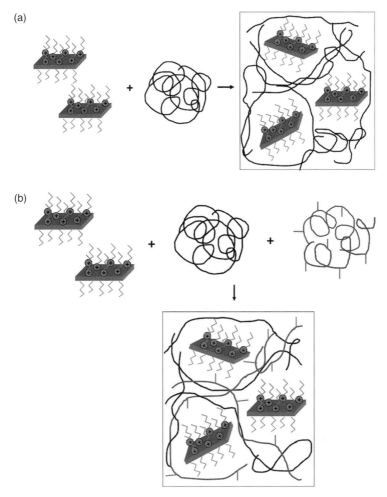

Figure 1.6 Representation of melt intercalation of (a) polymer with clay and (b) polymer in the presence of compatibilizer with clay.

1.2 Evaluation of thermal behavior

As has been mentioned, evaluation of the thermal performance of the surface-modified filler, as well as of composite materials, is important, because it affects the properties of the final nanocomposites significantly. High-resolution (Hi-Res) thermogravimetric analysis (TGA), in which the heating rate is coupled to the mass loss, that is, the sample temperature is not raised until the mass loss at a particular temperature is completed, is the method most commonly used for analysis of the thermal performance of materials.

First, the use of TGA is beneficial in quantifying the amount of organic matter present in the filler interlayers, thus generating an idea of the extent of ion exchange on the surface. If the amount of organic matter does not correspond to a satisfactory extent of

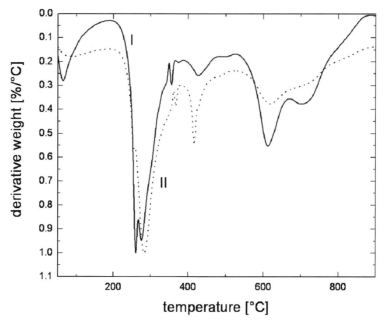

Figure 1.7 TGA thermograms of montmorillonite with a cation-exchange capacity of 880 μ.eq/g modified with (I) octadecyltrimethylammonium and (II) dioctadecyldimethylammonium.

cation exchange, the ion-exchange reaction has to be repeated. Thus, TGA provides an efficient tool for determining the organophilization of the filler surface. One has to be careful in treating the total weight loss achieved in a TGA analysis. Mass loss due to high-temperature dehydroxylation of the mineral has to be subtracted from the total mass loss to obtain information on the mass loss corresponding to the organic layer. The weight loss between 50 and 150 °C, corresponding to the evaporation of physisorbed water and solvent molecules, should also be subtracted from the total weight loss. The total exchanged moles of ammonium cations per gram of clay, φ_{amm}, is then calculated using the expression

$$\varphi_{amm} = W_{corr}/[(1 - W_{corr}) * M_{amm}],$$

where W_{corr} is the corrected mass loss, corresponding to organic weight loss owing only to alkylammonium ions, and M_{amm} is the molecular mass of the organic cation exchanged on the surface. Figure 1.7 shows the TGA thermograms of montmorillonite (with a cation-exchange capacity of 880 μ.eq/g) with octadecyltrimethylammonium and dioctade-cyldimethylammonium modifications. The increase in the overall organic matter attached to the filler can be observed from the increasing chain density of the surface modification. Increasing chain density also led to better thermal stability of the modified filler on the surface, in that the peak degradation temperature for the dioctadecyldimethylammonium-modified montmorillonite was higher than that for the octadecyltrimethylammonium-modified montmorillonite.

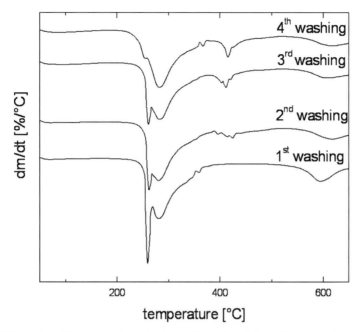

Figure 1.8 Effect of washing protocols on the thermal behavior of the surface-treated montmorillonite modified with dioctadecyldimethylammonium. Reproduced from [32] with permission from Wiley.

Second, high-resolution TGA also helps to ascertain the presence (or absence) of excess surface modification in the interlayers. Excess surface modification is not ionically bound to the surfaces of platelets, but is present as a pseudo-bilayer in the interlayers. The presence of such an excess of surface modification is important to ascertain, and such excess needs to be cleaned off before the filler is compounded with polymer. The reason is the low thermal stability of such pseudo-bilayers as compared with surface-bound molecules. The lower thermal degradation is much more problematic when the filler has to be compounded with the polymer at high temperatures. This low-temperature degradation of the unbound surface modification molecules can cause unwanted interactions with the polymer, leading to a reduction in the molecular weight as well as a deterioration of the interface between the polymer and filler phases [30, 31]. It is also worth noting that it is only high-resolution TGA that is able to detect the presence of such an excess, because other methods like X-ray diffraction are blind to it. Figure 1.8 shows the thermograms of surface-treated montmorillonite modified with dioctadecyldimethylammonium [32]. It is obvious that in the initial washing stages, there was excess surface modification that degraded at a lower temperature, represented by a sharp degradation peak. The montmorillonite after the fourth washing was observed to be relatively free from this excess, as the low temperature degradation peak is almost eliminated. As reported earlier, X-ray diffraction carried out for all four montmorillonite samples obtained after subsequent washing cycles had similar basal plane–spacing

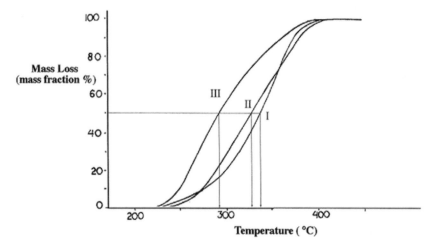

Figure 1.9 TGA analysis of (I) linear PMMA–clay nanocomposite, (II) cross-linked PMMA–clay nanocomposite, and (III) pure PMMA. The lines are drawn at 50% mass loss for comparison. Reproduced from [34] with permission from Elsevier.

values, indicating the importance of high–resolution TGA in ascertaining the surface cleanliness and hence the thermal stability of the modified clays. It also indicates that one should be careful while modifying the surfaces of montmorillonites, as more than one washing step may be required; otherwise the thermal stability of the filler may be hindered. Commercially treated montmorillonites have also been observed occasionally to contain an excess of modification molecules. It was reported recently for epoxy nanocomposites that by removing the excess modification molecules from the filler surface, the properties of the nanocomposites could be significantly improved [33].

In the case of nanocomposites, too, the use of TGA is important for ascertaining the thermal synergy between the organic and inorganic components of the system. Generally, it is observed that the thermal behavior of the nanocomposite is better than that of both the filler and the pure polymer matrix, indicating synergistic improvement in the thermal behavior. Figure 1.9 shows the thermal behavior of pure poly(methyl methacrylate) polymer and its composites (linear and cross-linked). The improvement in the thermal behavior after incorporation of filler is obvious from the higher temperature for 50% mass loss. The thermal stability of the nanocomposites is also dealt with in Section 1.3 by providing further examples.

1.3 Thermal stability of modified clays and nanocomposites

As mentioned in the earlier section, ammonium-based modifications of the surfaces of montmorillonites have been commonly employed. Thermogravimetric analysis has also been used to quantify the extent of surface modification, that is, the amount of

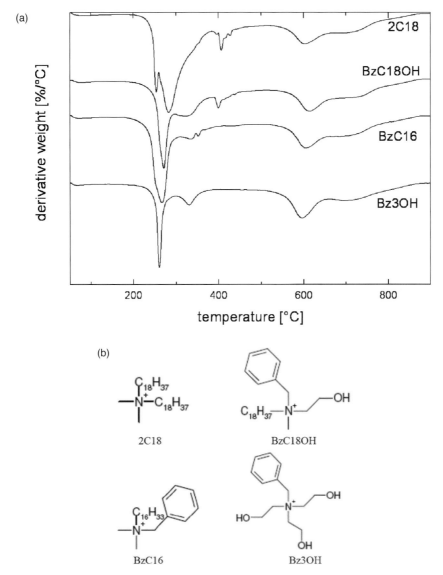

Figure 1.10 (a) TGA thermograms of the modified fillers with various ammonium ions and (b) chemical structures of the ammonium ions exchanged on the filler surface. Reproduced from [35] with permission from American Chemical Society.

organic matter and the presence or absence of excess surface modification molecules on the surfaces of montmorillonite platelets. Figure 1.10a shows the thermograms of montmorillonites with other ammonium-based modifications [35]. The ammonium ions used were benzyldibutyl(2-hydroxyethyl)ammonium chloride (Bz1OH), benzylbis(2-hydroxyethyl)butylammonium chloride (Bz2OH), benzyltriethanolammonium chloride

(Bz3OH), benzyl(2-hydroxyethyl)methyloctadecylammonium chloride (BzC18OH), dioctadecyldimethylammonium chloride (2C18), and benzyldimethylhexadecylammonium
chloride (BzC16). The chemical structures of the ammonium ions used for the surface
exchange are demonstrated in Figure 1.10b [35]. The absence of any low–temperature
degradation peak in the temperature region 200–250 °C in the thermograms of the modified montmorillonites (except the montmorillonite modified with dioctadecyldimethylammonium) confirms the absence of any pseudo-bilayer or excess surface modification in
the interlayers. In the case of montmorillonite modified with dioctadecyldimethylammonium ions, a sharp peak indicating low-temperature degradation corresponding to a small
amount of excess surface modification molecules is observed. Montmorillonites modified with dioctadecyldimethylammonium, trioctadecylmethylammonium, or tetraoctadecylammonium ions have commonly been observed to have the problem of formation of
pseudo-bilayers by the surface modification molecules, and extensive washing protocols
are necessary to completely wash off the unbound modifier molecules.

   As observed in the earlier section, the ammonium modifications have generally caused
the onset of thermal degradation at temperatures similar to the ones used for melt-
compounding of the polymers with the fillers. The degradation of a small extent of surface
modification can have detrimental effects on the composite microstructure and properties; therefore, it is of the utmost importance to retain the thermal stability of the surface
modification during the high-temperature compounding process. To achieve this, a number
of thermally stable modifications have been developed, based on imidazolium, phosphonium, pyridinium, and similar ions. Figure 1.11a shows the example of an imidazolium
salt (1-decyl-2-methyl-3-octadecylimidazolium bromide) [36]. The thermal behavior of the
montmorillonite with this organic modification is compared with that of montmorillonite
modified with dioctadecyldimethylammonium in Figure 1.11b. The imidazolium-modified
montmorillonite is more thermally stable, as the onset of degradation in this case is 50 °C
higher than for the ammonium-modified counterpart. The peak degradation temperature in
the case of imidazolium-modified clay is also roughly 35 °C higher than for the ammonium-
modified clay. The ammonium-modified clay also was observed to have a sharp degradation
peak around the point of onset of degradation, indicating the presence of pseudo-bilayers
affecting the thermal stability of the modified clay. The dynamic thermal behavior of both
imidazolium- and ammonium-modified montmorillonites was compared [36]. For this analysis, both the imidazolium- and ammonium-modified clays were first stabilized at 140 °C
and then subsequently heated to 180 °C. The clays were then kept under isothermal conditions at 180 °C for 90 min. The ammonium-modified montmorillonite was observed to
have greater weight loss than the imidazolium-modified montmorillonite in these 90 min.
This greater weight loss is a result of early decomposition of the surface modification
attached to the filler surface. It can be argued that the molecular weight of the ammonium
modification is higher than that of the imidazolium modification, thus justifying greater
organic matter loss, and also that the amount of weight loss in the two clays is not very
different, but it has to be considered that the decomposition of even a small number of
weaker C–N bonds in the organic modification in the earlier stages of compounding with

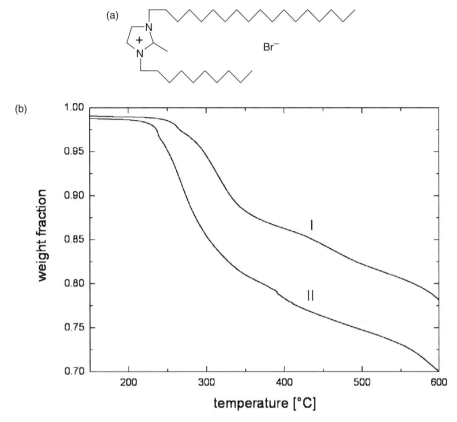

Figure 1.11 (a) Chemical structure of 1-decyl-2-methyl-3-octadecylimidazolium bromide and (b) comparison of the thermal behavior of imidazolium-modified montmorillonite (I) with dioctadecyldimethylammonium-modified montmorillonite (II). Reproduced from [36] with permission from Elsevier.

the polymer can cause a significant change in the structure and properties of the modifying molecules, thus affecting the microstructure and properties of the resulting composites. The properties of the composites were accordingly affected, with the composites containing imidazolium-modified clay showing much better barrier properties than the composites containing ammonium-modified clay.

Awad *et al.* reported extensive studies on the various imidazolium salts and montmorillonites modified with these salts [37]. The imidazolium salts included 1,2-dimethyl-3-propylimidazolium, 1-butyl-2,3-dimethylimidazolium, 1-decyl-2,3-dimethylimidazolium, 1,2-dimethyl-3-hexadecylimidazolium, 1,2-dimethyl-3-eicosylimidazolium, 1,2-dimethyl-3-ethylbenzene imidazolium, and 1-ethyl-3-methylimidazolium. The authors investigated the effect of counterion, alkyl chain length, and structural isomerism on the thermal stability of the imidazolium salts and compared their behavior with that of the conventional quaternary ammonium ions using a number of characterization techniques. It was reported that

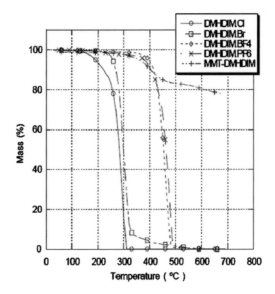

Figure 1.12 Thermogravimetric analysis of dimethylhexadecylimidazolium salts with different anions (Cl⁻, Br⁻, BF^{4-}, PF^{6-}). Montmorillonite modified with dimethylhexadecylimidazolium salt is also been shown. Reproduced from [37] with permission from Elsevier.

the type of anions in the imidazolium salts had an effect on their thermal stability. Halide-based anions were much less thermally stable than phosphate- and borate-based anions, as shown in Figure 1.12, where the thermal behavior of dimethylhexadecylimidazolium salt with different anions (Cl⁻, Br⁻, BF^{4-}, PF^{6-}) is demonstrated. The isomeric structure of the alkyl side group was also observed to affect the thermal stability. Methyl substitution in the 2 position, between two N atoms, was observed to enhance the thermal stability. The length of the chain attached to the nitrogen atom was also observed to be inversely proportional to thermal stability, as increasing the chain length increased the amount of organic matter in the interlayers, and as a result, the thermal performance deteriorated. The environment in which the measurement is made also affects the thermal stability of the organic modifications. Measurements carried out under nitrogen show much better thermal stability for the imidazolium ions than measurements performed in air. Table 1.1 also details the onset and peak degradation temperatures of various ammonium and imidazolium salts for comparison [37].

Phosphonium-based modifications have also found significant applications. One study of phosphonium-based modifications was reported by Avalos *et al.* [38]. The phosphonium ions used for ion exchange on the surface of filler platelets were triphenylvinylbenzylphos-phonium chloride (TVBPCl) and tetraoctylphosphonium bromide (TOPBr). Figure 1.13 shows the cumulative and differential thermal degradation plots for montmorillonites mod-ified with phosphonium salts. In the case of tetraoctylphosphonium-modified montmo-rillonite, the decomposition was observed to start at 300 °C and the peak degradation

Table 1.1 *Onset and peak degradation temperatures of various imidazolium and ammonium salts*

Sample	TGA under N_2 (°C)		TGA under air (°C)	
	T_{onset}	T_{peak}	T_{onset}	T_{peak}
DMDODA-Br	225	236	185	228
DMPIM-Cl	260	297	259	299
BDMIM-Cl	257	294	255	294
BMIM-Cl	234	285	232	284
DDMIM-Cl	239	287	237	284
HDMIM-Cl	230	292	229	287
DMHDIM-Cl	239	292	239	286
DMHDIM-Br	253	301	250	304
DMEiIM-Br	259	308	260	315
DMEtBIM-Br	275	339	275	330
DMPIM-BF_4	390	461	360	434
BDMIM-BF_4	405	475	347	428
DMiBIM-BF_4	350	429	347	404
DDMIM-BF_4	400	469	342	425
DMHDIM-BF_4	400	464	278	391
A1DMFM-BF_4	332	410	323	392
DMEiIM-BF_4	390	454	271	406
BDMIM-PF_6	425	499	358	396
DMiBIM-PF_6	382	439	357	424
DDMIM-PF_6	420	469	327	376
DMHDIM-PF_6	400	478	308	370
DMEtBIM-PF_6	386	464	334	397
EtMIM-$N(SO_2CF3)_2$	410	479	401	477

Note: Reproduced from [37] with permission from Elsevier.

temperature was higher than 425 °C. This shows the superior thermal behavior of phosphonium salts over ammonium salts. In the case of triphenylvinylbenzylphosphonium-modified montmorillonite, the peak degradation temperature was higher than 450 °C, though the onset of degradation was observed at a temperature lower than for tetraoctylphosphonium-modified montmorillonite.

The improvement in the thermal performance of the polymers after the incorporation of filler was demonstrated in Figure 1.9 for poly(methyl methacrylate). Figure 1.14 further shows the thermal behavior of polypropylene nanocomposites containing different volume fractions of montmorillonite modified with dioctadecyldimethylammonium. Curve 1 represents the thermal degradation of pure polymer. Curve 2 represents the composite with 1 vol% clay. The thermal stability of the polymer was significantly enhanced. Further

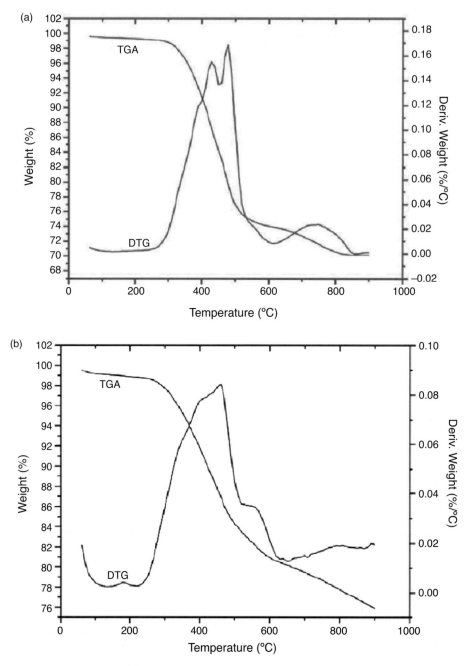

Figure 1.13 TGA and DTG plots of (a) tetraoctylphosphonium bromide (TOPBr)- and (b) triph-enylvinylbenzylphosphonium chloride (TVBPCl)-modified montmorillonites. Reproduced from [38] with permission from Elsevier.

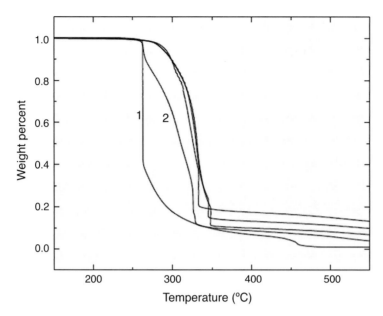

Figure 1.14 TGA thermograms of pure polypropylene (curve 1) and 2C18·M880 montmorillonite/ polypropylene nanocomposites containing different volume fractions. Curve 2 corresponds to composite with 1 vol% filler. Curves of the composites containing higher volume fractions are indistinguishable. Reproduced from [39] with permission from Sage Publishers.

addition of clay was subsequently observed to generate much more thermally stable behavior in the composites. The thermal behavior of the composites containing filler volume fractions of 2 vol% or more was observed to have synergistic improvement. The onset temperature for degradation in these composites was much higher than for the modified clay as well as for the pure polymer. The thermograms of composites with 2 vol% filler or more were practically indistinguishable, indicating that 2 vol% filler content was necessary to obtain the required thermal reinforcement.

In an interesting study, polybenzimidazole has also been used to generate thermally stable nanocomposites in which the clay modification was still ammonium-based [40]. Figure 1.15 represents a synthetic scheme for the preparation of sulfonated polybenzimidazole. The composites were synthesized by using different amounts of clay; their thermal behavior is reported in Figure 1.16. The thermograms of the pure polymer organically modified montmorillonite have also been compared with those of the composites. The thermal degradation of the composites is observed at much higher temperatures than that of the modified clay. Thus, in this system, the resulting polymer is very thermally stable, but the thermal stability is also contributed to by the polymer phase as well as the crystalline inorganic filler phase.

In another study, pyridine- and quinoline-containing salts were used to modify the montmorillonite organically. It was observed in thermogravimetric analysis that the quinolinium

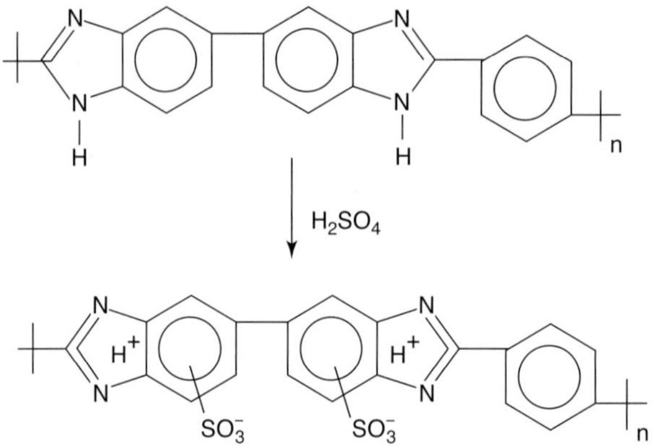

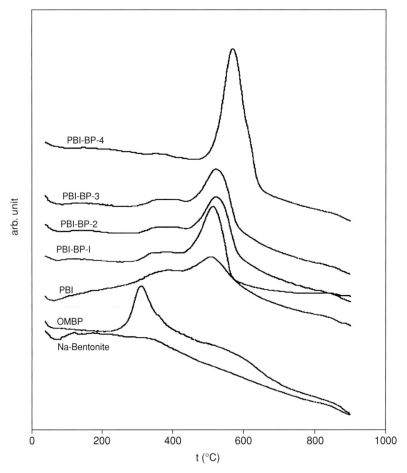

Figure 1.15 Synthesis of sulfonated polybenzimidazole. Reproduced from [40] with permission from Elsevier.

Figure 1.16 Differential thermal degradation curves for pristine bentonite, organically modified bentonite (OMBP), polybenzimidazole, and polybenzimidazole–clay hybrid materials. Reproduced from [40] with permission from Elsevier.

salt was more thermally stable than the pyridinium salt. An initial degradation temperature of 367 °C was observed to quinolinium-based clay, whereas this temperature was 295 °C in the case of pyridinium-based clay. The char yields in these clays were 75% and 70%, respectively. The polystyrene nanocomposites generated by the incorporation of varying amounts of the modified montmorillonites were observed to have an increase in the temperature at 50% degradation as compared to the pure polymer, although the initial degradation temperature in the composites was observed to decrease. Figure 1.17 shows the thermal behavior of the generated nanocomposites in comparison with that of the pure polymer.

Yei *et al.* [42] reported the synthesis of thermally stable nanocomposites by emulsion polymerization. In this mode of generation of polymer nanocomposites, it is possible to eliminate the thermal degradation of the organic modification as well as of polymer that occurs during the high-temperature compounding. In the case of emulsion polymerization, the polymerization is carried out in the aqueous phase, the heat is constantly dissipated, and the polymer nanocomposites of the same polymer can be generated at much lower temperatures. The montmorillonite was modified with cetylpyridinium chloride (CPC) and nanocomposites were generated using this mineral. The authors also reported the generation of inclusion complexes of CPC with cyclodextrin, which were then exchanged on the surfaces of the montmorillonite platelets.

The authors observed that the inclusion complex of CPC with cyclodextrin decomposed at higher temperatures then the pure CPC. A decomposition temperature of 284 °C was observed for this complex, compared to 220 °C for pure CPC. This confirmed the attainment of better thermal stability of the pyridinium modification by complexing with cyclodextrin. The thermal behavior of the polystyrene nanocomposites synthesized using both CPC and the inclusion complex of CPC with cyclodextrin has been demonstrated in Figure 1.18 and is compared with that of the pure polymer. Both the nanocomposites had better thermal stability than the pure polymer; however, the thermal performance of inclusion complex–intercalated clay nanocomposites was better than that of the CPC-intercalated clay nanocomposites. The temperature at 5% weight loss was observed to be 33 °C higher for the inclusion complex–intercalated clay nanocomposites as compared to the pure polymer, whereas this increase was 18 °C for the CPC–intercalated clay nanocomposites.

Figure 1.14 reported the thermal behavior of polypropylene nanocomposites with ammonium-modified montmorillonites. More thermally stable imidazolium-based modifications were exchanged on the surfaces of platelets and the composites were synthesized with varying filler content [36]. Figure 1.19a depicts the thermal behavior of the composites in comparison with that of the pure polymer. As also observed in Figure 1.14, the inclusion of only 1 vol% of the filler was responsible for significant enhancement in the thermal stability of the filler and the performance of the 1 vol% imidazolium-based filler was also much better than that of the 1 vol% ammonium-based filler in polymer matrix. The 4 vol% composite was observed to markedly enhance the thermal stability of the composite, as both the onset of degradation temperature and the peak degradation temperature were shifted to higher values.

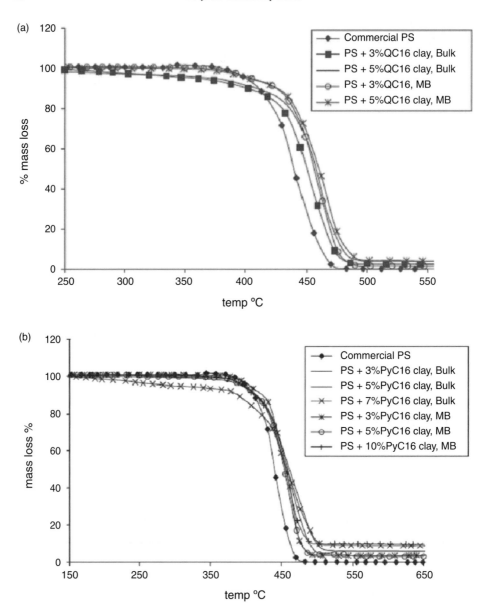

Figure 1.17 Thermal degradation behavior of nanocomposites containing (a) quinolinium- and (b) pyridinium-modified montmorillonites in different amounts. Reproduced from [41] with permission from Elsevier.

Another important factor to take into account is the effect on thermal stability of the composites formed by the addition of low–molecular weight compatibilizers. These compatibilizers are often added to polyolefin matrices such as polyethylene and polypropylene in order to enhance the compatibility of the organic and inorganic phases. Because of

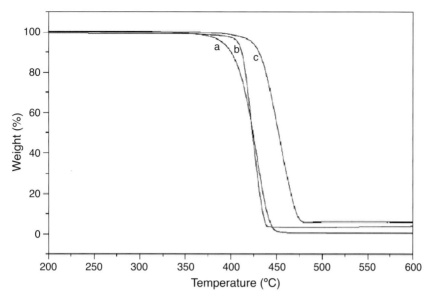

Figure 1.18 TGA thermograms (under nitrogen) of (a) pure PS, (b) the nanocomposite containing CPC as filler surface modification, and (c) the nanocomposite containing the inclusion complex of CPC with cyclodextrin as filler surface modification. Reproduced from [42] with permission from Elsevier.

their amphiphilic nature as well as their low molecular weight, they can easily intercalate the clay interlayers, thus facilitating the intercalation of pure polymer chains in the interlayers as well. However, the low molecular weight of the compatibilizers can lead to deterioration of the mechanical properties owing to plasticization and can also have a negative impact on the thermal stability of the composites. Therefore, it is important to analyze the thermal performance of the composites in order to quantify the effect of potential low-temperature degradation of compatibilizer on the thermal properties of the resulting composites. Figure 1.19b demonstrates the thermal behavior of the pure polymer and of the pure compatibilizer as well as of a nanocomposite containing 3 vol% of the imidazolium-treated montmorillonite and 2 wt% of the polypropylene-graft-maleic anhydride compatibilizer [43]. Though the compatibilizer had low-temperature degradation similar to that of the polymer, the composite was observed to have much better thermal stability than the constituents, indicating that the compatibilizer addition at 2 wt% levels did not affect the thermal performance of the composites.

Interestingly, it has recently been reported [44] that it is also possible to improve the thermal stability of the modified montmorillonite even when the ionic modification attached to the surface is ammonium based (Figure 1.20). By this technique, in which the physical adsorption of high–molecular weight poly(vinylpyrrolidone) (PVP) was carried out on the surface of montmorillonite premodified with dioctadecyldimethylammonium, the thermal stability of the clay improved with increasing adsorption of the polymer in the interlayers.

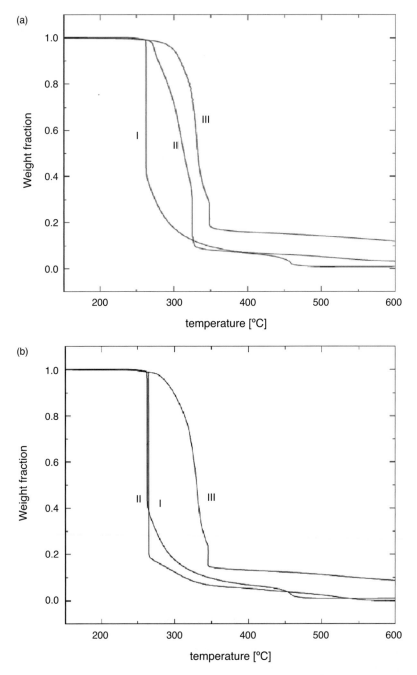

Figure 1.19 TGA thermograms of the polypropylene nanocomposites with imidazolium-modified montmorillonite. (a) Composites synthesized without compatibilizer: (I) pure polymer, (II) 1 vol% clay composites, and (III) 4 vol% clay composite; (b) composites synthesized with PP-g-MA compatibilizer: (I) pure polymer, (II) pure compatibilizer, and (III) 3 vol% imidazolium–clay composite with 2 wt% compatibilizer. Reproduced from [36] and [43] with permission from Elsevier.

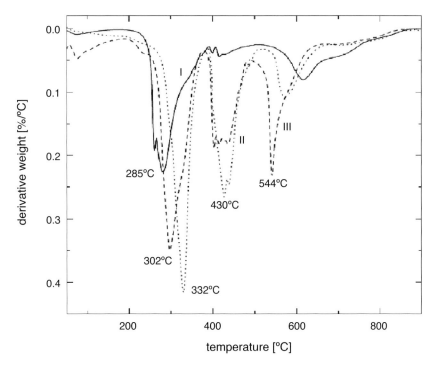

Figure 1.20 Thermograms of organically modified montmorillonite (dioctadecyldimethylammonium) before adsorption (I), after adsorption with smaller amounts of PVP (II), and after adsorption with larger amounts of PVP (III). Reproduced from [44] with permission from Elsevier.

The polymer chains were reported to form hydrogen bonds with the –OH groups present at the edges of the platelets or in the crystal structure of the fillers, as well as with the water molecules preadsorbed in the clay interlayers. The absence of any low–temperature degradation corresponding to the presence of free unbound polymer also confirmed that the excess polymer chains added during adsorption could be washed off. Thus, this further opens functional routes to achieving thermal stability with the preexisting systems.

References

1. S. W. Bailey, *Reviews in Mineralogy* (Blacksburg: Virginia Polytechnic Institute and State University, 1984).
2. G. W. Brindley and G. Brown, *Crystal Structures of Clay Minerals and Their X-Ray Identification* (London: Mineralogical Society, 1980).
3. L. W. Carter, J. G. Hendricks, and D. S. Bolley, *Elastomer Reinforced with Modified Clay* (assigned to National Lead Co.), United States Patent 2,531,396 (1950).
4. D. J. Greenland, Adsorption of poly(vinyl alcohols) by montmorillonite. *Journal of Colloid Science*, 18 (1963), 647–64.

5. K. Yano, A. Usuki, A. Okada, T. Kurauchi, and O. Kamigaito, Synthesis and properties of polyimide–clay hybrid. *Journal of Polymer Science, Part A: Polymer Chemistry*, 31 (1993), 2493–8.
6. Y. Kojima, K. Fukumori, A. Usuki, O. Okada, and T. Kurauchi, Gas permeabilities in rubber–clay hybrid. *Journal of Materials Science Letters*, 12 (1993), 889–90.
7. P. B. Messersmith and E. P. Giannelis, Synthesis and barrier properties of poly(ε-caprolactone)–layered silicate nanocomposites. *Journal of Polymer Science, Part A: Polymer Chemistry*, 33 (1995), 1047–57.
8. T. Lan, P. D. Kaviratna, and T. J. Pinnavaia, On the nature of polyimide–clay hybrid composites. *Chemistry of Materials*, 6 (1994), 573–5.
9. I.-J. Chin, T. Thurn-Albrecht, H.-C. Kim, T. P. Russell, and J. Wang, On exfoliation of montmorillonite in epoxy. *Polymer*, 42 (2001), 5947–52.
10. P. C. LeBaron, Z. Wang, and T. J. Pinnavaia, Polymer layered silicate nanocomposites: An overview. *Applied Clay Science*, 15 (1999), 11–29.
11. S. K. Lim, J. W. Kim, I.-J. Chin, Y. K. Kwon, and H. J. Choi, Preparation and interaction characteristics of organically modified montmorillonite nanocomposite with miscible polymer blend of poly(ethylene oxide) and poly(methyl methacrylate). *Chemistry of Materials*, 14 (2002), 1989–94.
12. Z. Wang and T. J. Pinnavaia, Nanolayer reinforcement of elastomeric polyurethane. *Chemistry of Materials*, 10 (1998), 3769–71.
13. K. Yano, A. Usuki, and A. Okada, Synthesis and properties of polyimide–clay hybrid films. *Journal of Polymer Science, Part A: Polymer Chemistry*, 35 (1997), 2289–94.
14. H. Shi, T. Lan, and T. J. Pinnavaia, Interfacial effects on the reinforcement properties of polymer–organoclay nanocomposites. *Chemistry of Materials*, 8 (1996), 1584–7.
15. E. P. Giannelis, Polymer layered silicate nanocomposites. *Advanced Materials*, 8 (1996), 29–35.
16. J. W. Gilman, C. L. Jackson, A. B. Morgan, R. Harris, Jr., E. Manias, E. P. Giannelis, M. Wuthenow, D. Hilton, and S. H. Phillips, Flammability properties of polymer–layered-silicate nanocomposites: Polypropylene and polystyrene nanocomposites. *Chemistry of Materials*, 12 (2000), 1866–73.
17. H. Heinz, H. J. Castelijns, and U. W. Suter, Structure and phase transitions of alkyl chains on mica. *Journal of the American Chemical Society*, 125 (2003), 9500–9510.
18. B. K. G. Theng, *The Chemistry of Clay–Organic Reactions* (New York: Wiley, 1974).
19. K. Jasmund and G. Lagaly, *Tonminerale und Tone Struktur* (Darmstadt: Steinkopff, 1993).
20. M. A. Osman, M. Ploetze, and U. W. Suter, Surface treatment of clay minerals: Thermal stability, basal plane spacing and surface coverage. *Journal of Materials Chemistry*, 13 (2003), 2359–66.
21. T. J. Pinnavaia, Intercalated clay catalysts. *Science*, 220 (1983), 365–71.
22. R. Krishnamoorti and E. P. Giannelis, Rheology of end tethered polymer layered silicate nanocomposites. *Macromolecules*, 30 (1997), 4097–4102.
23. H. Fischer, Polymer nanocomposites: From fundamental research to specific applications. *Materials Science and Engineering C*, 23 (2003), 763–72.
24. S. C. Tjong and Y. Z. Meng, Preparation and characterization of melt-compounded polyethylene/vermiculite nanocomposites. *Journal of Polymer Science, Part B: Polymer Physics*, 41 (2003), 1476–84.

25. J. Xu, R. K. Y. Li, Y. Xu, L. Li, and Y. Z. Meng, Preparation of poly(propylene carbonate)/organo-vermiculite nanocomposites via direct melt intercalation. *European Polymer Journal*, 41 (2005), 881–8.

26. M. A. Osman, M. Ploetze, and P. Skrabal, Structure and properties of alkylammonium monolayers self-assembled on montmorillonite platelets. *Journal of Physical Chemistry B*, 108 (2004), 2580–88.

27. H. Heinz, R. A. Vaia, R. Krishnamoorti, and B. L. Farmer, Self-assembly of alkylammonium chains on montmorillonite: Effect of chain length, headgroup structure, and cation exchange capacity. *Chemistry of Materials*, 19 (2007), 59–68.

28. R. A. Vaia, H. Ishii, and E. P. Giannelis, Synthesis and properties of two-dimensional nanostructures by direct intercalation of polymer melts in layered silicates. *Chemistry of Materials*, 5 (1993), 1694–6.

29. V. Mehrotra and E. P. Giannelis, Conducting molecular multilayers: Intercalation of conjugated polymers in layered media. *Materials Research Society Symposium Proceedings*, 171 (1990), 39–44.

30. V. Mittal, Polymer nanocomposites: Synthesis, microstructure and properties. In: *Optimization of Polymer Nanocomposite Properties*, ed. V. Mittal (Weinheim: Wiley–VCH, 2009), pp. 1–19.

31. V. Mittal, Role of clean clay surface on composite properties. In: *Polymer Nanocomposites: Advances in Filler Surface Modifications*, ed. V. Mittal (New York: Nova Science Publishers, 2009), pp. 143–52.

32. M. A. Osman, V. Mittal, and U. W. Suter, Poly(propylene)-layered silicate nanocomposites: Gas permeation properties and clay exfoliation. *Macromolecular Chemistry and Physics*, 208 (2007), 68–75.

33. V. Mittal, Effect of the presence of excess ammonium ions on the clay surface on permeation properties of epoxy nanocomposites. *Journal of Materials Science*, 43 (2008), 4972–8.

34. J. W. Gilman, Flammability and thermal stability studies of polymer layered-silicate clay/nanocomposites, 1. *Applied Clay Science*, 15 (1999), 31–49.

35. M. A. Osman, V. Mittal, M. Morbidelli, and U. W. Suter, Epoxy-layered silicate nanocomposites and their gas permeation properties. *Macromolecules*, 37 (2004), 7250–57.

36. V. Mittal, Gas permeation and mechanical properties of polypropylene nanocomposites with thermally-stable imidazolium modified clay. *European Polymer Journal*, 43 (2007), 3727–36.

37. W. H. Awad, J. W. Gilman, M. Nyden, R. H. Harris, Jr., T. E. Sutto, J. Callahan, P. C. Trulove, H. C. DeLong, and D. M. Fox, Thermal degradation studies of alkyl-imidazolium salts and their application in nanocomposites. *Thermochimica Acta*, 409 (2004), 3–11.

38. F. Avalos, J. C. Ortiz, R. Zitzumbo, M. A. López-Manchado, R. Verdejo, and M. Arroyo, Phosphonium salt intercalated montmorillonites. *Applied Clay Science*, 43 (2009), 27–32.

39. V. Mittal, Polypropylene-layered silicate nanocomposites: Filler matrix interactions and mechanical properties. *Journal of Thermoplastic Composites Materials*, 20 (2007), 575–99.

40. A. M. Gultek, G. Icduygu, and T. Seckin, Preparation and characterization of polybenzimidazole–clay hybrid materials. *Materials Science and Engineering B*, 107 (2004), 166–71.

41. G. Chigwada, D. Wang, and C. A. Wilkie, Polystyrene nanocomposites based on quinolinium and pyridinium surfactants. *Polymer Degradation and Stability*, 91 (2006), 848–55.
42. D.-R. Yei, S.-W. Kuo, H.-K. Fu, and F.-C. Chang, Enhanced thermal properties of PS nanocomposites formed from montmorillonite treated with a surfactant/cyclodextrin inclusion complex. *Polymer*, 46 (2005), 741–50.
43. V. Mittal, Polypropylene nanocomposites with thermally-stable imidazolium modified clay: Mechanical modeling and effect of compatibilizer. *Journal of Thermoplastic Composite Materials*, 22 (2009), 453–74.
44. V. Mittal and V. Herle, Physical adsorption of organic molecules on the surface of layered silicate clay platelets: A thermogravimetric study. *Journal of Colloid and Interface Science*, 327 (2008), 295–301.

2

Mechanism of thermal degradation of layered silicates modified with ammonium and other thermally stable salts

K. PIELICHOWSKI,[a] A. LESZCZYŃSKA,[a] AND J. NJUGUNA[b]

[a]Department of Chemistry and Technology of Polymers, Cracow University of Technology
[b]Department of Sustainable Systems, Cranfield University

2.1 Introduction

The development of polymer/clay nanocomposites as commercial materials faces the problem of limited miscibility of inorganic hydrophilic layered silicates and organic hydrophobic polymers. Intensive studies have led to various strategies, including the use of surface-active organic compounds, chemical modification of the polymer matrix, and application of macromolecular compatibilizers that produce a desired improvement of miscibility and therefore facilitate the formation of nanostructure. The application of organically modified clays provides certain properties to nanocomposite materials superior to those of systems containing sodium montmorillonite [1]. However, ammonium salts, which are most frequently applied [2, 3], suffer from thermal degradation during the fabrication and further processing of nanocomposites. This leads to changes in the surface properties of clays resulting in alteration of nanocomposite structure and related properties [4, 5] and facilitates the occurrence of some unwanted side reactions and the contamination of polymeric material with the products of thermal degradation of an organic modifier, which may be responsible for enhanced thermal degradation of the polymer matrix [6], accelerated aging, color formation [7, 8], plasticization effects [9], and so forth. The need to improve the thermal stability of organoclays applied in the preparation of polymeric nanocomposites has motivated the search for an organic modifier combining high thermal stability with high efficiency in facilitating dispersion of a nanofiller in a polymer matrix.

2.2 Changes in properties of organically modified layered silicates as a result of thermal degradation

Changes in the surface properties of cationic complexes, in which the surface metal cations of natural clays have been exchanged with an ammonium cationic surfactant, could be simply confirmed by mixing them with polar solvents and observing their dispersibility. It should be noted that quantitative measurements of surface properties may be obtained by contact-angle measurements [4, 10]. This method allowed measuring the change of surface polarity of Cloisite 30B (a dimethylbenzyl tallow ammonium chloride–modified montmorillonite) subjected to elevated temperatures for the duration of typical

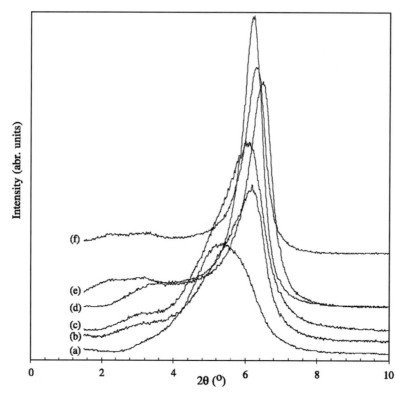

Figure 2.1 XRD scans of heat-treated Cloisite 30B surfaces. (a) Clay as received, (b) treated at 220 °C for 7 min, (c) treated at 220 °C for 7 min and 250 °C for 14 min, (d) treated at 220 °C for 7 min and 250 °C for 17 min, (e) treated at 250 °C for 60 min, and (f) treated at 250 °C for 120 min. Reprinted from Ref. [4], with permission from Elsevier.

polyamide–clay mixing. The polarity was reduced drastically after the clay surface was treated for more than 7 min at 220 °C or 7 min at 250 °C. After further heating for 1 h at 250 °C, the polarity of Cloisite 30B clay approached that of polypropylene (PP). Though the polarity of the clay surface decreased substantially with thermal decomposition, the XRD analysis of these surfaces, as shown in Figure 2.1, did not follow any such trend. The X-Ray diffractograms showed that heating clay at high temperatures resulted in a slight decrease in gallery spacing, as revealed from the shifting of the peaks to slightly higher values of 2θ. The study concluded that of XRD, FTIR, and contact-angle measurements, the last is much more sensitive to the effects of thermal decomposition of alkylammonium ions.

2.3 Mechanism of thermal degradation of organic modifiers and organically modified layered silicates

Prior to discussing the mechanism of thermal degradation of a hybrid system composed of a mineral deposit and an organic modifier, it is advisable to consider the thermal stability and degradation mechanism of neat components.

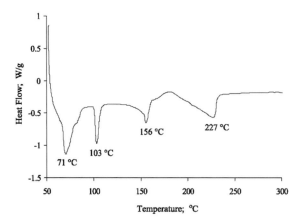

Figure 2.2 DSC scan of *n*-hexadecylammonium chloride. Reprinted from [9] with permission from Elsevier.

Unmodified sodium montmorillonite (Na$^+$MMT) displays good thermal stability in the temperature range 20–500 °C, evolving only moderate quantities of physically adsorbed water at temperatures up to 120 °C [11, 12] and the water from hydrated ions in the temperature range from 85 to 183 °C [13]. The dehydroxylation of the crystal lattice of MMT was observed at temperatures above 500 °C, at which most commercially available polymers have already degraded.

2.3.1 Ammonium salts and ammonium-modified layered silicates

Pristine ammonium salts show relatively low initial temperatures of degradation with respect to polymer processing temperatures. For example, alkylammonium chloride salt, n-C$_{16}$H$_{33}$NH$_3^+$Cl$^-$, subjected to DSC, underwent four endothermic transitions due to, respectively, melting of an impurity or free amines, melting of the chloride salt, thermal dissociation of the ammonium chloride, and subsequent degradation of the hydrocarbon chain [9]. The thermal dissociation, which leads to formation of n-C$_{16}$H$_{33}$NH$_2$ and HCl, started at 130 °C, showed a maximum at 156 °C, and continued with decreasing intensity until 162 °C, with some loss in weight of the specimen, purportedly due to hydrogen chloride emission (Figure 2.2).

The endothermic peak with a maximum at 227 °C was due to degradation of the hydrocarbon chain in the *n*-hexadecylammonium chloride.

There are two generally accepted mechanisms of degradation of neat ammonium salts – nucleophilic substitution and elimination reaction [14, 15]. In the former, nucleophilic attack on the R$_4$N$^+$ moiety by, for example, chloride ion leads to the formation of RCl and R$_3$N, which is essentially the reverse reaction of most quaternary ammonium syntheses (Scheme 2.1).

In cases where the quaternary ammonium incorporates different alkyl substituents, the least sterically hindered (i.e., methyl) and/or other electrophilic alkyl groups (i.e., benzyl) are generally the most susceptible to nucleophilic attack.

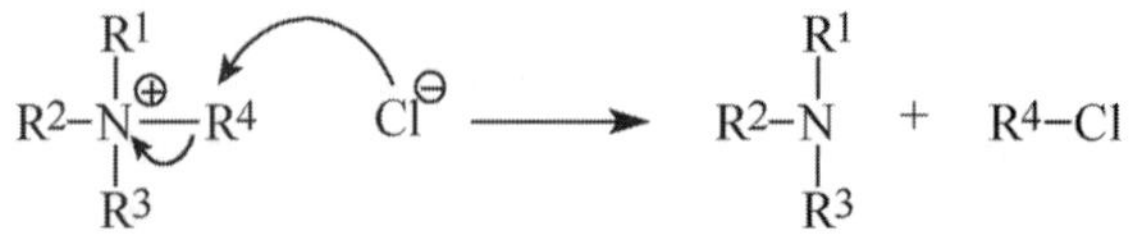

Scheme 2.1 Nucleophilic substitution leading to the decomposition of an ammonium surfactant [16].

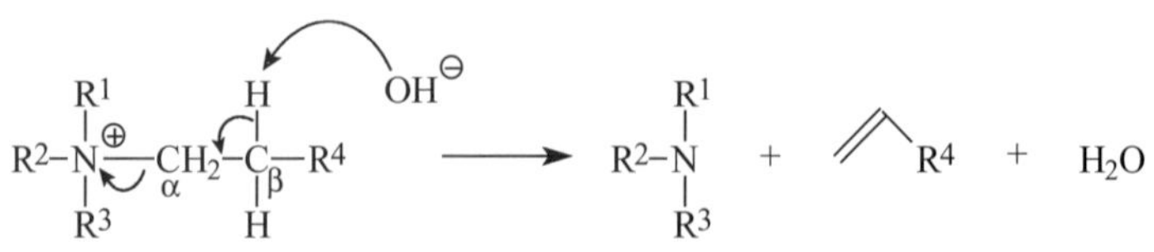

Scheme 2.2 Representative example of a Hofmann-type elimination reaction [16].

The Hofmann elimination reaction, on the other hand, is a process where a quaternary ammonium salt is decomposed into an olefin and a tertiary amine *via* exposure to basic conditions (*e.g.*, silver oxide and water) (Scheme 2.2).

The mechanism of this reaction is believed to be a bimolecular elimination, where a base, such as hydroxide, abstracts a hydrogen atom from the β-carbon of the quaternary ammonium salt [16].

Recently, by applying NMR analysis, Cui *et al.* found tertiary amines in nonvolatile residues and chloroalkanes in the volatile products of degradation of $M_2(HT)_1B_1N^+Cl^-$ at 250 °C for 5 min. Those results suggested that pure ammonium surfactants decompose primarily *via* nucleophilic attack of anions, such as the chloride anion [16].

To establish the mechanism of thermal decomposition of organically modified layered silicates, various analytical and thermoanalytical methods have been applied, including thermogravimetric analysis (TGA), pyrolysis techniques, gas chromatography (GC), and mass spectrometry (MS) [12, 15, 17, 18].

Intercalation of ammonium surfactant in the clay galleries increases its decomposition temperature [19–21] but, generally, the initial decomposition temperatures for commercially available organoclays are relatively low. For instance, the degradation temperatures of Cloisite clays, obtained by Cervantez-Uc *et al.* [17], are collected in Table 2.1.

Comparative studies of as-received organically modified layered silicates (OLS) and their residues after thermal degradation, carried out by FTIR, have shown changes in the absorption region of C–H bonds, mainly [11], as illustrated in Figure 2.3.

The thermal degradation of organoclays is mostly associated with the thermal decomposition of organic moieties, with minor changes in the crystal lattice of MMT. Changes in conformation and loss of organic matter from organoclays, resulting in collapse of organoclay layers, were also confirmed by WAXD and SAXS methods [22, 23].

The degradation of organoclays is believed to be a multistep process. Generally, evolution of water, originating mainly from the mineral component, was observed in the initial step of mass loss, whereas organic moieties were evolved in successive stages of degradation because of decomposition of surfactant. The different locations of organic molecules

Table 2.1 *The onset decomposition temperature, maximum mass loss rate temperature, and residual mass of the Cloisites*

Sample code	Onset decomposition temperature from TG (°C)	Maximum mass loss rate temperature from DTG (°C)	Residual mass (%)
Cloisite Na+	–	–	95
Cloisite 10A	160	245, 310, 395	66
Cloisite 15A	192	331, 447	60
Cloisite 20A	198	336, 451	63
Cloisite 25A	192	330, 390	74
Cloisite 30B	174	298, 427	73
Cloisite 93A	212	347	76

Note: Reprinted from Ref. [17] with permission from Elsevier.

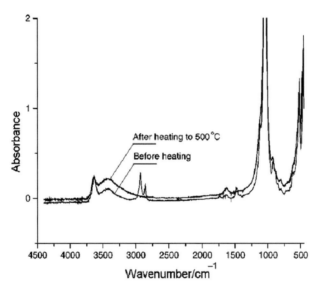

Figure 2.3 FTIR spectra of Na$^+$ montmorillonite before and after heating to 500 °C. Adapted from Ref. [11], with permission from Springer.

(inside or outside the clay galleries) and types of surfactant bonding to the mineral surface (physical or chemical) were considered as causes of the complex decomposition mechanism of organic modifier. For example, Ni *et al.* distinguished four steps of mass loss for cetyltrimethylammonium bromide–exchanged MMT by means of thermogravimetric analysis, which were assigned to the following processes: the first step from the ambient to 95 °C temperature range was attributed to the desorption of water; the second step,

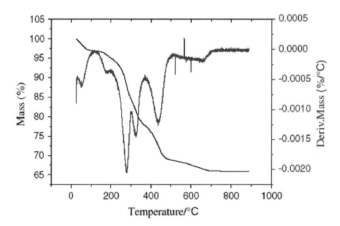

Figure 2.4 TG and DTG profiles of montmorillonite-CTAB. Reprinted from Ref. [13], with permission from Springer.

occurring from 125 to 185 °C – the very small peak – was assigned to the loss of hydration water from the Na^+ environment; the third step in the temperature of 190–500 °C should be identified as the decomposition of the surfactant; the fourth mass loss step between 600 and 710 °C was attributed to the loss of structural hydroxyl groups from the MMT lattice [13]. Remarkably, there were three separate peaks in the third range, at 210–290, 300–355, and 380–480 °C, referred to the decomposition of the physically adsorbed surfactant, interlayer-adsorbed surfactant molecules, and intercalated surfactant cations, respectively (see Figure 2.4).

The presence of physically adsorbed ammonium salts, which did not undergo cation–exchange reactions and preserved their anions, could be confirmed by qualitative analysis of evolved products of degradation [11]. Moreover, a clear indication of free surfactant in commercial organoclays was the presence of chlorine and sulfur, which were detected in organoclays by means of EDX, as well as the presence of two FTIR bands at 1376 and $1340\,cm^{-1}$, which were assigned to the covalent sulfate absorption region (R–O–SO_2–O–R) in a manner similar to that reported by Dyer [24].

Gao *et al.* [11] found that the gases evolved at 200 °C and 220 °C from MMT that was exchanged with trimethyloctadecyl quaternary ammonium chloride were the same as for the neat ammonium compound. However, most of the species determined at 300 and 400 °C were short or long alkenes without the chloro group [11]. This showed that the first bond breakage in the organic compound may occur outside the clay sheets, between N and C bonds, at around 200 °C, and products formed there can evolve quickly without further degradation because of weak hindrance in the clay sheets. The decomposition of intercalated organic compounds, on the other hand, was more advanced. Hindered between the clay sheets, they evolved slowly, and further decomposition occurred mainly between C–C bonds. A mechanism of OLS decomposition was proposed assuming two areas of surfactant molecule concentrations (Figure 2.5).

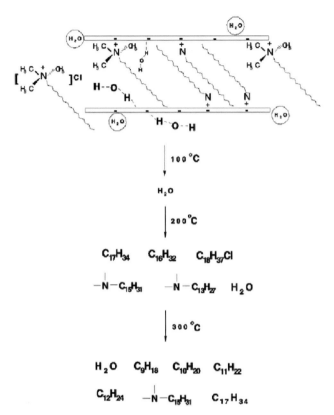

Figure 2.5 The proposed decomposition mechanism of MMT exchanged with trimethyloctadecyl quaternary ammonium chloride. Reprinted from Ref. [11], with permission from Springer.

It is noteworthy that DSC analysis confirmed that the molecular environment and thermal behavior of the surfactant within the montmorillonite galleries is different from that in the bulk state [25]. A slightly different scenario of OMMT degradation was proposed by Xie *et al*. [18]. In the initial stage of decomposition (around 200 °C), low–molecular weight organic compounds were released first, but the high–molecular weight organics were still trapped by the OLS matrix. As the temperature increased, the high–molecular weight species not only decomposed, but were also released, as they are from OLS (Table 2.2).

Interestingly, Xie *et al*. [18] and Bertini *et al*. [26] only observed long-chain alkanes and alkenes at 200 or 300 °C and relatively high residues (>95 wt%). Stoeffler and co-workers, who also investigated evolution of alkanes, postulated that the high molecular weight of evolved amines could prevent their vaporization, and therefore they underwent degradation by thermal cleavage of alkyl chains [27].

At higher temperatures, when the pyrolysis run was performed at 450 °C, the degradation proceeded to a greater extent according to the Hofmann elimination mechanism, because high yields of tertiary amines and olefins were achieved. Nucleophilic substitution S_N also

Table 2.2 *Organic species evolved in the temperature range 200–500 °C from MMT exchanged with trimethyloctadecyl quaternary ammonium chloride*

Sample ID	Temperature	
	300 °C	500 °C
SCPX 2048	1-Tridecanamine-N, N-dimethyl-$C_{15}H_{33}N$	1-Pentadecanamine-N, N-dimethyl-$C_{17}H_{37}N$
(MMT modified with trimethyloctadecyl quaternary ammonium chloride)	1-Hexadecene-$C_{16}H_{32}$	1-Decene-$C_{10}H_{20}$
	1-Heptadecene-$C_{17}H_{34}$	4-Nonene-2-methyl-$C_{10}H_{20}$
	Tetradecanal-$C_{14}H_{28}O$	1-Nonene-C_9H_{18}
	Octadecane-1-chloro-$C_{18}H_{37}Cl$	3-Undecene-(E)-$C_{11}H_{22}$
	1-Decene-$C_{10}H_{20}$	5-Dodecene-(Z)-$C_{12}H_{24}$
	1-Pentadecanamine-N, N-dimethyl-$C_{17}H_{37}N$	5-Octadecene-(E)-$C_{18}H_{36}$
	Pyrrolidine-N-(4-pentenyl)-$C_9H_{17}N$	E-15-heptadecenal-$C_{17}H_{32}O$
		1-Octadecanamine-$C_{18}H_{39}N$

Note: Reprinted from Ref. [18] with permission from Elsevier.

occurred, yielding chloroalkanes [26]. The intensity of chloroalkane emission gradually decreased as the time of extraction of OMMT with ethanol was increased. This provided a clear indication that the degradation of free and adsorbed ammonium salts in the presence of halide anions runs via nucleophilic substitution. The presence of chlorinated compounds in volatiles of commercial samples, which most likely arose from excess surfactant, was not detected for laboratory OLS [12]. Similar dependence of the thermal stability of organic cations on their bonding mode and location was observed [17, 18] and led to the conclusion that ammonium salts adsorbed onto clay surfaces have low thermal stability, apparently close to that of pure surfactants, and contribute to the decrease of the thermal stability of organoclays.

However, it was suggested that if the residual anions could be completely removed from the organoclays, the primary degradation pathway would switch to an elimination-type mechanism [16]. The Hofmann elimination of ammonium compounds was most probably the source of additional amounts of vinyl-type unsaturation found in melt-processed OMMT-PE relative to both the polymer control and the Na^+MMT-PE samples [28]. On the other hand, the presence of alkenes was also explained by three possible routes of decomposition: (i) pyrolysis of alkanes derived from the major component of the organic part, i.e., hydrogenated tallow (HT); (ii) pyrolysis of the tallow (unsaturated fatty acids used for the preparation of the quaternary ammonium salt); and (iii) decarboxylation of $RCOO^\bullet$ and $RCO^\bullet$ radicals [17, 29].

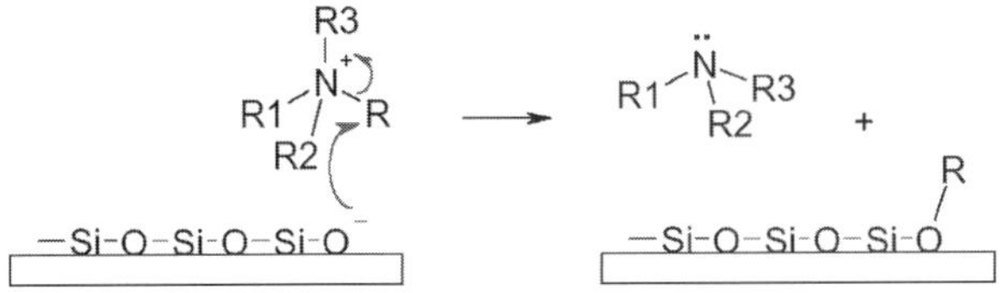

Scheme 2.3 S_N2 substitution of quaternary ammonium cation by oxygen anion of clay surface. Reprinted from Ref. [30], with permission.

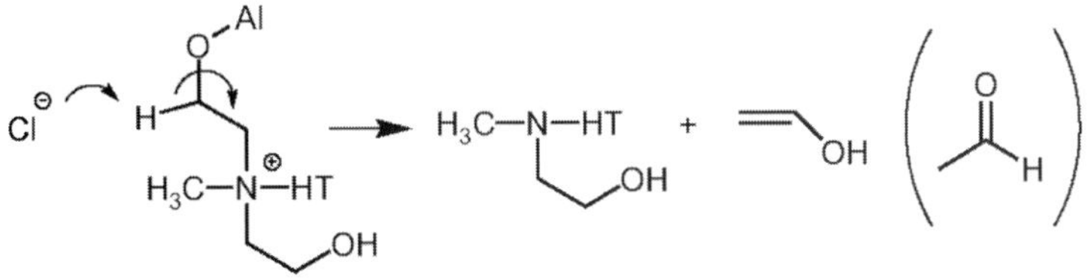

Scheme 2.4 Activation of a hydroxyethyl group with an Al site in an organoclay may facilitate a Hofmann-type elimination reaction. Reprinted from Ref. [16], with permission from Elsevier.

The evolution of tertiary amines, chloroalkanes, and alkenes, confirming simultaneous runs of nucleophilic substitution (S_N2) and – with lower yield – Hofmann elimination, was also observed by Galimberti et al. [30]. Interestingly, this work has suggested that clays modified with ammonium are susceptible to nucleophilic attack by organic compounds other than surfactant anions that are components of a nanocomposite system. This could explain the formation of long-chain carboxylic acid methyl esters through the nucleophilic substitution to the alkylammonium cation by the corresponding carboxylate anion (fatty acids were in fact used to assemble the master batch). The study further speculated that another possibility is that nucleophilic oxygen atoms in the periphery of the clay gallery can undergo substitution with a net increase of the hydrophobicity of the inorganic material (Scheme 2.3).

The concept of high-temperature reaction of carbon atoms present in the surfactant with oxygen atoms from the crystal structure of the MMT has also been formulated. Xie et al. [18] detected some CO_2 amounts, Edwards et al. [12] identified considerable amounts of alcohols and linear aldehydes from C16 to C18, and Cervantez-Uc et al. [17] confirmed the presence of carboxylic acids and aldehydes in the evolved products from thermal decomposition of OMMT under nonoxidative conditions. It was clear that aldehydes originated from tallow residue because the molecular weights of aldehydes agreed well with the size of these fragments [12]. Notably, a significant amount (~15%) of 2-chloroethanol and acetaldehyde was observed upon decomposition of Cloisite 30B, though these decomposition products were not detected in the respective neat surfactant [16]. Alternative decomposition mechanisms may therefore be operational for this organoclay, as presented in Scheme 2.4 – the presence of Lewis basic sites (e.g., Al) may activate the material toward elimination.

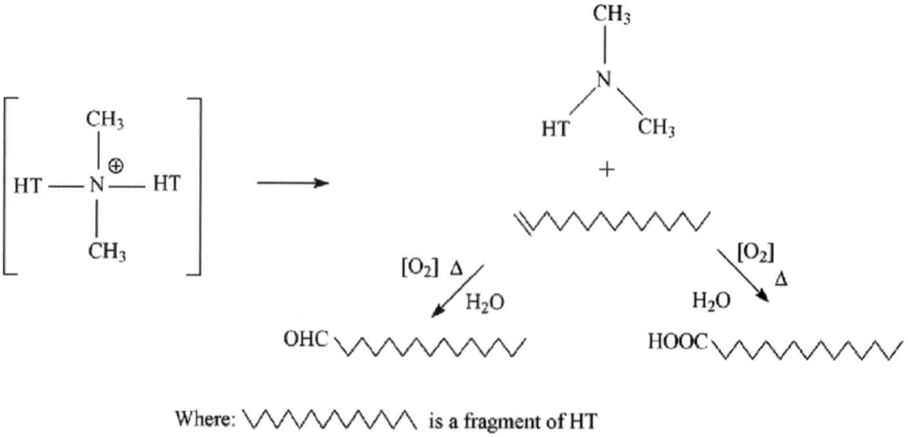

Where: ⋀⋀⋀⋀⋀⋀ is a fragment of HT

Scheme 2.5 Suggested reactions to explain the presence of aldehydes and carboxylic acids in the products evolved from the thermal degradation of Cloisites. Reprinted from Ref. [17], with permission from Elsevier.

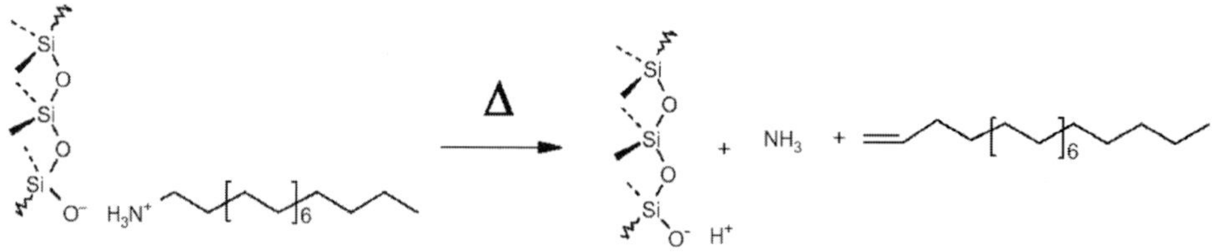

Scheme 2.6 Degradation of octadecylammonium ion inside the clays. Reprinted from Ref. [34], with permission from Elsevier.

The carbonyl compounds could possibly originate from the oxidation of residual organic species and from the presence of a small fraction of insoluble metal carbonate impurities [17]. Scheme 2.5 depicts another proposed mechanism of carbonyl compound formation involving oxidation of alkenes.

There currently exist a number of patents for the use of nonvolatile diluents (such as soybean oil) to reduce the viscosity of organoclay formulations [31], and carboxylic acids as alkyl tails in quaternary ammonium compounds [32] or as the corresponding counterions [33]. Therefore, the presence of carboxylic groups (and CO_2 after decomposition) might arise from raw material residue or from auxiliary compounds used during the industrial process of organoclay synthesis/modification.

Bellucci *et al.* have proposed that in the Hofmann elimination reaction amine, the corresponding olefin and acidic sites in the aluminosilicate are formed during OLS decomposition [34] (Scheme 2.6).

The formation of acidic sites is an important finding because it may have an influence on the course of polymer degradation reactions. By acceptance of single electrons from donor molecules with low ionization potential, leading to the formation of free radicals, acidic sites on MMT would accelerate the degradation of the polymer and promote cross-linking reactions [35]. Additionally, the olefin generated from the OMMT degradation may act as a sensitizer and promote coupling reactions. Meanwhile, the alkyl quaternary ammonium ions diffused throughout the polymeric matrix would also promote cross-linking reactions and improve the charred residue yield.

Additionally, $M-NH_4^+$ was proposed to be formed because of the set of reversible reactions [18]

$$M-R_3NH^+ + H_3O^+ \rightleftharpoons M-R_2NH_2^+ + RH_2O^+ \tag{2.1}$$

$$M-R_2NH_2^+ + H_3O^+ \rightleftharpoons M-RNH_3^+ + RH_2O^+ \tag{2.2}$$

$$M-RNH_3^+ + H_3O^+ \rightleftharpoons M-NH_4^+ + RH_2O^+. \tag{2.3}$$

During the gradual heating of organoclay complexes in oxidative atmospheres, oxidation of organic hydrogen and formation of water and charcoal are observed, leading to final oxidation in the range 400–750 °C and to the formation of CO_2 [36]. Interestingly, the formation of carbonyl compounds at 240 °C in air was detected after short durations (≥ 5 min) and progressive formation and evolution of different carbonyl compounds such as peroxy acids, esters, and lactones, as well as aldehydes and carboxylic acids, was observed with prolonged heating [37]. The use of an antioxidant system protected organoclays from oxidation as well as exerting a positive effect on the mechanical properties of the compatibilized nanocomposite blends [37].

The decomposition of ammonium compounds, as considered for neat organoclays, may not necessarily be the exact degradation process occurring in a polymer matrix. It may be altered by complex interactions between nanocomposite components and products of their thermal decomposition. The composition of commercial polymeric materials containing dispersed nanoparticles is quite complex, as it incorporates a set of additional additives, such as UV and thermal stabilizers, plasticizers, and dyes and other fillers. These additives may considerably influence the degradation mechanism [38, 39].

Additionally, the processing of a thermoplastic polymer may cause some decrease in its initial thermal stability. Because of chain scission under shear forces during processing, free radicals could be formed and alter the chemical structure of polymers. This process leads to the formation of weak sites at which the degradation starts at lower temperatures than in the case of unprocessed polymers, because of decreased activation energy of degradation. Organic moieties of organoclay subjected to shear forces and elevated temperatures under processing conditions undergo profound thermomechanical degradation. Indeed, the data available clearly reveal a larger surfactant loss from the clay galleries during melt-processing (estimated from WAXD diffraction patterns) than during thermogravimetric analysis [40].

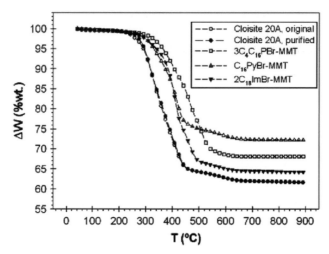

Figure 2.6 TG profiles of the various organoclays (argon, 10 °C/min). Reprinted from Ref. [27], with permission from Elsevier.

2.3.2 *Phosphonium salts and phosphonium-modified layered silicates*

Current efforts focus on selecting new clay modifiers that display significantly higher thermal stability than ammonium compounds, provide effective modification of the polymer/clay interface, and facilitate clay dispersion. In that line of thought, a group of cationic surfactants, including some ionic liquids, most frequently imidazolium, pyridinium, and phosphonium salts [41] with long hydrocarbon substituents, were investigated as potential surface-active compounds and therefore are discussed in detail in the follow-up.

The thermal stability of ammonium, phosphonium, imidazolium, and pyridinium compounds has been compared recently by Stoeffler *et al.* [27]. Although the surfactants differed in their alkyl substituents, the comparison displayed a general tendency of these types of organoclays to thermal stability. The neat surfactants decomposed in the following order: C_{16}PyBr (hexadecyl pyridinium bromide) $< 2C_{18}$ImBr (1,3-dioctadecyl imidazolium bromide) $< 3C_4C_{16}$PBr (tributylhexadecylphosphonium bromide), whereas modified clays took the order Cloisite 20A original (clay containing dimethyldioctadecylammonium chloride) $<$ Cloisite 20A purified $< C_{16}$PyBr-MMT$< 2C_{18}$ImBr-MMT $< 3C_4C_{16}$PBr-MMT. TG profiles of modified MMTs are shown in Figure 2.6.

It is postulated that thermal decomposition of tetra-alkylphosphonium salts proceeds through four types of reactions [42, 43]: (i) nucleophilic substitution at the α-carbon ($S_N(C)$); (ii) β-elimination; (iii) nucleophilic substitution at the phosphorus atom ($S_N(P)$); (iv) α-elimination. $S_N(C)$ nucleophilic substitution involves the attack of the halide on the α-carbon, leading to the formation of an alkyl halide and a tertiary phosphine:

$$R_3P^+-CH_2-CH_2-R' + X^- \rightarrow R_3P + X-CH_2-CH_2-R'. \tag{2.4}$$

β-elimination occurs in the presence of a basic anion (*e.g.*, Si–O$^-$), which extracts a hydrogen atom from a β-carbon of the quaternary phosphonium, leading to the formation of an alkene, a tertiary phosphine, and an acid:

$$R_3P^+ - CH_2 - CH_2 - R' + B^- \rightarrow R_3P + CH_2 = CH_2 - R' + BH. \qquad (2.5)$$

$S_N(P)$ nucleophilic substitution proceeds in the presence of a surface hydroxyl group that attacks the phosphorus atom, leading to the formation of a tertiary phosphine oxide and an alkane driven by the formation of a strong phosphoryl (P=O) bond:

$$R_3P^+ - CH_2 - CH_2 - R' + OH^- \rightarrow R_3P = O + CH_3 - CH_2 - R'. \qquad (2.6)$$

Finally, α-elimination takes place in the presence of a strong base, which extracts hydrogen from the α-carbon, resulting in the formation of an unsaturated quaternary phosphine and an acid:

$$R_3P^+ - CH_2 - CH_2 - R' + B^- \rightarrow R_3P = CH - CH_2 - R' + BH. \qquad (2.7)$$

Xie *et al.* [43] analyzed the thermal decomposition of tributyloctadecylphosphonium bromide during pyrolysis by applying GC and MS. They detected tributylphosphine, 1-bromoalkanes, and alkenes, indicating that the $S_N(C)$ and β-elimination mechanisms are favored for the pure surfactant. Both mechanisms generate trialkylphosphine, which can easily be fragmented in the MS source to produce hydrocarbons, because of the low energy of the C–P bond (264 kJ/mol) [27, 44]. Phosphonium salts behaved differently from ammonium salts because of the higher steric tolerance of the phosphorus atom and the participation of its low-lying *d*-orbitals in the processes of forming and breaking chemical bonds [43].

For purified tributyloctadecylphosphonium–modified montmorillonite ($3C_4C_{18}$PBr–MMT), tributylphosphine, alkenes, tributylphosphine oxide, and alkanes were detected, indicating the suggested occurrence of $S_N(P)$ and β-elimination mechanisms [27, 43]. Even though it was not possible to identify with certainty the mechanism involved in the thermal decomposition of $3C_4C_{16}$PBr–MMT, it is highly probable that it proceeded *via* a different pathway than pure surfactant, because the measured degradation temperature was lower than that for the pure surfactant.

The phosphine oxide has been formed during the modification of OMMT by the quaternary diphosphonium salt [MeOOCCH$_2$(Ph)$_2$PCH$_2$CH$_2$P(Ph)$_2$CH$_2$COOMe]Br$_2$, as confirmed by solid state ^{31}P NMR spectroscopy [45]. Another reaction occurring in the course of degradation of this phenyl-containing phosphonium–MMT, confirmed by GC–MS and ^{29}Si NMR results, was formation of Si–O–P–R species, with subsequent radical decomposition to yield biphenyl. This reaction is catalyzed by the hydroxyl groups of MMT and accompanied by structural changes in the MMT lattice. However, the thermal stability of phenylphosphonium–MMT was increased with respect to the free phosphonium salt, suggesting the formation of a nonvolatile Si–O–P system.

Calderon *et al.* found that the surfactant's isometry and molecular weight influenced the thermal stability of phosphonium-containing clays; that is, isometric phosphonium surfactants started to decompose at higher temperatures than compounds containing both short and long alkyl chains; surfactants with higher molecular weight were more thermally stable than lower–molecular weight compounds [46]. The earlier results of Patel *et al.* were in agreement with Calderon's generalization, because symmetric tetraphenylphosphonium MMT and phosphonium cations with longer alkyl chains exhibited the highest thermal stability [47]. Contrary to these findings, Patro *et al.* found that the T_{onset} of octadecylphosphonium cations was slightly lower than that of dodecyl-containing phosphonium modifiers [48]. It seems that different content of solvent, volatilized in the initial stage of decomposition, and different methodology applied for the determination of initial temperature of degradation could be responsible for the discrepancy of obtained results. Another important factor is the surfactant packing density estimated from the basal spacing of OMMT, as noted by Hedley *et al.* [49]. They reported that organic cation (hexadecyltributylphosphonium) with a loose arrangement was less thermally stable than tetrabutylphosphonium and butyltriphenylphosphonium, forming a monolayer in the gallery space of MMT.

2.3.3 Pyridinium salts and pyridinium-modified layered silicates

Alkylpyridinium salts have been reported to decompose via Hofmann elimination, similarly to quaternary alkylammonium salts [50]. For instance, the degradation of hexadecylpyridinium bromide proceeds through an elimination reaction, resulting in the formation of pyridine and hexadecene [27].

$$C_5H_5N^+-(CH_2)_{15}-CH_3 + C_5H_5N + CH_2=CH-(CH_2)_{13}-CH_3 + BH \qquad (2.8)$$

The qualitative MS data obtained for the thermal decomposition of hexadecylpyridinium-modified MMT did not show any change in the mechanism of degradation of the organic compound after intercalation. However, for all organoclays investigated in that work, MMT modified by hexadecylpyridinium bromide or tributylhexadecylphosphonium bromide, 1,3-dioctadecylimidazolium bromide–modified clays, and Cloisite 20A (MMT containing dimethyldioctadecylammonium chloride), the number of species detected by MS during the degradation of the intercalating agent was larger than for the corresponding surfactants. In addition to the specific groups evolved from the degradation of the cationic head and to short hydrocarbons (C4–C7) evolved from the alkyl chains, various products with higher molecular weight were formed, for example, tropylium cations [characteristic species – $C_7H_7^+$ ($m/e = 91$)], neutral acetylene molecules C_2H_2 ($m/e = 26$), and linear or cyclic hydrocarbons with multiple unsaturation (fragments with $m/e = 67, 81,$ and 95). The variety of products detected during the thermal decomposition of the organoclays is associated with numerous decomposition events observed in the TG thermograms, and is certainly associated with the confinement of the intercalating agent within the silicate layers: high–molecular weight compounds produced during primary decomposition steps might be trapped between the clay platelets and undergo secondary reactions, such as chain scission, oxidation, dehydrogenation, or cyclization [15, 27].

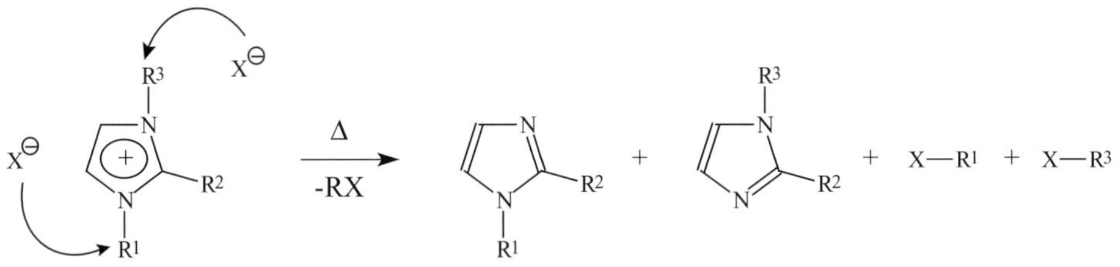

Scheme 2.7 Scheme of nucleophilic substitution reaction of an imidazolium compound [64].

2.3.4 Imidazolium salts and imidazolium-modified layered silicates

The application of imidazolium surfactants, which effectively modify interface in a number of polymer/clay systems and undergo degradation at temperatures significantly higher than ammonium compounds, provides an alternative to commonly applied ammonium modifiers [51, 52]. Highly thermally stable imidazolium surfactants have been applied for engineering and special polymers that were processed at high temperatures, such as poly(ethylene terephthalate) (PET) [53, 54], polycarbonate (PC) [55], acrylonitrile–butadiene–styrene terpolymer (ABS) [56], polypropylene (PP) [57–60], and poly(ethylene naphthalate) (PEN) [61]. It was confirmed that imidazolium ionic liquid, containing long alkyl chains, forms a stable complex with clays and undergoes only weak desorption, presumably owing to the longer alkyl side chain [62].

In comparison with ammonium and pyridinium salts, the thermal decomposition of alkylimidazolium salts is more difficult to predict because of the presence of two nitrogen atoms. Chan *et al.* [63] have studied the isothermal decomposition of 1,3-disubstituted imidazolium iodides in the temperature range 220–260 °C. The presence of alkylimidazoles and alkyl halides was detected in the decomposition products, indicating that the degradation proceeds mainly *via* a nucleophilic substitution reaction, most likely an S_N2 mechanism (Scheme 2.7).

For the sake of the degradation mechanism, it is clear that the thermal stability of imidazolium salts depends on the type of anions. In Awad *et al.*'s work, imidazolium salts took the order $PF_6^- > N(SO_2CF_3)^{2-} > BF_4^- > Cl^-$ with respect to thermal stability.

The nature of the alkyl substituents was found to rule the rate of cleavage for the N–C bonds [63, 64]. Methyl substitution in the 2 position (i.e., between the two N atoms) enhances the thermal stability. This may be due to the strong acidic character of the C-2 proton. It was observed that the thermal stability of imidazolium was also affected by the type of isomeric structure of the alkyl side group. This was evidenced by the observation that both 1-butyl-2,3-dimethylimidazolium tetrafluoroborate and 1-butyl-2,3-dimethylimidazolium hexafluorophosphate salts had higher onset decomposition temperatures than 1,2-dimethyl-3-isobutylimidazolium tetrafluoroborate and 1,2-dimethyl-3-isobutylimidazolium hexafluorophosphate salts. This reaction presumably proceeds via S_N1, as shown in Scheme 2.8.

Thermal stability of trialkylimidazolium salts decreases as the length of the alkyl group attached to the nitrogen increases [58, 64]. PS/imidazolium–MMT nanocomposites,

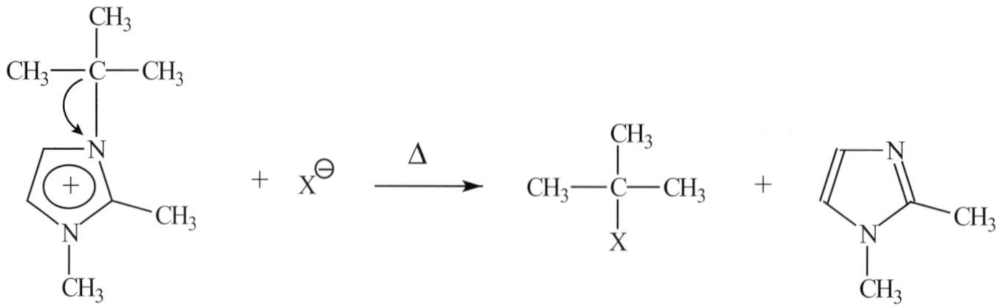

Scheme 2.8 $S_N 1$ reaction of an imidazolium compound [64].

however, showed a very good linear increase of the activation energy of degradation as a function of the number of carbon atoms in the surfactant alkyl chain [65].

The surrounding gas had a distinct effect on the thermal stability of tetrafluoroborate and hexafluorophosphate imidazolium salts, whereas the presence of oxygen appeared to have no effect on the decomposition temperatures of halide salts [64]. It was envisaged that the activation energy required for the thermal decomposition of the halide salts is lower than for their oxidative decomposition.

The incorporation of imidazolium chloride salts between clay layers drastically increased their initial decomposition temperature, whereas no significant improvement was observed in the thermal stability of the intercalated tetrafluoroborate and hexafluorophosphate salts [64]. Apparently this effect was connected with the removal of the anion and depended on its nucleophilicity. Gilman and co-workers reported that the replacement of sodium in natural MMT by 1-alkyl-2,3-dimethylimidazolium salts yielded organophilic MMT with a 100 °C improvement in thermal stability (in N_2) as compared to the alkylammonium-treated MMT [66]. In Langat *et al.*'s work the initial temperature of decomposition for clays modified with 1-hexadecyl-3-(10-hydroxydecyl)-2-methylimidazolium and 1-hexadecyl-2-methylimidazolium chloride was also very high – around 370 °C [67].

Awad *et al.* [64] did not notice any significant difference between the products evolved during the degradation of alkylimidazolium salts and of the montmorillonite clays modified by these salts. Although the end products of the degradation reactions for both groups of compounds were similar, the reaction pathways might be different because of the limitations that the interlayer space imposes on the orientation and packing arrangement of the surfactant molecules.

2.3.5 Silane-grafted organoclays

Another approach to enhancing the thermal stability of organoclays is to covalently graft the organic pendant group onto the backbone of the layered silicate [68–71]. The covalent bonding is expected to be much more stable than the ionic interactions in the organoclay, and therefore the thermal stability of the layered organosilicate should be significantly improved [72].

Silylation can serve as an example of the modification of montmorillonite, which offers covalent bonding of organic molecules to MMT layers. TG studies showed that the decomposition temperature of the organosilicate with a covalently grafted silylated chain can be as high as 430 °C for 7-octenylsiloxysilicate and *n*-octylsiloxysilicate powders [73]. Elsewhere, the hybrid material obtained by treating fully exchanged dimethyldistearylammonium–MMT with an alkylsilane compound gave stable siloxane bridges on the clay mineral surface, resulting in an organoclay of improved organophilicity and thermal stability [74]. The onset temperature for the alkylsilane-grafted ammonium-exchanged MMT was 207 °C and it was considerably higher than that for the equivalent (ungrafted) octadecyltrimethylammonium-modified clay (158 °C) [49, 75].

2.3.6 Surfactant mixtures as layered silicate modifiers

A significant increase of the initial temperature of decomposition was achieved for clay modified with a typical cationic ammonium surfactant and sodium dodecyl sulfonate [76]. This was attributed to coexistence of the two oppositely charged surfactants, which could form microstructures such as complexes, mixed micelles, bilayers, and tubules. The strong attractive interaction between oppositely charged head groups at the interface and hydrogen bonding caused an increase of the decomposition temperature of the corresponding surfactants.

2.4 Role of catalytic activity of clay minerals in the mechanism of organically modified layered silicate degradation

The enhanced degradation of polymer in the presence of OMMT is frequently explained in terms of the catalytic activity of acidic sites formed because of the Hofmann degradation of onium compounds [77, 78]. However, it should be emphasized that not only organomodifiers and products of their decomposition appear to have a catalytic effect on polymer degradation, but also acidic sites inherently present on the mineral surface and inclusions of transition metals should be considered as active catalytic sites – the latter could significantly enhance the degradation of the polymer matrix and all organic compounds in the nanocomposite material.

Several studies have reported that hydroxyl ends close to the surface show a typical Brønsted acid character (H^+ donor), whereas interlamellar hydroxyls display both Brønsted and Lewis (electron acceptor) acid characters [79–81]. Brønsted acidity essentially results from the dissociation of H_2O molecules, induced by the exchangeable cations with which they are associated [82]. In contrast, the dominant Lewis acid species are usually coordinatively unsaturated Al^{3+}, either exposed at the edges of the MMT lattice layers, or formed by dehydration of adsorbed oligomeric oxyaluminum cations. In the presence of water these Lewis acid species are hydrated and their acidity is masked. The relative amounts of these sites can vary with temperature [83]. At temperatures between 300 and 500 °C, Lewis acid sites are more stable than Brønsted ones.

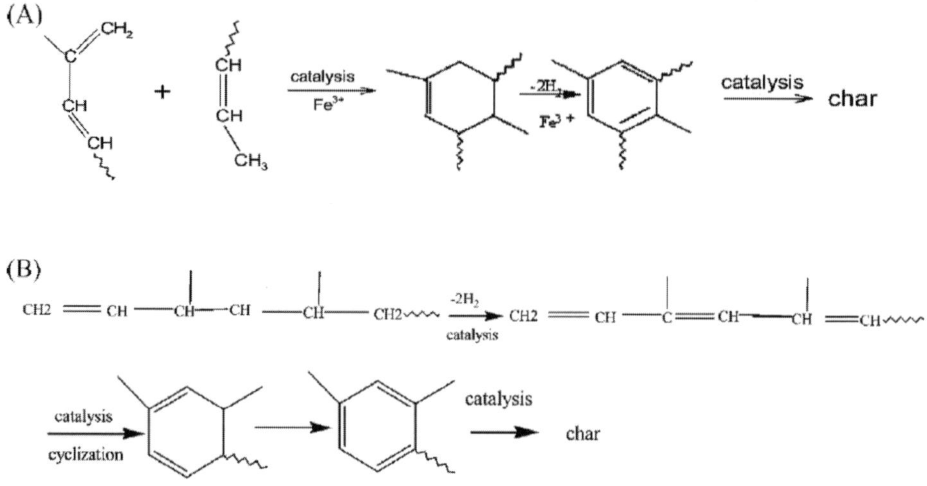

(A)

(B)

Scheme 2.9 Cyclization of olefins and catalyzed carbonization. Reprinted from Ref. [88], with permission from Wiley.

The catalytic effect of the natural clay was more evident for the degradation of phosphonium salt than for ammonium compounds [43]. The difference in the onset decomposition temperature of the phosphonium halides and clays modified by these surfactant reached 90 °C [43, 47].

Moreover, several works demonstrated that the metal ions, such as Fe^{3+} or Cu^{2+}, naturally present in clay minerals act as catalytic sites enhancing the degradation reactions [84, 85]. It was suggested that Fe^{3+} cations facilitated decomposition of hydroperoxides through a reversible oxidative–reductive catalytic process between Fe^{3+} and Fe^{2+} [86]. The reasons may be that the cations of some transition metals (e.g., Fe^{3+}) could lead to the formation of macroradicals, promote molecular cross linking, and increase the charred residues. The effect was attributed to the ability of the cation to form complexes in which the metal atoms were coordinately bonded, for example, to the carbonyl oxygen atom of the amide group [87]. From the EDX analysis, it was found that the content of naturally occurring iron in commercial Cloisite was at the 3.5 wt% level [17]. Kong *et al.* proposed a scheme of successive reactions leading to the formation of carbonaceous char in the course of degradation of hydrocarbon polymers in the presence of Fe-rich OMMT (Scheme 2.9) [88]. It is likely that alkyl tails of organic modifiers could undergo such reactions in the interlayer spaces of organoclays.

The Fe^{3+} ion was also suggested to act as the operative site for radical trapping during the thermal degradation of the nanocomposite fibers [89].

The presence of metal species in the montmorillonite structure may catalytically enable the oxidative cleavage of alkene substituents in alkylammonium compounds to produce aldehydes at elevated temperatures [15, 18].

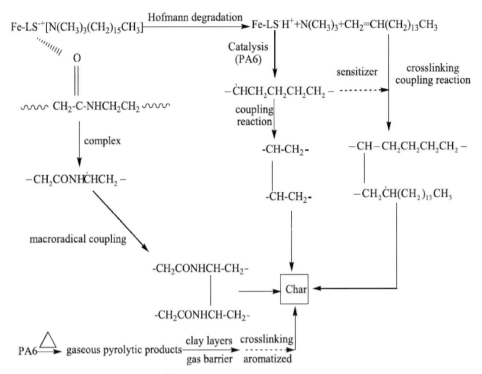

Scheme 2.10 Illustration of the possible catalyzing carbonization mechanism for PA6 nanocomposite fibers. Reprinted from Ref. [84], with permission from Elsevier.

Several studies, focusing on the thermal stability of polymer/surfactant blends (without layered silicate), demonstrated lowering of the initial decomposition temperature and/or acceleration of mass loss, thus confirming that the ammonium compound itself or products of its decomposition act as catalytic agents. Bordes *et al.* [90] tested the thermomechanical degradation of a melt-mixed blend of poly(hydroxybutyrate) (PHB) and ammonium salt. All investigated ammonium compounds enhanced the decrease of molecular mass of the polymer, but the least detrimental was oleyl bis(2-hydroxyethyl)methylammonium chloride, which was thus suggested as the most appropriate modifier for PHB. It has also been observed that the presence of hydroxyvalerate comonomer made the polymer more sensitive to degradation regardless of the surfactant type. No doubt, the organomodifier itself and the products of its degradation have the potential to influence the polymer decomposition process. Notably, for some polymer-based nanocomposites, the presence of reactive products of organoclay degradation may have advantageous effects such as enhanced cross linking [30] or charring during burning [84]. In the work of Cai and co-workers [84], the presence of organoclay degradation products was indicated as an important component initiating new chemical paths in a complex charring process, which is crucial in fire retardant formulations (Scheme 2.10).

It was also observed that pristine sodium and organically modified clays showed catalytic activity in photooxidation reactions, which seemed to be independent of the type of organic modifier and its thermal stability (ammonium or imidazolium) [91]. This effect was ascribed to the activity of catalytic sites on the MMT surface that accept single electrons from donor molecules, forming free radicals in the polymer matrix.

Finally, it should be stressed that the high interface area in nanocomposite materials enhances the total catalytic activity of clays [92]. Indeed, the higher the dispersion degree achieved, the stronger effects on thermal and photooxidative stability of nanocomposite material were observed [91].

2.5 Parameters influencing the thermal stability of organically modified layered silicates

2.5.1 Excess organic treatment

The identification of free adsorbed surfactant in OMMT clearly advanced knowledge on the degradation mechanism of OMMT and had important implications for its manufacture. Xi *et al.* displayed changes in the degradation route of organoclays with different amounts of ammonium cations with respect to the total cation exchange capacity (CEC) of MMT [93, 94]. As the concentration of ammonium ions reached 0.6 CEC of clay, two fractions of surfactant molecules were formed, differing in the type of bonding to MMT surface. Part of the provided surfactant molecules adhere to surface sites *via* electrostatic interactions and part are adsorbed onto the montmorillonite surface. The bonding of the latter was much weaker, rendering it more susceptible to thermal degradation (Figure 2.7).

If the concentration increased further, a third fraction of surfactant occurred that was weakly attached by van der Waals forces to a layer of sorbed surfactant cations and underwent thermal degradation at temperatures very similar to that for pure surfactant [95, 96]. Additionally, Osman *et al.* found that the thermal stability of alkylammonium self-assembled monolayers on montmorillonite, estimated by the time needed for 5% isothermal mass loss, was high enough to enable compounding it with several commercially available polymers [97].

Similar decomposition profiles with characteristic steps corresponding to the different packing densities and adsorption states of surfactants were observed for imidazolium [57]- and phosphonium [48]-modified organoclays. They reflect different modes of adsorption of organic cations onto existing hydroxyl groups with an acid character in the MMT [98]. Generally, the molecules adsorbed onto the Brønsted acid groups were thought to be less thermally resistive than those adsorbed onto the Lewis acid groups.

Excess organic treatment, which may occur in the typical synthesis of an organoclay, can remove or weaken some of the thermal, flammability, and mechanical property benefits provided by the nanocomposite structure and morphology. Adsorbed surfactant residue undergoes separate degradation at lower temperatures than ionically bonded residue. Therefore, repeated washing of OMMT precipitate was most frequently applied with the purpose of removing excess surfactant, and the reduction of the DTG degradation peak could

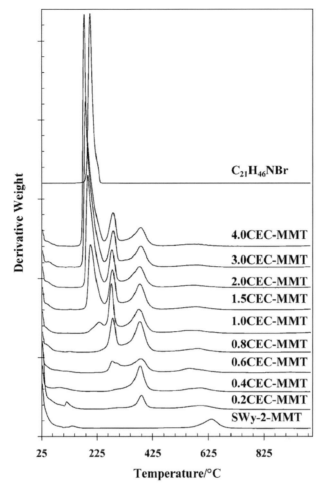

Figure 2.7 High-resolution DTG analysis of montmorillonite, surfactant, and surfactant–montmorillonite hybrids. Reprinted from Ref. [93], with permission from Elsevier.

serve as a measure of washing efficiency. However, despite repeated rinsing, some physically adsorbed surfactant can still remain in OMMT [19]. High efficiency of washing was observed for surfactants containing only alkyl tails, to which a water/ethanol solution was applied. A smaller portion of adsorbed surfactant was removed by washing with water alone and in the case of ammonium with benzyl substituent. Moreover, cations with longer alkyl chains, as well as with benzyl substituent, resisted the washing process much better [64]. Elsewhere, it has been found that washing with toluene is not effective, but ethanol removes a considerable amount of excess surfactant [26]. Davis and co-workers applied a washing procedure to remove bromide anions and found the efficiency to be solvent-dependent; sequential extraction, first with ethanol, then with tetrahydrofuran, gave

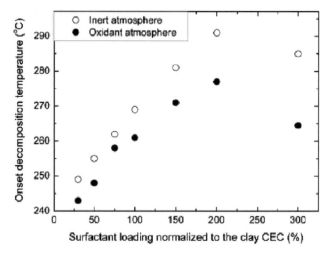

Figure 2.8 Effect of clay modification degree on the onset decomposition temperature (estimated at 2% weight loss) of PLLA hybrids. Reprinted from Ref. [102], with permission from Wiley.

the best results [99]. Morgan and Harris [100] and Bertini *et al.* [26] applied Soxhlet extraction to organoclays after synthesis and detected a decrease in the organic content of OLS and in interlayer distance. This operation slightly altered the thermal characteristics of OLS and the resulting polypropylene nanocomposites. Moreover, it also suppressed the plasticization effect, because nanocomposites comprising PP and extracted OLS gained higher values of modulus [100].

Washing the organoclay with alcohol proved to be an effective way to remove the excess surfactant from the clay galleries, but not the by-product of the ion-exchange process – NaCl [101]. It might be a source of chlorine in washed organoclay and contribute to the formation of alkyl chlorides.

The enlargement of the surfactant molecule seemed to render adsorption onto the interlayer more difficult [13]. The second reason for a low adsorption level might be the strong electrostatic interaction of the polar head group of the gemini surfactant – more surfactants are inclined to be adsorbed onto the external surface of the clay.

When analyzing the thermal stability of OMMT with different surfactant loadings, one may expect that the nanocomposites obtained with an excess of surfactant, with respect to the CEC of the clay, will not show enhanced thermal stability as compared with those having equivalent or even lower amounts of modifier. Unexpectedly, it has been reported that a moderate excess of surfactant was beneficial for poly(L-lactic acid) in terms of the degree of dispersion of the nanofiller, leading to better thermal and mechanical properties – Figure 2.8 [102]. An ideal balance between thermal and mechanical properties was obtained at surfactant quantity 1.5 times the clay CEC.

Moreover, an advantageous effect of excess surfactant on nanostructure formation has been reported in the literature [103–105]. Overexchange of the surfactant was even found necessary for intercalation of extruded PS into the silicate sheets [100]. Contrary to these reports, several findings presented difficulties in dispersing OMMT with excess modifier [106–109]. Surprisingly, in the experiment of VanderHart *et al.*, extensive decomposition of the organic ammonium cation did not result in poor mixing of clay in the PA-6 matrix; in fact, as judged by NMR, nanocomposites with the best dispersion of MMT also had the most profoundly degraded organic modifier [110]. The extent of surfactant degradation seemed not to be influenced by the total organoclay content in some nanocomposites, for example, those based on PE matrix [40].

Generally, divergent results were obtained even when the same type of polymer was tested. The exact effect of the amphiphilic cation concentration on the final nanocomposite structure and properties still remains unclear. However, it should be stressed that thermal stability of the organic modifier and nanostructure formation are mutually dependent parameters, and both are influenced by the type and concentration of organic modifiers of MMT [111, 112].

2.5.2 Purity of the mineral

The work by García-López and co-workers showed clearly that the thermal stability of OLS depends on the purity of the mineral [107]. The presence of other minerals naturally occurring in the bentonite deposits apart from montmorillonite, such as dolomite, illite, quartz, calcite, and plagioclase, could result in a greater quantity of surfactant being physically adsorbed onto the mineral surface and, due to weak links to the clay surfaces, undergoing thermal degradation at lower temperatures than the intercalated surfactant. Weight loss in the range of 200–250 °C, attributed to the adsorbed surfactant, was 22% in modified bentonite and only 8% in modified montmorillonite. It was found that the thermal treatment of modified clays at 300 °C under nitrogen eliminated the free surfactant, because the first and the second stage of mass loss almost disappeared from the TG profiles of thermally modified bentonite (sample BT-2 ashed in TGA 300 °C in Figure 2.9).

2.5.3 Chemical constitution of surfactant

The surfactant structural features were found to affect nanocomposite morphology and properties significantly. For example, in the MMT organophilization process for high–molecular weight PA-6-based nanocomposites, decreasing the number of long alkyl tails from two to one tallow, using methyl rather than hydroxyethyl groups, and using an equivalent amount of surfactant, as opposed to adding excess, lead to greater extent of silicate platelet exfoliation, increased moduli, higher yield strengths, and lower elongation at breaks [113]. These trends were opposite to what has been seen in LDPE, LLDPE and poly(ethylene-co-methacrylic acid)-based nanocomposites [105, 114]. On the other hand, no significant differences were found by Araújo *et al.* in the mechanical properties and crystallization parameters of PP/clay nanocomposites containing different types of quaternary ammonium salts [115]. Similarly, Cui *et al.* concluded that the differences in the

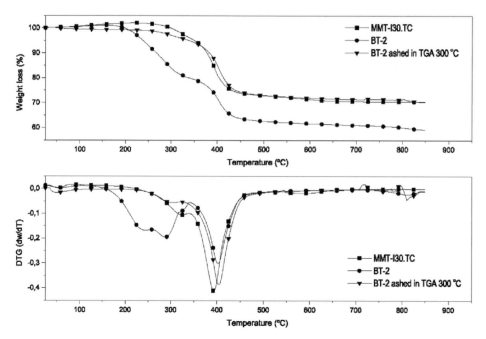

Figure 2.9 TG and DTG profiles of bentonites modified with octadecylammonium chloride: laboratory synthesized (BT-2), commercial grade (MMT-I30.TC), and calcined BT-2 bentonite. Reprinted
from Ref. [107], with permission from Elsevier.

thermal stability of organoclays do not appear to have a significant effect on the morphology
and properties of the nanocomposites formed [101].

The efficiency of a surfactant in platelet exfoliation and interfacial bonding should
be considered as the main criterion for its selection. However, it is advisable to select
a surfactant with respect to its thermal stability a well. The number and length of alkyl
chains, type of counterion, and class of amine (or ammonium salt) – primary, secondary,
or tertiary – were considered important parameters influencing the thermal stability of
nanocomposites.

2.5.3.1 Number and length of alkyl chains

Several research works reported that the thermal stability of organoclays increases with the
number of long alkyl tails in the ammonium molecule (Figure 2.10) [40, 97].

Similar results for isothermal degradation of different alkylammonium-modified clays
were obtained by Cui *et al.* [16]. However, the authors noted that isothermal percentage
mass loss data for samples with significantly different molecular weights do not precisely
reflect the extent of degradation (conversion). Comparable stability was displayed when
mass loss was related to moles of surfactant surfactants.

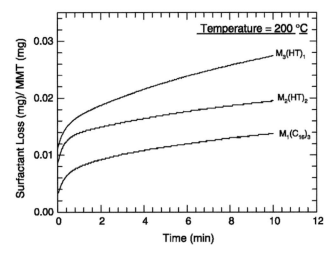

Figure 2.10 Isothermal TG results showing the mass loss for $M_3(HT)_1$, $M_2(HT)_2$, and $M_1(C_{16})_3$ organoclays at 200 °C under nitrogen. Reprinted from Ref. [40], with permission from Elsevier.

It has also been observed that the onset temperature of degradation of epoxy nanocomposites was lowered more strongly by ammonium salts with longer alkyl chains than by compounds with shorter alkyl substituents [116].

2.5.3.2 Presence of aromatic structures

The substituents of the ammonium cations may also influence the thermal stability of neat surfactants as well as of corresponding organoclays. In general, methyl and benzyl groups in these materials were susceptible at elevated temperatures to nucleophilic attack by the halide anions [16, 30].

On the other hand, clays modified with salts of aromatic amines, in which the N atom is directly linked to aromatic ring, exhibited higher thermal stability then those modified with alkylammonium compounds [117, 118]. The use of a combination of aromatic and aliphatic ammonium compounds was an effective strategy for synthesis of clays with sufficient thermal stability for thermoset polyimide resin (PI) [117]. In that case, the aromatic component of OLS provided higher thermal stability and the aliphatic component promoted intercalation. The benefits of combining aromatic and aliphatic surfactants to treat clay are also reported by other authors [119]. It was shown that clay treated with a 1:1 molar mixture of methylene dianiline (MDA) and dodecylamine resulted in an improvement in thermooxidative stability of PI-based nanocomposites.

Contrary to this finding, the intercalated aliphatic phosphonium salt (tetraoctylphosphonium) was thermally more stable than the aromatic compound (triphenylvinylbenzylphosphonium) [82]. This was attributed to the steric hindrance of the aromatic groups bonded to the phosphorus and the lower energy required for large aromatic phosphonium cations to be eliminated from the MMT surface. Moreover, alkylphosphonium modifiers, investigated in this work, were more effective in increasing the basal spacing of OMMT and further promoting the exfoliation of montmorillonite particles in natural rubber.

2.5.3.3 *Presence of other functional groups*

Recently, more complex (and efficient) nanocomposite systems with reactive components have been investigated. The main idea is to introduce new reactive components in order to achieve better material performance and/or to bring synergistic effects to nanocomposites [119–123]. The chemical reaction between polymer and reactive clay treatments is a strong driving force for the intercalation/delamination of MMT, resulting in more efficient nano-structure formation. It is expected that a stronger influence of nanoadditive on polymer properties is achieved when a higher degree of dispersion is obtained. Improvement in mechanical and thermal properties is also due to strong tethering between the clay and the surrounding polymer, which originated from chemical reactions.

2.5.3.4 *Type of anion*

The anions associated with the ammonium cations play a significant role in the thermal stability of the surfactant salts and the surfactants in organoclay. Anions that are weak nucleophiles (*e.g.*, $MeSO_4^-$) can result in more thermally stable surfactants [101]. As-received organoclay made from alkylammonium methyl sulfide ($M_2(HT)_2^+\ MeSO_4^-$) is more stable than as-received organoclay made from alkylammonium chloride if the surfactant loading is the same (and relatively high), but the differences between them tend to diminish as the surfactant loading decreases. The unbound surfactant and the associated anion in the as-received organoclay cause the differences. After removal of the free surfactant and the associated anion, either Cl^- or $MeSO_4^+$, the two organoclays had exactly the same thermal behavior.

2.6 Conclusions

Proper understanding of mechanisms governing the degradation of organoclay and the identification of products evolved during this process is necessary for engineering the optimal composition of nanocomposites with regard to the polymer type and its specific application. A survey of recent literature on the OLS degradation mechanism provides a great deal of information, which can be useful in further technological development of polymer/clay nanocomposites.

It is likely that organic molecules may be located in and associated to the MMT galleries in at least three manners: ionically bonded, physically adsorbed in clay galleries, and/or confined in the interlayer environment. The intercalated organic cations counterbalancing the negative charge of MMT layers displayed the highest onset temperatures of degradation, significantly higher than that measured for pristine surfactant, whereas free molecules, caught in OLS, start to decompose at temperatures close to that of neat surfactant. Therefore, it is advisable to avoid the use of excessive amounts of surfactant in relation to the CEC of MMT during the synthesis of OLS, and to subject prepared OLS to careful washing. However, an advantageous effect of excess surfactant on nanostructure formation has been reported in literature.

Two mechanisms of degradation are proposed for organoclays with ammonium surfactants. First is the elimination-type mechanism (Hofmann elimination), which is assumed to be primary degradation pathway if the salt anions are completely removed from the organoclay, and second is nucleophilic substitution with an attack of residual anions.

For surfactants that can decompose according to a nucleophilic substitution mechanism, the type of anion and its nucleophilicity were found to influence the thermal stability of salts and surfactant–clay complexes strongly. In the presence of halide anions, which are strong nucleophilic agents, the degradation path was activated at a lower temperature than it was observed at for PF_6^-, $N(SO_2CF_3)^{2-}$, or BF_4 anions, for example.

The presence of free amines and alkenes, which are predominant products of OLS degradation and capable of reacting with numerous polymer matrices, may significantly decrease properties of nanocomposites, for example, of the polyurethane/clay system, as they undergo transurethanization reactions diminishing the molecular weight of the polymer. Moreover, the presence of leachable low–molecular weight organic compounds, released during processing, may prevent the use of nanocomposite materials in medical or food contact applications. On the other hand, the olefins and amines generated from the OMMT degradation may act as sensitizers that promote the coupling and cross-linking reactions and improve the charred residue yield. The latter effect plays an important role in lowering nanocomposite flammability.

Several organic compounds, especially imidazolium, pyridinium, and phosphonium salts, were shown to have noticeably higher thermal stability, and they could be applied as compatibilizers for polymer/clay nanocomposites, providing a good level of dispersion and improved properties. Additional functionalization of surfactants is most often aimed at improvement of miscibility or synergistic enhancement of nanocomposite properties; however, any change in the chemical constitution of an organomodifier involves simultaneous alteration of its thermal characteristic.

It seems that further research on polymer–clay nanocomposites would focus on the synthesis of organic modifiers with enhanced thermal stability and providing additional functionalities for formation of interfacial bonds and interaction through, for example, graft polymerization.

References

1. S. Kar, P. K. Maji, and A. K. Bhowmick, Chlorinated polyethylene nanocomposites: Thermal and mechanical behavior. *Journal of Materials Science*, 45 (2010), 64–73.
2. D. Carastan and N. Demarquette, Microstructure of nanocomposites of styrenic polymers. *Macromolecular Symposia*, 233 (2006), 152–60.
3. G. Liu, L. Zhang, D. Zhao, and X. Qu, Bulk polymerization of styrene in the presence of organomodified montmorillonite. *Journal of Applied Polymer Science*, 96 (2005), 1146–52.

56 *Mechanism of thermal degradation*

4. D. Dharaiya and S. C. Jana, Thermal decomposition of alkyl ammonium ions and its effects on surface polarity of organically treated nanoclay. *Polymer*, 46 (2005), 10139–47.
5. P. J. Yoon, D. L. Hunter, and D. R. Paul, Polycarbonate nanocomposites. Part 1. Effect of organoclay structure on morphology and properties. *Polymer*, 44 (2003), 5323–39.
6. O. Monticelli, Z. Musina, A. Frache, F. Bellucci, G. Camino, and S. Russo, Influence of compatibilizer degradation on formation and properties of PA6/organoclay nanocomposites. *Polymer Degradation and Stability*, 92 (2007), 370–78.
7. P. J. Yoon, D. L. Hunter, and D. R. Paul, Polycarbonate nanocomposites. Part 2. Degradation and color formation. *Polymer*, 44 (2003), 5341–54.
8. T. D. Fornes, P. J. Yoon, and D. R. Paul, Polymer matrix degradation and color formation in melt processed nylon 6/clay nanocomposites. *Polymer*, 44 (2003), 7545–56.
9. J. H. Park and S. C. Jana, Adverse effects of thermal dissociation of alkyl ammonium ions on nanoclay exfoliation in epoxy–clay systems. *Polymer*, 45 (2004), 7673–9.
10. L. Le Pluart, J. Duchet, H. Sautereau, and J. Gerard, Surface modifications of montmorillonite for tailored interfaces in nanocomposites. *Journal of Adhesion*, 78 (2002), 645–62.
11. Z. Gao, W. Xie, J. M. Hwu, L. Wells, and W.-P. Pan, The characterization of organic modified montmorillonite and its filled PMMA nanocomposite. *Journal of Thermal Analysis and Calorimetry*, 64 (2001), 467–75.
12. G. Edwards, P. Halley, G. Kerven, and D. Martin, Thermal stability analysis of organo-silicates, using solid phase microextraction techniques. *Thermochimica Acta*, 429 (2005), 13–18.
13. R. Ni, Y. Huang, and C. Yao, Thermogravimetric analysis of organoclays intercalated with the gemini surfactants. *Journal of Thermal Analysis and Calorimetry*, 96 (2009), 943–7.
14. R. T. Morrison and R. N. Boyd, *Organic Chemistry*, 6th ed. (London: Prentice-Hall International, 1992).
15. W. Xie, Z. Gao, W. P. Pan, D. Hunter, A. Singh, and R. Vaia, Thermal degradation chemistry of alkyl quaternary ammonium montmorillonite. *Chemistry of Materials*, 13 (2001), 2979–90.
16. L. Cui, D. M. Khramov, C. W. Bielawski, D. L. Hunter, P. J. Yoon, and D. R. Paul, Effect of organoclay purity and degradation on nanocomposite performance. Part 1. Surfactant degradation. *Polymer*, 49 (2008), 3751–61.
17. J. M. Cervantes-Uc, J. V. Cauich-Rodríguez, H. Vázquez-Torres, L. F. Garfias-Mesías, and D. R. Paul, Thermal degradation of commercially available organoclays studied by TGA-FTIR. *Thermochimica Acta*, 457 (2007), 92–102.
18. W. Xie, Z. Gao, K. Liu, W. P. Pan, R. Vaia, D. Hunter, and A. Singh, Thermal characterization of organically modified montmorillonite. *Thermochimica Acta*, 367–8 (2001), 339–50.
19. A. Vazquez, M. López, G. Kortaberria, L. Martín, and I. Mondragon, Modification of montmorillonite with cationic surfactants: Thermal and chemical analysis including CEC determination. *Applied Clay Science*, 41 (2008), 24–36.
20. J. Zhu, H. He, J. Guo, D. Yang, and X. Xie, Arrangement models of alkylammonium cations in the interlayer of HDTMA pillared montmorillonites. *Chinese Science Bulletin*, 48 (2003), 368–72.

21. G. P. Gillman and E. A. Sumpter, Modification to the compulsive exchange method for measuring exchange characteristics of soils. *Australian Journal of Soil Research*, 24 (1986), 61–6.
22. M. Gelfer, C. Burger, A. Fadeev, I. Sics, B. Chu, B. S. Hsiao, A. Heintz, K. Kojo, S. L. Hsu, M. Si, and A. Rafailovich, Thermally induced phase transitions and morphological changes in organoclays. *Langmuir*, 20 (2004), 3746–58.
23. F. Kooli and P. C. M. M. Magusin, Adsorption of cetyltrimethyl ammonium ions on an acid-activated smectite and their thermal stability. *Clay Minerals*, 40 (2005), 233–43.
24. J. R. Dyer, *Applications of Absorption Spectroscopy of Organic Compounds*, 1st ed. (Upper Saddle River, NJ: Prentice Hall, 1965).
25. H. P. He, Z. Ding, J. X. Zhu, P. Yuan, Y. F. Xi, D. Yang, and R. L. Frost, Thermal characterization of surfactant-modified montmorillonites. *Clays and Clay Minerals*, 53 (2005), 287–93.
26. F. Bertini, M. Canetti, G. Leone, and I. Tritto, Thermal behavior and pyrolysis products of modified organo-layered silicates as intermediates for in situ polymerization. *Journal of Analytical and Applied Pyrolysis*, 86 (2009), 74–81.
27. K. Stoeffler, P. G. Lafleur, and J. Denault, Thermal decomposition of various alkyl onium organoclays: Effect on polyethylene terephthalate nanocomposites' properties. *Polymer Degradation and Stability*, 93 (2008), 1332–50.
28. N. T. Dintcheva, S. Al-Malaika, and F. P. La Mantia, Effect of extrusion and photo-oxidation on polyethylene/clay nanocomposites. *Polymer Degradation and Stability*, 94 (2009), 1571–88.
29. J. W. Alencar, P. B. Alves, and A. A. Craveiro, Pyrolysis of tropical vegetable oils. *Journal of Agricultural and Food Chemistry*, 31 (1983), 1268–70.
30. M. Galimberti, M. Martino, M. Guenzi, G. Leonardi, and A. Citterio, Thermal stability of ammonium salts as compatibilizers in polymer/layered silicate nanocomposites. *e-Polymers*, (2009), 056.
31. C. Cody, B. Campbell, A. Chiavoni, and E. Magauran, *Organoclay Compositions*, U.S. Patent 5,634,969 (1997).
32. C. A. Cody and S. J. Kemnetz, *Improved Organophilic Clay Gellant and Processes for Preparing Organophilic Clay Gellants*, EU Patent Application 0312988 (1989).
33. W. S. Mardis and C. Malcolm, *Organophilic Organic-Clay Complexes*, UK Patent GB 2107693 (1983).
34. F. Bellucci, G. Camino, A. Frache, and A. Sarra, Catalytic charring–volatilization competition in organoclay nanocomposites. *Polymer Degradation and Stability*, 92 (2007), 425–36.
35. H. L. Qin, Z. G. Zhang, M. Feng, F. L. Gong, S. M. Zhang, and M. S. Yang, The influence of interlayer cations on the photo-oxidative degradation of polyethylene/montmorillonite composites. *Journal of Polymer Science Part B: Polymer Physics*, 42 (2004), 3006–12.
36. S. Yariv, The role of charcoal on DTA curves of organo-clay complexes: An overview. *Applied Clay Science*, 24 (2004), 225–36.
37. R. Scaffaro, M. C. Mistretta, and F. P. La Mantia, Compatibilized polyamide 6/polyethylene blend–clay nanocomposites: Effect of the degradation and stabilization of the clay modifier. *Polymer Degradation and Stability*, 93 (2008), 1267–74.
38. K. Pielichowski and J. Njuguna, *Thermal Degradation of Polymeric Materials* (Shawbury: Rapra, 2005).

39. K. Pielichowski and A. Leszczyńska, TG-FTIR study of the thermal degradation of polyoxymethylene (POM)/thermoplastic polyurethane (TPU) blends. *Journal of Thermal Analysis and Calorimetry*, 78 (2004), 631–7.
40. R. K. Shah and D. R. Paul, Organoclay degradation in melt processed polyethylene nanocomposites. *Polymer*, 47 (2006), 4075–84.
41. T. Takekoshi, F. F. Khouri, J. R. Campbell, T. C. Jordan, and K. H. Dai, *Layered Minerals and Compositions Comprising the Same*, U.S. Patent 5,707,439 (1998).
42. R. F. Hudson, *Structure and Mechanism in Organo-Phosphorus Chemistry* (New York: Academic Press, 1965).
43. W. Xie, R. Xie, W.-P. Pan, D. Hunter, B. Koene, L.-S. Tan, and R. Vaia, Thermal stability of quaternary phosphonium modified montmorillonites. *Chemistry of Materials*, 14 (2002), 4837–45.
44. R. C. Weast, *Handbook of Chemistry and Physics*, 65th ed. (Boca Raton, FL: CRC Press, 1984).
45. S. Semenzato, A. Lorenzetti, M. Modesti, E. Ugel, D. Hrelja, S. Besco, R. A. Michelin, A. Sassi, G. Facchin, F. Zorzi, and R. Bertani, A novel phosphorus polyurethane FOAM/montmorillonite nanocomposite: Preparation, characterization and thermal behaviour. *Applied Clay Science*, 44 (2009), 35–42.
46. J. U. Calderon, B. Lennox, and M. R. Kamal, Thermally stable phosphonium–montmorillonite organoclays. *Applied Clay Science*, 40 (2008), 90–98.
47. H. A. Patel, R. S. Somani, H. C. Bajaj, and R. V. Jasra, Preparation and characterization of phosphonium montmorillonite, with enhanced thermal stability. *Applied Clay Science*, 35 (2007), 194–200.
48. T. U. Patro, D. V. Khakhar, and A. Misra, Phosphonium-based layered silicate-poly(ethylene terephthalate) nanocomposites: Stability, thermal and mechanical properties. *Journal of Applied Polymer Science*, 113 (2009), 1720–32.
49. C. B. Hedley, G. Yuan, and B. K. G. Theng, Thermal analysis of montmorillonites modified with quaternary phosphonium and ammonium surfactants. *Applied Clay Science*, 35 (2007), 180–88.
50. L. Moens, D. M. Blake, D. L. Rudnicki, and M. J. Hale, Advanced thermal storage fluids for solar parabolic trough systems. *Journal of Solar Energy Engineering*, 125 (2003), 112–16.
51. C. L. Toh, L. F. Xi, S. K. Lau, K. P. Pramoda, Y. C. Chua, and X. H. Lu, Packing behaviors of structurally different polyhedral oligomeric silsesquioxane–imidazolium surfactants in clay. *Journal of Physical Chemistry B*, 114 (2010), 207–14.
52. H. L. Ngo, K. Le Compte, L. Hargens, and A. B. Mc Ewen, Thermal properties of imidazolium ionic liquids. *Thermochimica Acta*, 357–8 (2000), 97–102.
53. S. A. Monemian, V. Goodarzi, P. Zahedi, and M. T. Angaji, PET/imidazolium-based OMMT nanocomposites via in situ polymerization: Morphological, thermal, and nonisothermal crystallization studies. *Advances in Polymer Technology*, 26 (2007), 247–57.
54. C. H. Davis, L. J. Mathias, J. W. Gilman, J. R. Schiraldi, P. Trulove, and T. E. Sutto, Effects of melt-processing conditions on the quality of poly(ethylene terephthalate) montmorillonite clay nanocomposites. *Journal of Polymer Science Part B: Polymer Physics*, 40 (2002), 2661–6.
55. L. Cui, J. E. Bara, Y. Brun, Y. Yoo, P. J. Yoon, and D. R. Paul, Polyamide- and polycarbonate-based nanocomposites prepared from thermally stable imidazolium organoclay. *Polymer*, 50 (2009), 2492–502.

56. M. Modesti, S. Besco, A. Lorenzetti, M. Zammarano, V. Causin, C. Marega, J. W. Gilman, D. M. Fox, P. C. Trulove, H. C. De Long, and P. H. Maupin, Imidazolium–modified clay-based ABS nanocomposites: A comparison between melt-blending and solution-sonication processes. *Polymers for Advanced Technologies*, 19 (2008), 1576–83.

57. Y. S. Ding, X. M. Zhang, R. Y. Xiong, S. Y. Wu, M. Zha, and H. O. Tang, Effects of montmorillonite interlayer micro-circumstance on the PP melting intercalation. *European Polymer Journal*, 44 (2008), 24–31.

58. N. H. Kim, S. V. Malhotra, and M. Xanthos, Modification of cationic nanoclays with ionic liquids. *Microporous and Mesoporous Materials*, 96 (2006), 29–35.

59. A. He, H. Hu, Y. Huang, J.-Y. Dong, and C. C. Han, Isotactic Poly(propylene)/monoalkylimidazolium-modified montmorillonite nanocomposites: Preparation by intercalative polymerization and thermal stability study. *Macromolecular Rapid Communication*, 25 (2004), 2008–13.

60. V. Mittal, Gas permeation and mechanical properties of polypropylene nanocomposites with thermally-stable imidazolium modified clay. *European Polymer Journal*, 43 (2007), 3727–36.

61. Y. C. Chua, X. H. Lu, and Y. Wan, Polymorphism behavior of poly(ethylene naphthalate)/clay nanocomposites. *Journal of Polymer Science Part B: Polymer Physics*, 44 (2006), 1040–49.

62. M. Matzke, K. Thiele, A. Muller, and J. Filser, Sorption and desorption of imidazolium based ionic liquids in different soil types. *Chemosphere*, 74 (2009), 568–74.

63. B. K. M. Chan, N.-H. Chang, and M. R. Grimmett, The synthesis and thermolysis of imidazole quaternary salts. *Australian Journal of Chemistry*, 30 (1977), 2005–13.

64. W. H. Awad, J. W. Gilman, M. Nyden, R. H. Harris, T. E. Sutto, J. Callahan, P. C. Trulove, H. C. DeLong, and D. M. Fox, Thermal degradation studies of alkylimidazolium salts and their application in nanocomposites. *Thermochimica Acta*, 409 (2004), 3–11.

65. L. Abate, I. Blanco, F. A. Bottino, G. Di Pasquale, E. Fabbri, A. Orestano, and A. Pollicino, Kinetic study of the thermal degradation of PS/MMT nanocomposites prepared with imidazolium surfactants. *Journal of Thermal Analysis and Calorimetry*, 91 (2008), 681–6.

66. J. W. Gilman, W. H. Awad, R. D. Davis, J. Shields, R. H. Harris Jr., C. Davis, A. B. Morgan, T. E. Sutto, J. Callahan, P. C. Trulove, and H. C. Delong, Polymer/layered silicate nanocomposites from thermally stable trialkylimidazoliumtreated montmorillonite. *Chemistry of Materials*, 14 (2002), 3776–85.

67. J. Langat, S. Bellayer, P. Hudrlik, A. Hudrlik, P. H. Maupin, J. W. Gilman, Sr., and D. Raghavan, Synthesis of imidazolium salts and their application in epoxy montmorillonite nanocomposites. *Polymer*, 47 (2006), 6698–709.

68. E. Ruiz-Hitzky and J. M. Rojo, Intracrystalline grafting on layer silicic acids. *Nature*, 287 (1980), 28.

69. C. Chen, D. Katsoulis, and M. E. Kenney, *Sheet and Tube Organosilicon Polymers*, U.S. Patent 5627241 (1996).

70. C. Chen, D. Katsoulis, and M. E. Kenney, *Silicone Gels and Composites from Sheet and Tube Organofunctional Siloxane Polymers*, U.S. Patent 6013705 (2000).

71. T. C. Chao, D. Katsoulis, and M. E. Kenney, Synthesis and characterization of organosilicon sheet and tube polymers. *Chemistry of Materials*, 13 (2001), 4269.

72. W. S. Chow and S. S. Neoh, Dynamic mechanical, thermal, and morphological properties of silane-treated montmorillonite reinforced polycarbonate nanocomposites. *Journal of Applied Polymer Science*, 114 (2009), 3967–75.

73. C. Chen, D. Yebassa, and D. Raghavan, Synthesis, characterization, and mechanical properties evaluation of thermally stable apophyllite vinyl ester nanocomposites. *Polymers for Advanced Technologies*, 18 (2007), 574–81.

74. T. Abiko and M. Onikata. *Development of Highly Dispersed Nanocomposites Using Novel Modification of Bentonite Clays*. Presented at the Japan Society for Polymer Processing (JSPP) Symposium, Kanazawa, 3–4 November, 2003.

75. E. Ruiz-Hitzky and A. Van Meerbeek, Clay mineral and organoclay–polymer nanocomposites. In: *Handbook of Clay Science*, ed. F. Bergaya, B. K. G. Theng, and G. Lagal (Amsterdam: Elsevier, 2006), pp. 583–621.

76. D. Chen, J. X. Zhu, P. Yuan, S. J. Yang, T.-H. Chen, and H. P. He, Preparation and characterization of anion–cation surfactants modified montmorillonite. *Journal of Thermal Analysis Calorimetry*, 94 (2008), 841–8.

77. K. Pielichowski and A. Leszczyńska, Polyoxymethylene-based nanocomposites with montmorillonite: An introductory study. *Polimery*, 51 (2006), 60–66.

78. G. Camino, G. Tartaglione, A. Frache, C. Manferti, and G. Costa, Thermal and combustion behaviour of layered silicate–epoxy nanocomposites. *Polymer Degradation and Stability*, 90 (2005), 354–62.

79. J. Billingham, C. Breen, and J. Yarwood, In situ determination of Brønsted/Lewis acidity on cation-exchanged clay mineral surfaces by ATR-IR. *Clay Minerals*, 31 (1996), 513–22.

80. M. R. S. Kou, S. Mendioroz, and V. Munoz, Evaluation of the acidity of pillared montmorillonites by pyridine adsorption. *Clays and Clay Minerals*, 48 (2000), 528–36.

81. Q. B. Li, K. C. Hunter, and A. L. L. East, A theoretical comparison of Lewis acid vs Brønsted acid catalysis for *n*-hexane → propane + propene. *Journal of Physical Chemistry A*, 109 (2005), 6223–31.

82. F. Avalos, J. C. Ortiz, R. Zitzumbo, M. A. López-Manchado, R. Verdejo, and M. Arroyo, Phosphonium salt intercalated montmorillonites. *Applied Clay Science*, 43 (2009), 27–32.

83. H. Ming-Yuan, L. Zhonghui, and M. Enze, Acidic and hydrocarbon catalytic properties of pillared clay. *Catalysis Today*, 2 (1988), 321–38.

84. Y. Cai, F. Huang, Q. Wei, L. Song, Y. Hu, Y. Ye, Y. Xu, and W. Gao, Structure, morphology, thermal stability and carbonization mechanism studies of electrospun PA6/Fe-OMT nanocomposite fibers. *Polymer Degradation and Stability*, 93 (2008), 2180–85.

85. J. Liu, Y. Hu, S. F. Wang, L. Song, Z. Y. Chen, and W. C. Fan, Preparation and characterization of nylon 6/Cu^{2+}-exchanged and Fe^{3+}-exchanged montmorillonite nanocomposite. *Colloid and Polymer Science*, 282 (2004), 291–4.

86. N. S. Allen, M. J. Harrison, M. Ledward, and G. W. Fellows, Thermal and photochemical degradation of nylon 6,6 polymer. Part III – Influence of iron and metal deactivators. *Polymer Degradation and Stability*, 23 (1989), 165–74.

87. P. Dunn and G. F. Sansom, The stress cracking of polyamides by metal salts. Part II. Mechanism of cracking. *Journal of Applied Polymer Science*, 13 (1969), 1657–72.

88. Q. Kong, Y. Hu, L. Song, and C. Yi, Synergistic flammability and thermal stability of polypropylene/aluminum trihydroxide/Fe-montmorillonite nanocomposites. *Polymers for Advanced Technologies*, 20 (2009), 404–9.

89. J. Zhu, F. M. Uhl, A. B. Morgan, and C. A. Wilkie, Studies on the mechanism by which the formation of nanocomposites enhances thermal stability. *Chemistry of Materials*, 13 (2001), 4649–54.
90. P. Bordes, E. Hablot, E. Pollet, and L. Avérous, Effect of clay organomodifiers on degradation of polyhydroxyalkanoates. *Polymer Degradation and Stability*, 94 (2009), 789–96.
91. F. A. Bottino, G. Di Pasquale, E. Fabbri, A. Orestano, and A. Pollicino, Influence of montmorillonite nano-dispersion on polystyrene photo-oxidation. *Polymer Degradation and Stability*, 94 (2009), 369–74.
92. K. Pielichowski, A. Leszczyńska, and J. Njuguna, Mechanism of thermal stability enhancement in polymer nanocomposites. In: *Optimization of Polymer Nanocomposite Properties*, ed. V. Mittal (Weinheim: Wiley–VCH, 2010), pp. 195–210.
93. Y. Xi, Z. Ding, H. He, and R. L. Frost, Structure of organoclays – An X-Ray diffraction and thermogravimetric analysis study. *Journal of Colloid and Interface Science*, 277 (2004), 116–20.
94. Y. Xi, Q. Zhou, R. L. Frost, and H. He, Thermal stability of octadecyltrimethyl-ammonium bromide modified montmorillonite organoclay. *Journal of Colloid and Interface Science*, 311 (2007), 347–53.
95. R. Zhu, L. Zhu, J. Zhu, and L. Xu, Structure of cetyltrimethylammonium intercalated hydrobiotite. *Applied Clay Science*, 42 (2008), 224–31.
96. B. Zidelkheir and M. Abdelgoad, Effect of surfactant agent upon the structure of montmorillonite X-Ray diffraction and thermal analysis. *Journal of Thermal Analysis and Calorimetry*, 94 (2008), 181–7.
97. M. A. Osman, M. Ploetze, and U. W. Suter, Surface treatment of clay minerals – Thermal stability, basal-plane spacing and surface coverage. *Journal of Materials Chemistry*, 13 (2003), 2359–66.
98. F. Avalos, J. C. Ortiz, R. Zitzumbo, M. A. López-Manchado, R. Verdejo, and M. Arroyo, Effect of montmorillonite intercalant structure on the cure parameters of natural rubber. *European Polymer Journal*, 44 (2008), 3108–15.
99. R. D. Davis, J. W. Gilman, T. W. Sutto, J. H. Callahan, P. C. Trulove, and H. C. De Long, Improved thermal stability of organically modified layered silicates. *Clays and Clay Minerals*, 52 (2004), 171–9.
100. A. B. Morgan and J. D. Harris, Effects of organoclay Soxhlet extraction on mechanical properties, flammability properties and organoclay dispersion of polypropylene nanocomposites. *Polymer*, 44 (2003), 2313–20.
101. L. Cui, D. L. Hunter, P. J. Yoon, and D. R. Paul, Effect of organoclay purity and degradation on nanocomposite performance. Part 2. Morphology and properties of nanocomposites. *Polymer*, 49 (2008), 3762–9.
102. S. I. Marras, A. Tsimpliaraki, I. Zuburtikudis, and C. Panayiotou, Morphological, thermal, and mechanical characteristics of polymer/layered silicate nanocomposites: The role of filler modification level. *Polymer Engineering and Science*, 49 (2009), 1206–17.
103. K. R. Ratinac, R. G. Gilbert, L. Ye, A. S. Jones, and S. P. Ringer, The effects of processing and organoclay properties on the structure of poly(methyl methacrylate)–clay nanocomposites. *Polymer*, 47 (2006), 6337.
104. T. Mandalia and F. Bergaya, Organo clay mineral–melted polyolefin nanocomposites: Effect of surfactant/CEC ratio. *Journal of Chemistry and Physics of Solids*, 67 (2006), 836.

105. R. K. Shah, D. L. Hunter, and D. R. Paul, Nanocomposites from poly(ethylene-co-methacrylic acid) ionomers: Effect of surfactant structure on morphology and properties. *Polymer*, 46 (2005), 2646–62.
106. G. Panek, S. Schleidt, Q. Mao, M. Wolkenhauer, H. W. Spiess, and G. Jeschke, Heterogeneity of the surfactant layer in organically modified silicates and polymer/layered silicate composites. *Macromolecules*, 39 (2006), 2191.
107. D. Garcia-Lopez, I. Gobernado-Mitre, J. F. Fernandez, J. C. Merino, and J. M. Pastor, Influence of clay modification process in PA6-layered silicate nanocomposite properties. *Polymer*, 46 (2005), 2758–65.
108. S. S. Lee, C. S. Lee, M. H. Kim, S. Y. Kwak, M. Park, S. Lim, C. R. Choe, and J. Kim, Specific interaction governing the melt intercalation of clay with poly(styrene-co-acrylonitrile) copolymers. *Journal of Polymer Science Part B: Polymer Physics*, 39 (2001), 2430.
109. Z. F. Zhao, T. Tang, Y. X. Qin, and B. T. Huang, Effects of surfactant loadings on the dispersion of clays in maleated polypropylene. *Langmuir*, 19 (2003), 7157.
110. D. L. VanderHart, A. Asano, and J. W. Gilman, Solid-state NMR investigation of paramagnetic nylon-6 clay nanocomposites. 2. Measurement of clay dispersion, crystal stratification, and stability of organic modifiers. *Chemistry of Materials*, 13 (2001), 3796–809.
111. A. Leszczyńska, J. Njuguna, K. Pielichowski, and J. R. Banerjee, Polymer/montmorillonite nanocomposites with improved thermal properties. Part I. Factors influencing thermal stability and mechanisms of thermal stability improvement. *Thermochimica Acta*, 453 (2007), 75–96.
112. A. Leszczyńska, J. Njuguna, K. Pielichowski, and J. R. Banerjee, Polymer/montmorillonite nanocomposites with improved thermal properties. Part II. Thermal stability of montmorillonite nanocomposites based on different polymeric matrixes. *Thermochimica Acta*, 454 (2007), 1–22.
113. T. D. Fornes, P. J. Yoon, D. L. Hunter, H. Keskkula, and D. R. Paul, Effect of organoclay structure on nylon 6 nanocomposite morphology and properties. *Polymer*, 43 (2002), 5915–33.
114. S. Hotta and D. R. Paul, Nanocomposites formed from linear low density polyethylene and organoclays. *Polymer*, 45 (2004), 7639–54.
115. E. M. Araújo, R. Barbosa, A. D. Oliveira, C. R. S. Morais, T. J. A. deMélo, and A. G. Souza, Thermal and mechanical properties of pe/organoclay nanocomposites. *Journal of Thermal Analysis and Calorimetry*, 87 (2007), 811–14.
116. A. Gu and G. Liang, Thermal degradation behaviour and kinetic analysis of epoxy/montmorillonite nanocomposites. *Polymer Degradation and Stability*, 80 (2003), 383–91.
117. M. J. Gintert, S. C. Jana, and S. G. Miller, An optimum organic treatment of nanoclay for PMR-15 nanocomposites. *Polymer*, 48 (2007), 7573–81.
118. Z.-M. Liang, J. Yin, and H.-J. Xu, Polyimide/montmorillonite nanocomposites based on thermally stable, rigid-rod aromatic amine modifiers. *Polymer*, 44 (2003), 1391–9.
119. S. Campbell and D. Scheiman, Orientation of aromatic ion exchange diamines and the effect on melt viscosity and thermal stability of PMR-15/silicate nanocomposites. *High Performance Polymers*, 14 (2002), 17–30.
120. A. Pattanayak and S. Jana, Synthesis of thermoplastic polyurethane nanocomposites of reactive nanoclay by bulk polymerization methods. *Polymer*, 46 (2005), 3275–88.
121. F. Cao and S. Jana, Nanoclay-tethered shape memory polyurethane nanocomposites. *Polymer*, 48 (2007), 3790–800.

122. S. Ahmed and F. Jones, A review of particulate reinforcement theories for polymer composites. *Journal of Materials Science*, 25 (1990), 4933–42.
123. J. Xiong, Y. Liu, X. Yang, and X. Wang, Thermal and mechanical properties of polyurethane/montmorillonite nanocomposites based on a novel reactive modifier. *Polymer Degradation and Stability*, 86 (2004), 549–55.

3

Thermal stability of polystyrene nanocomposites from improved thermally stable organoclays

MUSA R. KAMAL AND JORGE URIBE-CALDERON

Department of Chemical Engineering, McGill University

3.1 Introduction

Polymer/clay nanocomposites exhibit remarkable improvement in material properties relative to unfilled polymers or conventional composites. These improvements can include increased tensile modulus, mechanical strength, and heat resistance and reduced gas permeability and flammability [1]. There are various methods of preparing polymer/clay nanocomposites: (i) in situ polymerization, (ii) solution intercalation, (iii) melt intercalation, and (iv) in situ template synthesis [2].

Nanoclays are difficult to disperse in polymer matrices, because of the strong attractive forces among the clay platelets and the commonly hydrophobic nature of polymers. Thus, it is necessary to modify pristine nanoclays in order to (i) render them compatible with most polymers and (ii) enlarge the basal spacing of clay to favor polymer intercalation. Several approaches are used to modify clays and clay minerals. They include adsorption, ion exchange (with inorganic cations and organic cations), binding with inorganic and organic anions (mainly at the edges), grafting of organic compounds, reaction with acids, pillaring by different types of poly(hydroxo metal) cations, intraparticle and interparticle polymerization, dehydroxylation and calcination, delamination and reaggregation of smectites, lyophilization, and exposure to ultrasound and plasma [3]. Ion exchange with organic cations (surfactants) is well established as the preferred method for modifying clay. The resulting material is usually named organoclay.

Surfactants for clay modification usually include long aliphatic chains in their molecular structure. The aliphatic chains arrange themselves, depending on the size and concentration of surfactant molecules, into monolayers, bilayers, pseudo-trimolecular layers, or inclined paraffin structures. The molecular arrangement determines the final interlayer distance [4, 5]. By choosing the appropriate type and concentration of surfactants, it is possible to enhance polymer–clay compatibility and to promote delamination of the nanoclay under suitable synthetic or processing conditions. The optimum enhancement in properties is achieved when the clay is exfoliated and homogeneously distributed in the polymer matrix. Traditionally, nanoclays have been treated with amine, ammonium, sulfonium, phosphonium, and imidazolium surfactants. In some cases, the modifiers are oligomers or polymers.

In this chapter, we shall deal mainly with montmorillonite (MMT) clay and with organo-clays and polymer/clay nanocomposites containing polystyrene (PS) as the matrix and incorporating MMT as the clay.

3.2 Thermal stability

The thermal stability of materials can be defined as their resistance to mass loss and/or chemical change or degradation at higher temperatures. In the case of organoclays, the main concern is the mass loss that results from degradation of the organic modifier, primarily during melt-processing of the nanocomposite.

Understanding the relationship between the molecular structure and the thermal stability (decomposition temperature and rate) of the organoclays and the subsequent influence on the stability of the polymer host is critical. Several analytical techniques have been used to determine the thermal stability of different organoclays and to indentify the decomposition products: conventional and high-resolution thermogravimetric analysis (TGA) coupled with Fourier transform infrared spectroscopy (FTIR) and mass spectrometry (MS), pyrolysis/gas chromatography (GC)-MS, and solid phase microextraction (SPME) [6–12].

Thermal stability of the organoclay is usually determined by TGA, which is a common and well-established technique. In TGA, weight loss of the sample, due to the forma-tion of volatile products during thermal degradation of the organic modifier, is monitored as a function of temperature. Two different degradation processes can be distinguished: nonoxidative and oxidative degradations. The former is observed when the experiment is conducted under an inert gas flow (nitrogen, argon, or helium). In the latter case, air or oxy-gen flow is employed. TGA analysis can be carried out in the isothermal or nonisothermal (scanning) mode.

3.2.1 Thermal stability of organoclays

There are different possible arrangements for the attachment of surfactants to montmo-rillonite (MMT): (i) surfactant cations are intercalated into the interlayer spaces through cation exchange and bound to surface sites via electrostatic interaction; (ii) surfactant cations and/or molecules are physically adsorbed onto the external surfaces of the parti-cles; and (iii) surfactant molecules are located within the interlayer spaces. Organoclays with surfactants bound to the clay surface exhibit the best thermal stability [13]. As will be discussed later, organoclays with unbound organic modifier (i.e., in excess) are more susceptible to degradation, which could compromise the properties of the material [12]. Thus, it is desirable to wash the organoclay with appropriate solvents to remove free and excess surfactant from clay galleries [14].

Thermal decomposition of ammonium salts generally follows either a Hofmann elimina-tion reaction or an SN2 nucleophilic substitution [7, 15] (Figure 3.1). Hofmann elimination occurs in the presence of basic anions, such as hydroxyl groups, which extract hydro-gen from the alkyl chain of the quaternary ammonium, yielding an olefinic and tertiary

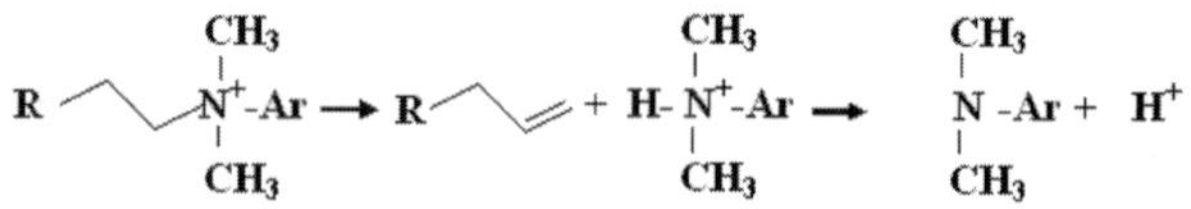

$$R \overset{CH_3}{\underset{CH_3}{\diagup\diagdown N^+ \text{-Ar}}} \longrightarrow R\diagup\diagdown + H\text{-}\overset{CH_3}{\underset{CH_3}{N^+\text{-Ar}}} \longrightarrow \overset{CH_3}{\underset{CH_3}{N\text{-Ar}}} + H^+$$

Figure 3.1 Example of Hofman elimination reaction.

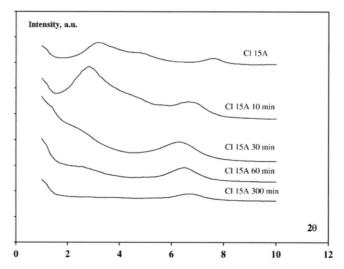

Figure 3.2 WAXS diffraction pattern of ammonium organoclay (Cl 15A: Cloisite 15A) exposed to air at 240 °C. Reproduced with permission from [18].

amino group. In the case of polystyrene resins, the olefinic group from the Hofmann reaction may react with oxygen to generate free radicals and attack the polystyrene backbone [16]. Radicals from organoclay decomposition could cause polymer degradation, leading to possible deterioration of composite properties [12]. Detailed studies concerning the chemical species generated during thermal degradation of ammonium, phosphonium, and imidazolium organoclays have been reported [10, 11].

The degradation and/or release of the intercalated surfactants in the galleries can promote the collapse of the interlayer spacing and changes in the surface characteristics. The thermal stability of organoclays depends on the type and concentration of surfactants and the interlayer structure [17].

Oxygen can accelerate the thermal degradation of surfactants. Oxidized degradation products of the modifier (i.e., α-olefins transforming into various carboxyl compounds) can increase the basal spacing of organoclays during the early stages of isothermal decomposition. However, progressive gallery collapse has been observed, as the decomposition products migrate toward the surface and eventually volatilize [18] (Figure 3.2).

Surface characteristic of organoclays are affected by the decomposition of surfactant molecules. The surface energy of clay changes with the thermal decomposition of the surfactant, which depends on the decomposition environment (inert or oxidative). The migration of nonpolar compounds that are produced during thermal decomposition of alkyl surfactants to the surface, and the further elimination of the surfactant, provoke changes in the surface characteristics of the clay. Contact angle measurements provide a simple technique for estimating the surface energy of clay. Thus, the migration of α-olefin compounds to the surface can be considered responsible for the increase of the contact angle in the early stages of decomposition. In a nonoxidative atmosphere, the increase of the contact angle (decrease of polarity) may be attributed to modifier migration together with α-olefin formation and migration [18]. Changes in surface polarity of the clay mineral during nanocomposite melt-compounding have a direct influence on the organoclay–polymer affinity, which has a significant impact on the dispersion of the clay mineral particles in the polymer matrix [19].

3.2.2 Thermal stability of PS resin

The thermal degradation pathway of pure PS involves chain scission followed by depolymerization. The resulting products are styrene monomer, dimer, and trimer through an intrachain reaction. However, the presence of clay causes changes in the degradation pathway of polystyrene. In the thermal decomposition of PS nanocomposites, the presence of products from interchain reactions is significant, because the radicals have more opportunity for transfer [20] (Figure 3.3).

The inclusion of nanoclay in the polymer matrix improves thermal stability, because clay can act as a heat sink and mass transport barrier for the volatile products generated during decomposition. Thus, clay can shift the sample decomposition to higher temperature in the early stages of thermal decomposition. On the other hand, when heat is supplied by an external heat source, silicate layers can accumulate heat that serves as an internal heat source to accelerate the decomposition process at higher temperatures [1]. Additionally, the barrier effect of the clay layers in the nanocomposites can increase the possibility of radical transfer reactions to produce tertiary radicals and head-to-head structures, and to cause hydrogen abstraction from the condensed species with double bonds and the formation of char [20, 21]. Char protects the bulk of the sample from heat and decreases the rate of mass loss during thermal decomposition, thus providing improved flame resistance. However, acidic sites present on MMT layers after removal of the modifier molecules can accelerate polymer decomposition. Thus, the final effect of clay on the thermal stability of nanocomposites depends on the intrinsic properties of the polymer matrix, and more specifically, on the routes of polymer degradation [21–24].

3.3 Polystyrene/clay nanocomposites from thermally stable organoclays

Various compounds are used to modify clays in order to produce organoclays that satisfy the requirements of nanocomposite synthesis. Although many modifiers serve mainly as surfactants that expand the gallery spaces between clay platelets and enhance interaction with the polymer matrix, others play an important role by initiating polymerization

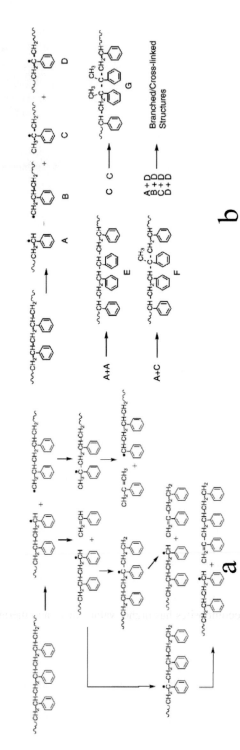

Figure 3.3 Thermal decomposition of (a) neat PS and (b) PS/clay nanocomposites (reproduced from [20] with permission).

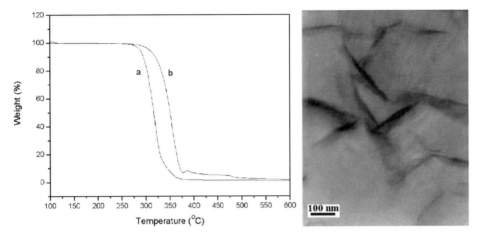

Figure 3.4 TGA curves of (a) pure APP and (b) pure APB and TEM image of APB-treated nanocomposite at high magnifications. Reproduced from [25] with permission.

reactions, as in the case of in situ nanocomposite preparation. The various types of modifiers are discussed in the following sections.

3.3.1 Polystyrene/clay nanocomposites from unreactive thermally stable organoclays

3.3.1.1 Ammonium/amine surfactants

Quaternary alkylammonium salts or alkylamines are the cationic surfactants most commonly used in the modification of layered silicates. They are synthesized by complete alkylation of ammonia or amines. Many efforts have been made to produce ammonium surfactants to improve the affinity between the clay mineral and the polymer [3]. Ammonium organoclays undergo thermal degradation at temperatures below or comparable to the melt-processing temperatures of many polymers.

One way to improve the thermal stability of surfactant molecules is the introduction of aromatic rings into the structure. For example, ammonium 4-(4-adamantylphenoxy)-1-butanamine (APB) salts exhibit higher thermal stability than allyltriphenylphosphonium chloride (APP) (Figure 3.4, Table 3.1). It has been reported that PS nanocomposites prepared by emulsion polymerization, using ammonium surfactants, exhibit exfoliated structures. The PS in such nanocomposites has higher values of T_g and thermal decomposition temperature than are observed for virgin PS [25].

Phenylacetophenone dimethylhexadecylammonium (BPNC16) salt (Table 3.1) was used to treat montmorillonite (MMT) [26]. The long alkyl chain was used to obtain organoclays with large basal spacing. BPNC16 clay was used to prepare nanocomposites with polystyrene (PS), acrylonitrile–butadiene–styrene (ABS), and high–impact polystyrene (HIPS) matrices by in situ bulk polymerization or melt-blending processes. The nanocomposites exhibited improved thermal stability and flame retardancy.

Table 3.1 *Chemical structure of ammonium surfactants with improved thermal stability*

Surfactant	Chemical structure	Reference
Ammonium4-(4-adamantylphenoxy)-1-butanamine (APB)		25
Phenylacetophenone dimethylhexadecyl ammonium (BPNC16) salt		26
Bromoalkyl carbazoles (5AC salt)		27
Alkyl carbazole salt (10AC salt)		27
Dialkyl carbazole salt (10ACDD salt)		27
Quinolinium (QC16) salt		28
Cetylpyridinium chloride/α-cyclodextrin		29

Table 3.1 (*cont.*)

Surfactant	Chemical structure	Reference
Aminopropylisobutyl polyhedral oligomeric silsesquioxane (POSS)	R = i-butyl	31
Polyhedral oligomeric silsesquioxane (C$_{20}$-POSS)	CH$_3$(CH$_2$)$_{17}$—N, C$_{20}$-POSS	32, 33
Protonated amino-propylisobutylti-tanosilsesquioxane (Ti-NH$_3$POSS)	R = tBu, R' = iPrO	30

Carbazole units and quinoline with long alkyl chains have been also used to modify clay for the preparation of PS nanocomposites by both bulk polymerization and melt-blending [27, 28] (Table 3.1). Poor clay dispersion was observed when the organoclay contained more than two alkyl chains. Bulk polymerization yielded nanocomposites with better dispersion and reduced flammability, compared to the melt-blending process.

The combination of cetylpyridinium chloride surfactant and cyclodextrine yields a thermally stable organoclay (Table 3.1). Usually, pyridium salts degrade at temperatures above 200 °C. However, the pyridium/cyclodextrine complex degrades at substantially higher temperatures (Figure 3.5). The basal spacing in pristine clay changed from 1.43 to 2.27 nm upon insertion of pyridium [29]. The d-spacing of the clay with intercalated pyridium/cyclodextrine complex became 5.12 nm. The large expansion of the basal spacing occurred because the linear aliphatic chain within the complex could not bend inside the galleries of the clay. Clay exfoliation was observed, and improved thermal properties were obtained in the cetylpyridinium/cyclodextrine complex–PS nanocomposites prepared by emulsion polymerization.

Polyhedral oligomeric silsesquioxanes (POSSs) could be used as building blocks for multifunctional composite solids and for hierarchical inorganic/organic hybrid architectures. These compounds may include inert hydrocarbon moieties, reactive groups, and

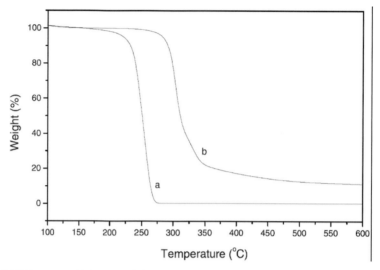

Figure 3.5 TGA curves of (a) pure pyridium and (b) the pyridium/cyclodextrine complex. Reproduced from [29] with permission.

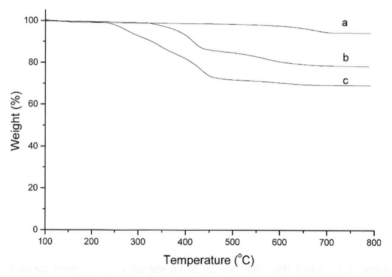

Figure 3.6 TGA traces of (a) pure clay, (b) clay intercalated with the POSS, and (c) clay intercalated with CPC. Reproduced from [31] with permission.

metal centers (catalysts), all in the same molecule. Table 3.1 shows two different POSS surfactants used to treat clay [30, 31].

Figure 3.6 shows the thermal stability of organoclay prepared with aminopropylisobutyl polyhedral oligomeric silsesquioxane (POSS) and cetylpyridinium chloride (CPC) [31].

The thermal stability of the POSS organoclay is substantially higher than that of the pyridinium clay. Although the basal spacing of the POSS-modified clay was smaller than that of the CPC-modified clay, better intercalation/exfoliation was observed in PS nanocomposites containing the POSS-treated clay, obtained by emulsion polymerization. Moreover, the presence of POSS enhanced the thermal stability of the polystyrene. Polyhedral oligomeric silsesquioxane (C20–POSS) surfactants containing a long alkyl chain were used to treat MMT [32, 33] (Table 3.1). The inclusion of the long alkyl chain reduced the thermal stability of C20–POSS. However, PS nanocomposites, prepared via emulsion polymerization, exhibited exfoliated structures and improved thermal properties.

The use of bifunctional protonated titanium-containing aminopropylisobutyl POSS (Ti–NH₃POSS) to intercalate synthetic sodium saponite leads to significant thermal stabilization of PS nanocomposites [30] (Table 3.1). The catalytic activity of the Ti centers leads to the formation of stable charring products by secondary reactions. This causes significant improvement of the thermal stability and flame retardancy of the nanocomposite.

3.3.1.2 Phosphonium

Phosphonium compounds are widely used as flame retardants/stabilizers and offer additional opportunities for improvement of polymer–layered silicate nanocomposites. Some of the phosphonium modifiers considered in the literature for polymer/clay nanocomposite synthesis include triphenyldodecylphosphonium bromide, P-C12; tributyltetradecylphosphonium bromide, PC-14; tributylhexadecylphosphonium bromide, P-C16; tributyloctadecylphosphonium bromide, P-C18; tetraphenylphosphonium bromide, P-4Ph; and tetraoctylphosphonium bromide, P-4C8 (Table 3.2). Figure 3.7 shows the derivative TGA curves for the phosphonium organoclays [15]. Most of these organoclays begin to undergo thermal decomposition above 250 °C, which makes them suitable for the melt-processing of nanocomposites.

n-Hexadecyltriphenylphosphonium and stearyltributylphosphonium bromide were used to modify MMT and synthetic mica–montmorillonite [34, 35]. Nanocomposites obtained by bulk polymerization of PS and incorporating the resulting organoclays yielded intercalated/exfoliated structures. TGA/FTIR analysis indicated that both phosphonium organoclays degrade by the Hofmann elimination mechanism. In general, PS thermal stability and fire-retarding properties improved with clay content. The presence of structural iron (natural clay), rather than that present as an impurity, significantly raised the onset temperature of thermal degradation in polymer–clay nanocomposites.

Nanocomposites were prepared by incorporating MMT-[2-(dimethylamino)-ethyl]triphenylphosphonium bromide in PS during free radical polymerization (Table 3.2) [36, 37]. Both the dielectric constant and dielectric loss were lowered for nanocomposites. The decrease of the dielectric constant was correlated with the extent of exfoliation of clay. MMT-[2-(dimethylamino)ethyl]triphenylphosphonium bromide was used to synthesize poly(styrene-co-acrylonitrile) nanocomposites in a typical free radical polymerization

Table 3.2 *Phosphonium surfactants*

Surfactant	Chemical structure	Reference
Triphenyldodecylphosphonium		15
Tributyltetradecylphosphonium		15, 44
Tributylhexadecylphosphonium		15
Tributyloctadecylphosphonium		15
Tetraphenylphosphonium		15
Tetraoctylphosphonium		15, 44
Hexadecyltriphenylphosphonium		34, 35, 39, 42
[2-(Dimethylamino)ethyl] triphenylphosphonium		36, 37, 38
Allyltriphenylphosphonium chloride (APP)		43
Trihexyltetradecylphosphonium		44
Tetra-*n*-butylphosphonium		44

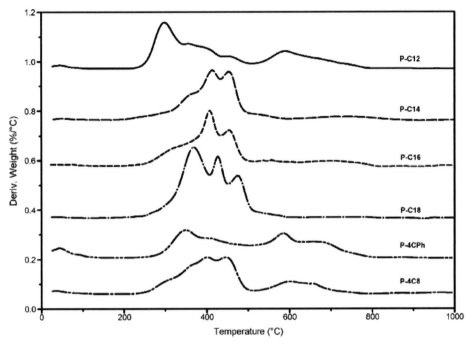

Figure 3.7 DTG curves from TGA of quaternary phosphonium montmorillonites (P-C12, P-C14, P-C16, P-C18, P-4Ph, and P-4C8). Reproduced from [15] with permission.

medium [38]. Clay exfoliation was obtained even at 10 wt% clay content, even though molecular weight decreased with clay content. The thermal stability and barrier properties were considerably improved.

Sodium fluorinated synthetic mica was used to obtain organoclays [39]. Mica was treated with triphenyl-*n*-hexadecylphosphonium to produce PS nanocomposites by melt-compounding. The results indicated the advantages of synthetic versus natural clays with regard to flammability of nanocomposites. Synthetic clays have an advantage in color, purity, and batch-to-batch consistency when compared with natural clays.

Nanocomposites of syndiotactic polystyrene (sPS) employing MMT–hexadecyltributylphosphonium [40, 41] and high-impact polystyrene (HIPS)/MMT–hexadecyltriphenylphosphonium [42] were prepared by melt-blending and in situ coordination–insertion polymerization. Partially exfoliated or intercalated materials were obtained in all cases, and a decrease of crystallinity of sPS was observed. However, the presence of clay did not have a strong influence on the sPS thermal transitions. Thermal decomposition of the material was slowed and mechanical properties were improved in the presence of low organoclay content. Intercalated HIPS nanocomposites were obtained, with improved thermal and flame retardant properties compared to pure HIPS (Figure 3.8).

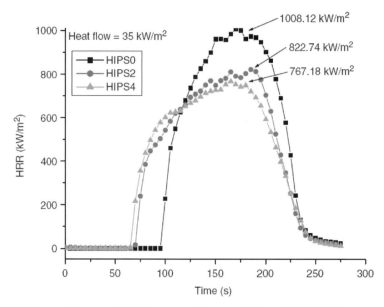

Figure 3.8 The heat release rate (HRR) curves of the pure HIPS and HIPS/OMT nanocomposites. Reproduced from [42] with permission.

Exfoliated PS nanocomposites were obtained by emulsion polymerization in the presence of a reactive phosphonium surfactant (APP) (Table 3.2) [43]. The thermal decomposition temperature and T_g of the PS component in the nanocomposite were higher than those of the virgin PS.

Several phosphonium organoclays were used in the preparation of PS/MMT nanocomposites [44] by melt-compounding. The phosphonium salts included tributyltetrade-cylphosphonium chloride–Ph1, trihexyltetradecylphosphonium chloride–Ph2, tetra-*n*-octylphosphonium bromide–Ph3, and tetra-*n*-butylphosphonium chloride–Ph4 (Table 3.2). The thermal stability of the phosphonium organoclays was significantly higher than that of ammonium organoclays in both scanning and isothermal (220 °C) TGA experiments (Figure 3.9). The quality of the clay dispersion depended on several parameters, such as the molecular weight of the polymer and the surface properties of modified clays. The improvements in thermal, mechanical, and barrier properties were explained in terms of the quality of dispersion and the surface energy characteristic of the organoclays. The quality of clay dispersion in the PS resin was correlated with the initial basal spacing of the organoclay and the value of the Hamaker constant of the modified clay (A_{11}). The nanocomposite tensile modulus was related to the work of adhesion at the interface and other surface properties [45].

3.3.1.3 Imidazolium

Imidazolium surfactants have been used to enhance the thermal stability of organoclays. Trialkylimidazolium salt derivatives were prepared with propyl, butyl, decyl, and hexadecyl

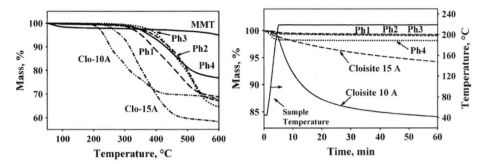

Figure 3.9 Temperature-scanning TGA (left) and isothermal TGA curves (right) for phosphonium and ammonium organoclays (referred to as Clo-10A: Cloisite 10A and Clo-15A: Cloisite 15A) under nitrogen [44].

chains attached to the imidazolium through one of the nitrogens (Table 3.3) [46–48]. The clays used were MMT and fluoro mica. The organoclays were used to prepare PS nanocomposites by melt-compounding or solution-blending. Intercalated/exfoliated structures were obtained.

The thermally stable reactive surfactant imidazolium salts (C_{12}, C_{16}, and C_{18}) were used to modify montmorillonite (Table 3.3) to prepare PS nanocomposites via in situ bulk polymerization [49–53]. In spite of the long alkyl chains in the compounds, the basal spacing of modified MMT was small. Nonetheless, the nanocomposites exhibited a higher thermal degradation temperature than the ammonium counterparts. The presence of the reactive group in the imidazolium cations promoted clay dispersion, thus yielding partially exfoliated PS, even when the surfactant contained a relatively short alkyl chain (Figure 3.10). Photo-oxidation experiments by UV exposure showed that the rate of photo-oxidation of the PS nanocomposite was higher than that of pristine PS. Nanocomposites with a higher degree of exfoliation showed a shorter photodegradation induction period and degraded more quickly than less exfoliated nanocomposites (PS/C12/MMT 3%). It is possible that the active catalytic sites present in MMT can accept single electrons from donor molecules (PS matrix), forming free radicals that accelerate the normal photo-oxidative process. Additionally, a rheological technique was used to differentiate (quantify) the degree of exfoliation/dispersion in samples of polystyrene PS–clay nanocomposites. PS nanocomposites demonstrated a change of pattern in dynamic mechanical spectrum as a function of the degree of exfoliation.

To increase the basal spacing of imidazolium organoclays, monoalkyl- and dialkylimidazolium surfactants were used to modify MMT (Table 3.3), yielding a basal spacing of 2.27 nm in the organoclays [54]. Syndiotactic polystyrene (sPS)/imidazolium–MMT nanocomposites were prepared under static melt-intercalation conditions, in the absence of high shear rates or solvents. TGA data showed that the onset and maximum decomposition temperatures of the imidazolium–MMTs were 100 °C higher than those of alkylammonium clays. The thermal stability of sPS nanocomposites was improved, in comparison with that

Table 3.3 *Imidazolium surfactants*

Surfactant	Chemical structure	Reference
Trialkylimidazolium		46, 47, 48
Reactive trialkylimidazolium		49, 50, 51, 52, 53
Monoalkyl- and dialkylimidazolium surfactants		54
1-Hexadecyl-1H-benzimidazole		55
1,3-Dihexadecyl-3H-benzimidazol-1-ium bromide		
2-Methyl-1-hexadecyl-1H-benzimidazole		
2-Methyl-1,3-dihexadecyl-3H-benzimidazol-1-ium bromide		

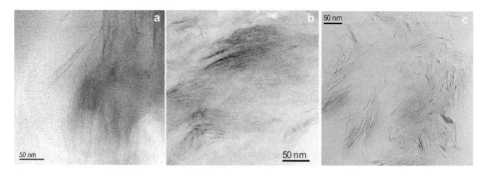

Figure 3.10 TEM microphotographs of PS/C12/MMT 3% (a), PS/C16/MMT 3% (b), and PS/C18/MMT 3% (c). Reproduced from [51] with permission.

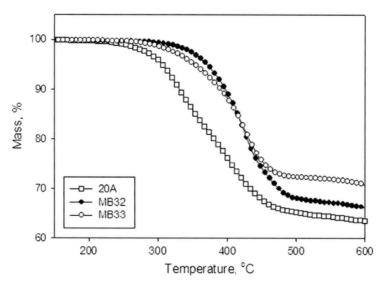

Figure 3.11 TGA curves for the MB32, MB33, and 20A (Cloisite 20A) organoclays. Reproduced from [55] with permission.

of neat sPS. Moreover, the β-crystal form of sPS became dominant. This crystallization response was attributed to the heterogeneous nucleation action by the inorganic fillers.

Nanocomposites of PS, ABS, and HIPS were prepared with benzimidazolium (1-hexadecyl-1H-benzimidazole,(BZ32), 1,3-dihexadecyl-3H-benzimidazol-1-iumbromide, 2-methyl-1-hexadecyl-1H-benzimidazole, and (BZ33) 2-methyl-1,3-dihexadecyl-3H-benzimidazol-1-ium bromide) modified clays by melt-blending (Table 3.3) [55]. Both organically modified clays exhibited higher thermal stability than the conventional ammonium-based organoclay (Figure 3.11). The thermal stability and fire-retarding properties of the nanocomposites were enhanced to about the same levels as seen for other organically modified commercially available clays. The chemical structure seemed to allow better diffusion of the polymer into the clay galleries and, therefore, better dispersion. PS, ABS, and HIPS nanocomposites from both clays produced mostly intercalated nanocomposites and showed good mesoscale dispersion. The fire-retarding properties of the nanocomposites were improved, especially in terms of the reduction in the peak heat-release rate

Awad *et al.* [56] reported on the thermal degradation of a series of alkylimidazolium molten salts. Elemental analysis, TGA, and thermal desorption mass spectroscopy (TDMS) were used to characterize the degradation process. A correlation was observed between the chain lengths of the alkyl groups and the thermo-oxidative stability: as the chain length increased from propyl, butyl, decyl, hexadecyl, and octadecyl to eicosyl, the stability decreased. Analysis of the decomposition products by FTIR provided information about the decomposition products. It suggested that the thermal decomposition of imidazolium salts followed an SN2 process (Figure 3.12).

Table 3.4 *Miscellaneous surfactants with improved thermal stability*

Surfactant	Chemical Structure	Reference
Triphenylhexadecylstibonium trifluoromethylsulfonate	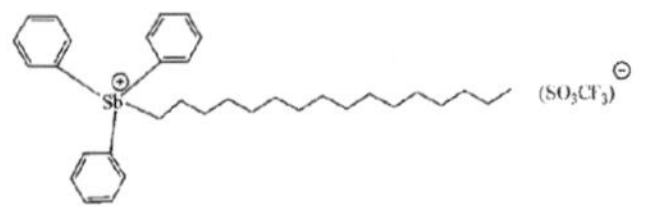	57
3-(Trimethoxysilyl)propyl methacrylate (MPTMS)	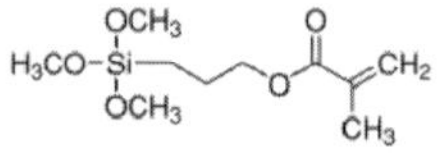	58, 59
Styryltropylium	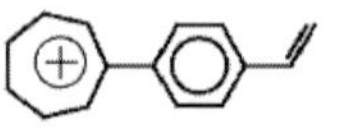	60
γ-Methacryloxypropyltrimethoxysilane– Halloysite nanotubes	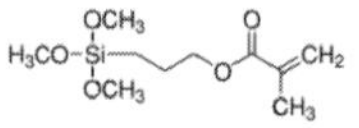	61

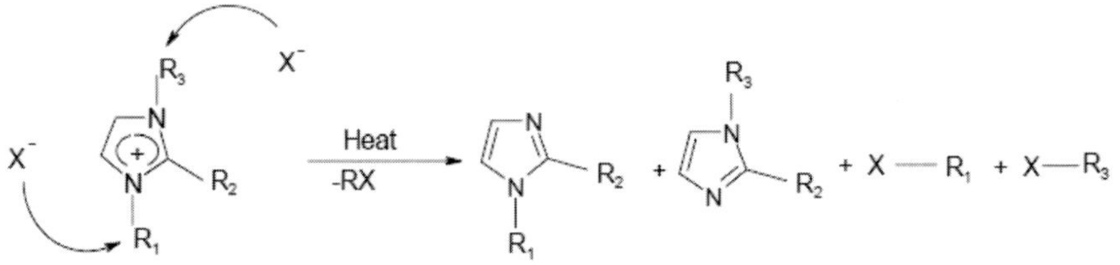

Figure 3.12 SN2 process for imidazolium quaternary salts. Reproduced from [56] with permission.

3.3.1.4 Miscellaneous

Other thermally stable surfactants have been used in the preparation of organoclays to produce PS nanocomposites [57]. Triphenylhexadecylstibonium trifluoromethylsulfonate (Table 3.4) was prepared with sodium montmorillonite to prepare polystyrene nanocomposites by bulk polymerization. The organoclay was not uniformly distributed throughout the polymer matrix, but there was evidence of polymer intercalation and a small amount of clay exfoliation (Figure 3.13). The nanocomposite showed enhanced thermal stability.

Fully exfoliated PS/clay nanocomposites were prepared via free radical polymerization in dispersion [58, 59]. Thermally stable organoclay was obtained by modifying MMT with 3-(trimethoxysilyl)propyl methacrylate (MPTMS) (Table 3.4, Figure 3.14). Analyses by XRD and TEM revealed that nanocomposites with low clay loadings exhibited exfoliated structures, whereas intercalated structures were obtained at higher clay loadings. Another silane-type organoclay was prepared by modification of vermiculite with the

Figure 3.13 TEM image of SbC16 PS nanocomposite (scale bar 50 nm). Reproduced from [57] with permission.

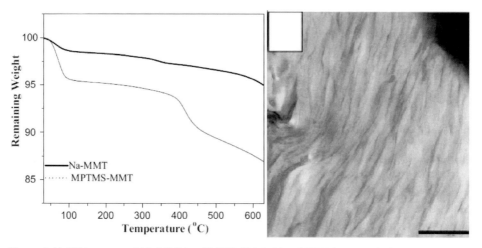

Figure 3.14 TGA curves of Na-MMT and MPTMS-MMT and TEM images of PS nanocomposites containing 1% clay loading (scale bar = 100 nm). Reproduced from [58] with permission.

γ-methacryloxypropyltrimethoxysilane. PS nanocomposites were prepared by bulk free radical copolymerization. The thermal stability of the nanocomposites was better than that of the pure polymeric materials or those without the modifier.

Styryltropylium (a carbo-cation) was used to modify clay for producing PS nanocomposites by in situ emulsion polymerization (Table 3.4) [60]. The resulting nanocomposites exhibited a mixture of intercalated and exfoliated structures. The nanocomposites exhibited improved thermal stability and fire retardancy. Halloysite nanotubes (HNT) are a kind of aluminosilicate clay with a hollow nanotubular structure (about 20–50 nm in diameter and several hundred nanometers in length) [61]. HNT was modified with γ-methacryloxypropyltrimethoxysilane to produce a nanoclay (Table 3.4). HNT/PS nanocomposites were prepared by in situ bulk polymerization. The thermal stability of the HNT/PS nanocomposites was better than that of the pure polystyrene.

3.3.2 Polystyrene/clay nanocomposites from reactive thermally stable organoclays

The use of organoclays containing reactive species as surfactants is an interesting approach to the synthesis of PS nanocomposites. In this case, surface-initiated polymerization (SIP) is promoted by the action of the initiators anchored on the clay surface. Fan *et al.* [62] designed two initiators with quaternized amine end-groups for cation exchange with MMT. One initiator molecule had cationic groups at both chain ends (bicationic free radical initiator); whereas the second initiator had a cationic group at one end (monocationic free radical initiator, Table 3.5). Figure 3.15 indicates the thermal stability of both the bicationic free radical initiator and the resulting organoclay. SIP product from the bicationic initiator retained some intercalated structure, whereas a highly exfoliated structure was achieved with the monocationic initiator. It is possible that the spatial arrangement of the bicationic initiator in the clay limits monomer diffusion during polymerization. Consequently, the monocationic initiator yielded a higher–molecular weight polymer.

Two polymerizable cationic surfactants were synthesized to produce thermally stable organoclays: (11-acryloyloxyundecyl)dimethyl(2-hydroxyethyl)ammonium bromide (called hydroxyethyl surfmer), and (11-acryloyloxyundecyl)dimethylethylammonium bromide (called ethyl surfmer (Table 3.5) [63]. PS nanocomposites were produced by bulk polymerization and by free radical polymerization using these organoclays. Exfoliated structures were obtained with the ethyl surfmer-modified clay, whereas a mixed exfoliated/intercalated structure was obtained using the hydroxyethyl surfmer-modified clay. The nanocomposites exhibited enhanced thermal stability and an increase in the glass transition temperature, in addition to improved mechanical properties relative to polystyrene. However, intercalated structures were obtained when nanocomposites were prepared in solution, because of competition between the solvent molecules and monomer in penetrating the clay galleries. Enhanced thermal stability was also obtained in the solution polymerization case.

Table 3.5 *Reactive surfactants*

Reactive surfactant	Chemical structure	Reference
2,2′-Azobis(isobutyramidine hydrochloride) (AIBA) type		62
((11-Acryloyloxyundecyl)dimethyl(2-hydroxyethyl)ammonium bromide: hydroxyethyl surfmer		62
(11-Acryloyloxyundecyl)dimethylethylammonium bromide:ethyl surfmer		62
N,*N*-Dimethyl-*N*-(4-{[(phenylcarbonothioyl)thio]-methyl}benzyl)ethanammonium bromide (PCDBAB)		64
N-[4-({[(Dodecylthio)carbonothioyl]thio}methyl)benzyl]-*N*,*N*-dimethylethanammonium bromide (DCTBAB)		64
2,2-Azobis(2-(1-(2-hydroxyethyl)-2-imidazolin-2-yl)propane) dihydrochloride monohydrate (VA060)		65

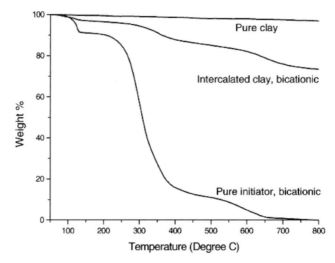

Figure 3.15 TGA traces of pure clay, clay intercalated with bicationic initiator, and pure bicationic initiator. Reproduced from [62] with permission.

Other thermally stable organoclays for atom transfer radical polymerization were considered: DCTBAB and PCDBAB (Table 3.5) [64]. Organoclays were used to prepare nanocomposites by polymerization of styrene in bulk. The thermal stability of the nanocomposites decreased as the clay loading increased, because of the competing effects of the number of attached chains relative to the unattached chains, level of clay exfoliation, and clay distribution. The overall thermal stability improved slightly.

Other thermally stable organoclays were synthesized: MMT-2,20-azobis(2-(1-(2-hydroxyethyl)-2-imidazolin-2-yl)propane) dihydrochloride monohydrate (VA060) [65] (Table 3.5, Figure 3.16). PS–clay nanocomposites were prepared by in situ bulk free radical polymerization. Some chain transfer (RAFT) agents were used in the process. The resulting nanocomposites exhibited intercalated/exfoliated morphologies (Figure 3.16). The RAFT agents promoted controlled polymerization and the production of polymers with narrow polydispersities. Improvements in the thermal stability were observed in all cases.

3.3.3 *Polystyrene/clay nanocomposites from polymer/oligomer modified organoclays*

Surface modification by polymers is an effective method of enhancing polymer–clay compatibility. There are two mechanisms for surface modification of clay minerals with polymers: (i) physical adsorption and (ii) chemical grafting of functional polymers onto the clay surface. Physical adsorption modifies the surface properties of clay, but the structure of the clay mineral is preserved. However, the forces between the adsorbed molecules and the clay mineral might be weak, causing desorption. Grafting of functional polymers onto the surfaces of clay minerals provides stronger clay–polymer interaction and permits control

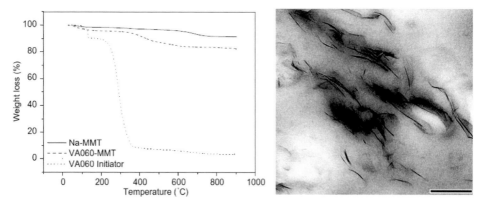

Figure 3.16 TGA curves for MMT, VA60 surfactant, and MMT-VA60 organoclays (left) and TEM images of PS-VA060–1 (scale bar = 200 nm) (right). Reproduced from [65] with permission.

and tuning of the properties of clay mineral surfaces. Surface modification by grafting usually involves condensation of functionalized polymers with reactive groups on the solid substrate. This method does not yield extremely dense polymer brushes, because chemical adsorption of the first fraction of chains hinders the diffusion of other chains to the surface for further attachments [66].

A typical polymer-modified organoclay for the preparation of PS nanocomposites could be a PS polymer with some cationic sites to be attached to the clay surface. Polystyryl quaternary ammonium salts (PSQAS) were synthesized via three different polymerization routes: anionic polymerization (AP), free radical co-polymerization (CP), and chloromethylation of polystyrene (CM) [67] (Table 3.6). TGA results indicated that CP-PSQAS and CM-PSQAS were thermally stable, because the C–C bonds in aromatics are much stronger than those in aliphatic compounds. The polystyryl-modified clays exhibited intercalated or partially exfoliated structures.

Su *et al.* [68] studied the influence of different amines and phosphines on the thermal stability of clays organically modified with styrene oligomers. PS-modified organoclays were used to prepare PS and PMMA nanocomposites by melt intercalation. The synthesized surfactants included 1,2-dimethyl-3-polystyrylimidazolium chloride (CDMID), *N,N,N*-trimethylpolystyrylammonium chloride (CTMA), *N,N*-dimethyl-*N*-hexadecylpolystyrylammonium chloride (CDMH), *N,N*-dimethyl-*N*-benzylpolystyrylammonium chloride (CDMBA), and triphenylpolystyrylphosphonium chloride (CTPP) (see Table 3.6). The increasing order of thermal stability of the oligomeric surfactants, based on the 10% mass loss, was CTPP > CDMID > CDMH > CTMA > CDMBA. As expected, phosphonium and imidazolium were more thermally stable than ammonium salts. All of these organoclays, except CDMID, yielded exfoliated polystyrene and poly(methyl methacrylate) nanocomposites.

Thermal stability of polystyrene nanocomposites

Table 3.6 *Oligomeric surfactants*

Surfactant	Molecular structure	Reference
Polystyryl quaternary ammonium salts (anionic polymerization: AP-PSQAS)		67
Polystyryl quaternary ammonium (free radical copolymerization: CP-PSQAS)		67
Polystyryl quaternary ammonium (chloromethylation: CM-PSQAS)		67
1,2-Dimethyl-3-polystyrylimidazolium chloride (CDMID)		68
N,N,N-Trimethylpolystyrylammonium chloride (CTMA)		68
N,N-Dimethyl-N-hexadecylpolystyrylammonium chloride (CDMH)		69
N,N-Dimethyl-N-benzylpolystyrylammonium chloride (CDMBA)		69
Triphenylpolystyrylphosphonium chloride (CTPP)		69
Copolymer of styrene and vinylbenzyl chloride (COPS)		69, 70, 71
Copolymer of methyl methacrylate and vinylbenzyl chloride (MAPS)		69

Table 3.6 (*cont.*)

Surfactant	Molecular structure	Reference
Polybutadiene-grafted	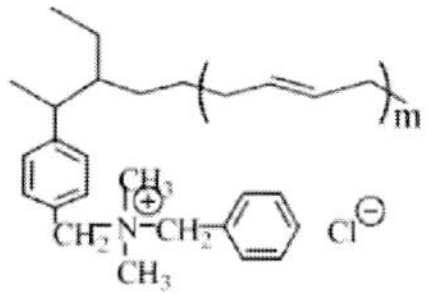	73, 74
Poly(styrene-b-4-vinylpyridine) (SVP)	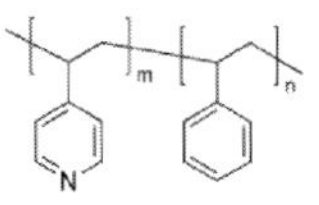	75
Diphenyl-4-vinylphenylphosphate (DPVPP)	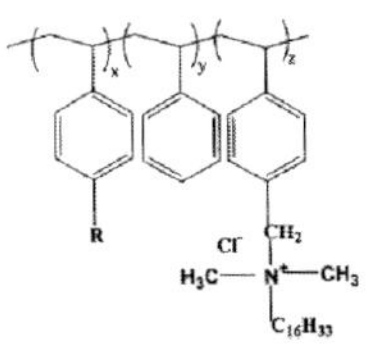	76
Terpolymer (vinylbenzyl chloride-styrene-dibromostyrene)	R = diphenyl 4-vinylbenzyl phosphate 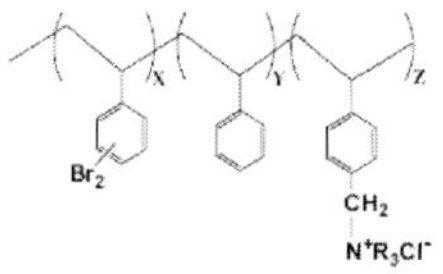	77
Triclay (vinylbenzyl chloride–styrene–lauryl acrylate copolymer)	R = methyl or ethyl or hexadecyl NA	78, 79, 80
Poly(styrene-co-acrylonitrile) ammonium salt	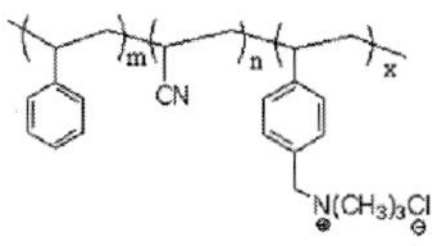	81
Terpolymer (maleic anhydride–styrene–vinylbenzyltrimethyl), MAST	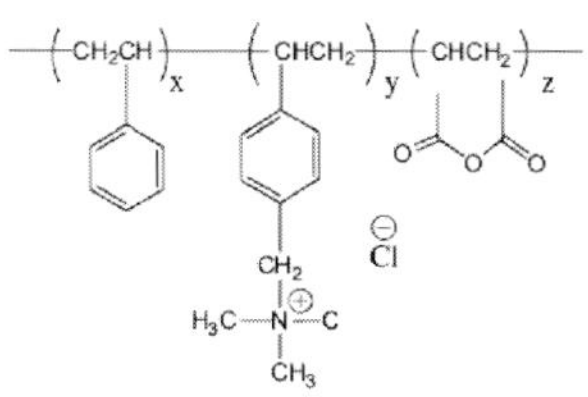	82

(*cont.*)

Table 3.6 (*cont.*)

Surfactant	Molecular structure	Reference
Poly(2-methyacryloyloxyethyl hexadecyldimethyl ammonium bromide) PMMA12		83

Wilkie and co-workers [69, 70] synthesized two organically modified clays to produce nanocomposites of PS, HIPS, and ABS terpolymer. They used the following copolymers to modify clay: vinylbenzyl chloride (COPS) and methyl methacrylate and vinylbenzyl chloride (MAPS). The cation head for clay modification with these compounds was ammonium. After melt-blending, styrene copolymer–modified clays yielded exfoliated nanocomposites, whereas the methacrylate copolymer clays yielded a mixture of immiscible and intercalated nanocomposites. In general, all nanocomposites exhibited improved thermal stability and mechanical properties, in addition to improvements in flame retardancy, depending on the quality of clay dispersion.

Polystyrene nanocomposites, incorporating clay intercalated with a copolymer of styrene and vinylbenzyltrimethylammonium chloride (COPS), were prepared using melt-compounding or the cosolvent method [71] (Table 3.6). COPS nanocomposites yielded interlayer spacings in the range 7–8 nm. The COPS phase appeared as large, immiscible domains within the PS matrix. In other work, PS nanocomposites were prepared by solution-blending, using organophilically modified silicates (natural and synthetic) with the ammonium salt of poly(styrene-ran-(4-vinylbenzylchloride)) [72]. The rheological properties of nanocomposites changed to solid-like behavior as the organoclay content increased. Nanocomposites from synthetic clay and from organoclays with short aspect ratios exhibited better clay dispersion than the natural clay series.

A butadiene-modified clay was prepared to produce PS, HIPS, ABS terpolymer, PMMA, polypropylene, and polyethylene nanocomposites by melt- or solution blending [73, 74]. The butadiene surfactant was obtained from the reaction of vinylbenzyl chloride–grafted polybutadiene with a tertiary amine (Table 3.6). All the composites were immiscible micro-composites.

Sen *et al.* [75] prepared PS nanocomposites by free radical polymerization in the presence of organically modified MMT with low–molecular weight quarternized poly(styrene-b-4-vinylpyridine) (SVP) (Table 3.6). Clay modification was carried out in different compositions of THF and water. Copolymer intercalation and thermal stability of the resulting organoclays depended on the THF/water proportion. Greater distances were obtained with higher THF proportions. Exfoliated nanocomposite structures were obtained when the MMT modification was conducted in the presence of 50 or 66% THF in the solution. Exfoliated nanocomposites showed high thermal stability and the best dynamic mechanical responses.

Oligomers of styrene, vinylbenzyl chloride, and diphenylvinylphenylphosphate (DPVBP) or diphenylvinylbenzylphosphate (DPVBP) were reacted with an amine and then ion-exchanged onto clay [76] (Table 3.6). The clays showed good thermal stability (degradation onset temperature in the range 330–340 °C). The clays were melt-blended with PS to produce intercalated nanocomposites. The use of aromatic phosphate compounds provided additional fire retardancy in both the condensed and vapor phases. However, the large amount of phosphate caused some loss in mechanical properties. Similarly, oligomers containing vinylbenzyl chloride, styrene, and dibromostyrene in different proportions were used to prepare nanocomposites both by bulk polymerization and by melt-blending [77]. The presence of dibromostyrene promoted the flame retardancy of polystyrene nanocomposites, compared to both the virgin polymer and polystyrene nanocomposites prepared from non–halogen containing organically modified clays. All the organoclays showed excellent thermal stability, and the bromide content did not have a great effect on thermal stability. The reduction in peak heat release rate was more significant for bulk polymerized samples than for melt-blended samples.

Oligomeric organoclay, called Triclay (Table 3.6), was obtained by modifying MMT with the ammonium salt of vinylbenzyl chloride (VBC), styrene (St), and lauryl acrylate copolymer [78]. Polymers used to produce nanocomposites were PS, HIPS, styrene–acrylonitrile copolymer (SAN), and ABS. Polymer polarity determined the quality of clay dispersion: intercalated nanocomposites were formed for PS and HIPS, whereas delaminated nanocomposites were formed for SAN and ABS. Thermal stability and fire retardancy were improved, but the effect of clay exfoliation was not evident. Young's modulus was improved, whereas tensile strength was unchanged, but elongation at break was reduced substantially in most nanocomposites. Similar results were obtained with a modified Triclay composition used to explore the influence of higher VBC and lower organic content in the clay [79, 80]. Chen *et al.* [81] produced organoclays incorporating oligomeric poly(styrene-co-acrylonitrile) quaternary ammonium salt prepared by free radical polymerization of a mixture of styrene, acrylonitrile, and vinyl benzyl chloride, with different proportions of acrylonitrile. The thermal stability of these organoclays would be suitable for the production of nanocomposites containing polymers that require high processing temperatures.

Nanocomposites of PS, HIPS, ABS terpolymer, polypropylene, and polyethylene were prepared by melt-blending using a methyl methacrylate oligomerically modified clay [82] (Table 3.6). The modified clay produced nanocomposites with PS and HIPS, but microcomposites were obtained with ABS, polypropylene, and polyethylene, depending on the polymer–clay interactions. The results suggested that clay dispersion was good in all cases. Maleic anhydride (MA) grafted-PS (MAST) clay nanocomposite exhibited tactoids and well-dispersed clay [83] (Figure 3.17). Flame retardancy was slightly improved.

Table 3.7 *List of phosphonium salts used in MMT modification [44]*

Organoclay/surfactant	Chemical name	MW, g/mol	Melting point, °C
Ph1: Cyphos IL 167[a]	Tributyl-tetradecyl-P$^+$Cl$^-$	434	45
Ph2: Cyphos IL 101[a]	Trihexyl-tetradecyl-P$^+$Cl$^-$	487	50
Ph3: Cyphos IL 166[a]	Tetra-*n*-octyl P$^+$Br$^-$	532	42
Ph4: Cyphos IL 164[a]	Tetra-*n*-butyl-P$^+$Cl$^-$	294	82

[a] Cyphos is a trademark owned by Cytec, Inc.

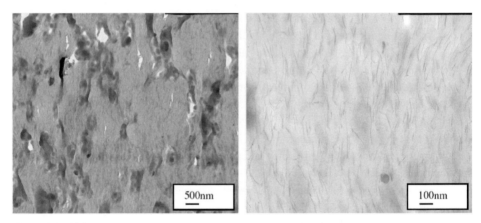

Figure 3.17 TEM image at low (left) and high (right) magnification for HIPS melt-blended with PMMA 12 clay. Reproduced from [82] with permission.

3.4 Factors influencing the selection of clay modifiers

The development of a complete formulation and processing system for the production of nanocomposites or thermally stable clays and nanocomposites must resolve a number of important issues, beyond the thermal stability of the clay modifier. We will describe some aspects of the approach followed by our group to achieve this goal. The main objective of the case study to be described was to evaluate a number of phosphonium compounds for the production of thermally stable organoclays that would be suitable for preparing satisfactory PS nanocomposites by melt-compounding.

In the first step, commercially available phosphonium salts, listed in Table 3.7, were used to modify montmorillonite. Standard techniques were used with water-soluble surfactants, whereas a two-phase reaction was used with surfactants exhibiting low water solubility. The thermal stability of the organoclays and the quality of the PS nanocomposites incorporating these organoclays were compared to the behavior exhibited by a conventional ammonium organoclay.

Table 3.8 *Summary of basal spacing (nm) of organoclays and PS nanocomposites*

Organoclay	Initial d_{001}	PS1510–d_{001}	PS1220–d_{001}
Ph1	2.32	2.32	2.38 (+0.06)
Ph2	2.52	2.32 (−0.20)	2.32 (−0.20)
Ph3	2.52	2.26 (−0.26)	2.45 (−0.07)
Ph4	1.84	1.84	1.84
Cloisite 10A	2.05	1.47 (−0.58)	1.47 (−.058)

Notes: Cloisite refers to Cloisite 10A. PS1510 and PS1220 are low– and high–molecular weight resins, respectively. Reproduced from [84].

Thermal stability of modified clay depends on the degree of cation modification. As the concentration of cation surfactants exceeds the natural CEC of the clay, the cations in excess are no longer attached to the clay surface and are susceptible to early thermal degradation. In general, phosphonium organoclays exhibited higher thermal stability than ammonium organoclays (Figure 3.9) [44]. The extent of the improvement depended on the molecular weight of the surfactant, with higher–molecular weight materials exhibiting higher thermal stability. The basal spacing in the organoclays correlated with molecular weight (Table 3.8) [44]. It was generally wider for higher molecular weight surfactants. Basal spacing also increased with the amount of surfactant used, up to the CEC equivalent of the clay. However, it should be emphasized that steric effects are important. The thermal stability of the PS/organoclay nanocomposites incorporating the various nanoclays is reflected in Figure 3.18 [84]. It is clear that all of the phosphonium organoclays yielded nanocomposites with thermal stability higher than that of the ammonium organoclay.

Various characteristics of the final nanocomposite, including the basal spacing in the clay, the quality of adhesion at the clay–polymer interface, and the ultimate mechanical, barrier, and other properties of the nanocomposite, are strongly influenced by the surface energy characteristics of the organoclay and the interfacial interactions between the organoclay and the polymer matrix. Thus, the surface energy characteristics of the various organoclays were determined, along with the interfacial energy characteristics of each organoclay with polystyrene, both at room temperature and at the processing temperature. The measured surface energy values were employed to determine the relevant values of the thermodynamic work of adhesion and the Hamaker constant for the various PS–organoclay pairs [45]. The quality of clay dispersion in PS resins was correlated with the initial basal spacing of organoclay and the values of the Hamaker constant for the polymer–organoclay system (A_{131}) and the clay (A_{11}) at the processing temperature (Figure 3.19) [45].

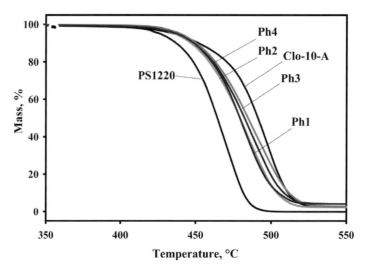

Figure 3.18 TGA curves for phosphonium- and ammonium-treated clay nanocomposites [44].

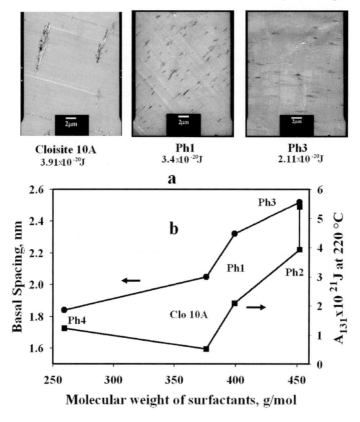

Figure 3.19 (a) TEM pictures of PS nanocomposites prepared with different organoclays (the clay content is 2% in all cases). (b) Effect of molecular weight of surfactant on the basal spacing and A_{131} (clay–polymer–clay) [45].

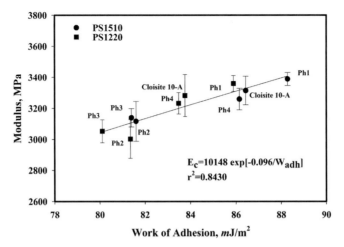

Figure 3.20 Nanocomposite moduli as a function of thermodynamic work of adhesion at 5% clay content [84].

Figure 3.20 [84] shows a good correlation between the tensile moduli of the various nanocomposites and the work of adhesion for these systems. It confirms that the overall performance of nanocomposites is influenced, not only by the quality of filler dispersion and/or polymer intercalation/exfoliation, but also by the quality (strength) of adhesion at the clay–polymer interface.

3.5 Concluding remarks

The thermal stability of organoclays is an important factor in the production of polymer/clay nanocomposites, especially in the case of melt-compounding. Depending on the polymer and clay systems of interest and the method used for nanocomposite formation, a variety of approaches may be used to impart thermal stability to the organoclay. Recent work in the field has made it possible to obtain thermally stable surfactants/modifiers that not only enhance the thermal stability of the organoclay and nanocomposite, but also enhance the quality of dispersion and the extent of intercalation/exfoliation. It is suggested that, through consideration of surface energy and interfacial interactions, further advances in modifier/surfactant development will lead to significant improvements in polymer/clay nanocomposite performance.

References

1. S. Sinha Ray and M. Okamoto, Polymer/layered silicate nanocomposites: A review from preparation to processing. *Progress in Polymer Science*, 28 (2003), 1539–1641.
2. S. Pavlidou and C. D. Papaspyrides, A review on polymer–layered silicate nanocomposites. *Progress in Polymer Science*, 33 (2008), 1119–98.

3. L. B. de Paiva, A. R. Morales, and F. R. Valenzuela Díaz, Organoclays: Properties, preparation and applications. *Applied Clay Science*, 42 (2008), 8–24.
4. G. Lagaly, Interaction of alkylamines with different types of layered compounds. *Solid State Ionics*, 22 (1986), 43–51.
5. G. Lagaly and K. Beneke, Intercalation and exchange reactions of clay minerals and non-clay layer compounds. *Colloid and Polymer Science*, 269 (1991), 1198–1211.
6. Y. Xi, Q. Zhou, R. L. Frost, and H. He, Thermal stability of octadecyltrimethylammonium bromide modified montmorillonite organoclay. *Journal of Colloid and Interface Science*, 311 (2007), 347–53.
7. W. Xie, Z. Gao, W.-P. Pan, D. Hunter, A. Singh, and R. Vaia, Thermal degradation chemistry of alkyl quaternary ammonium montmorillonite. *Chemistry of Materials*, 13 (2001), 2979–90.
8. W. Xie, Z. Gao, K. Liu, W.-P. Pan, R. Vaia, D. Hunter, and A. Singh, Thermal characterization of organically modified montmorillonite. *Thermochimica Acta*, 367–8 (2001), 339–50.
9. F. Avalos, J. C. Ortiz, R. Zitzumbo, M. A. López-Manchado, R. Verdejo, and M. Arroyo, Phosphonium salt intercalated montmorillonites. *Applied Clay Science*, 43 (2009), 27–32.
10. K. Stoeffler, P. G. Lafleur, and J. Denault, Thermal decomposition of various alkyl onium organoclays: Effect on polyethylene terephthalate nanocomposites' properties. *Polymer Degradation and Stability*, 93 (2008), 1332–50.
11. J. M. Cervantes-Uc, J. V. Cauich-Rodrıguez, H. Vazquez-Torres, L. F. Garfias-Mesıas, and D. R. Paul, Thermal degradation of commercially available organoclays studied by TGA–FTIR. *Thermochimica Acta*, 457 (2007), 92–102.
12. G. Edwards, P. Halley, G. Kerven, and D. Martina, Thermal stability analysis of organo-silicates, using solid phase microextraction techniques. *Thermochimica Acta*, 429 (2005), 13–18.
13. Y. Xi, R. Frost, H. He, T. Kloprogge, and T. Bostrom, Modification of Wyoming montmorillonite surfaces using a cationic surfactant. *Langmuir*, 21 (2005), 8675–80.
14. L. Cui, D. M. Khramov, C. W. Bielawski, D. L. Hunter, P. J. Yoon, and D. R. Paul, Effect of organoclay purity and degradation on nanocomposite performance. Part 1. Surfactant degradation. *Polymer*, 49 (2008), 3751–61.
15. W. Xie, R. Xie, W.-P. Pan, D. Hunter, B. Koene, L.-S. Tan, and R. Vaia, Thermal stability of quaternary phosphonium modified montmorillonites. *Chemistry of Materials*, 14 (2002), 4837–45.
16. N. Nassar, L. A. Utracki, and M. R. Kamal, Melt intercalation in montmorillonite/polystyrene nanocomposites. *International Polymer Processing*, 20 (2005), 423–31.
17. J. W. Lee, Y. T. Lim, and O. O. Park, Thermal characteristics of organoclay and their effects upon the formation of polypropylene/organoclay nanocomposites. *Polymer Bulletin*, 45 (2000), 191–8.
18. R. Scaffaro, M. C. Mistretta, F. P. La Mantia, and A. Frache, Effect of heating of organo-montmorillonites under different atmospheres. *Applied Clay Science*, 45 (2009), 185–93.
19. D. Dharaiya and S. C. Jana, Thermal decomposition of alkyl ammonium ions and its effects on surface polarity of organically treated nanoclay. *Polymer*, 46 (2005), 10139–47.

20. B. N. Jang and C. A. Wilkie, The thermal degradation of polystyrene nanocomposite. *Polymer*, 46 (2005), 2933–42.

21. K. Chen and S. Vyazovkin, Mechanistic differences in degradation of polystyrene and polystyrene–clay nanocomposite: Thermal and thermo-oxidative degradation. *Macromolecular Chemistry and Physics*, 207 (2006), 587–95.

21. A. Leszczynska, J. Njuguna, K. Pielichowski, and J. R. Banerjee, Polymer/montmorillonite nanocomposites with improved thermal properties. Part I. Factors influencing thermal stability and mechanisms of thermal stability improvement. *Thermochimica Acta*, 453 (2007), 75–96.

22. A. Leszczynska, J. Njuguna, K. Pielichowski, and J. R, Banerjee. Polymer/montmorillonite nanocomposites with improved thermal properties. Part II. Thermal stability of montmorillonite nanocomposites based on different polymeric matrixes. *Thermochimica Acta*, 454 (2007), 1–22.

23. J. W. Gilman, R. H. Harris, J. R. Shields, T. Kashiwagi, and A. B. Morgan, A study of the flammability reduction mechanism of polystyrene-layered silicate nanocomposite: Layered silicate reinforced carbonaceous char. *Polymers for Advanced Technologies*, 17 (2006), 263–71.

24. A. P. Kumar, D. Depan, N. S. Tomer, and R. P. Singha, Nanoscale particles for polymer degradation and stabilization – Trend and future perspectives. *Progress in Polymer Science*, 34 (2009), 479–515.

25. D.-R. Yei, H.-K. Fu, Y.-H. Chang, S.-W. Kuo, J.-M. Huang, and F.-C. Chang, Thermal properties of polystyrene nanocomposites formed from rigid intercalation agent-treated montmorillonite. *Journal of Polymer Science, Part B: Polymer Physics*, 45 (2007), 1781–7.

26. G. Chigwada, D. Wang, D. D. Jiang, and C. A. Wilkie, Styrenic nanocomposites prepared using a novel biphenyl-containing modified clay. *Polymer Degradation and Stability*, 91 (2006), 755–62.

27. G. Chigwada, D. D. Jiang, and C. A. Wilkie, Polystyrene nanocomposites based on carbazole-containing surfactants. *Thermochimica Acta*, 436 (2005), 113–21.

28. G. Chigwada, D. Wang, and C. A. Wilkie, Polystyrene nanocomposites based on quinolinium and pyridinium surfactants. *Polymer Degradation and Stability*, 91 (2006), 848–55.

29. D.-R. Yei, S.-W. Kuo, H.-K. Fu, and F.-C. Chang, Enhanced thermal properties of PS nanocomposites formed from montmorillonite treated with a surfactant/cyclodextrin inclusion complex. *Polymer*, 46 (2005), 741–50.

30. F. Carniato, C. Bisio, G. Gatti, E. Boccaleri, L. Bertinetti, S. Coluccia, O. Monticelli, and L. Marchese, Titanosilsesquioxanes embedded in synthetic clay as a hybrid material for polymer science. *Angewandte Chemie International Edition*, 48 (2009), 6059–61.

31. D.-R. Yei, S.-W. Kuo, Y.-C. Su, and F.-C. Chang, Enhanced thermal properties of PS nanocomposites formed from inorganic POSS-treated montmorillonite. *Polymer*, 45 (2004), 2633–40.

32. H.-K. Fu, S.-W. Kuo, D.-R. Yeh, and F.-C. Chang, Properties enhancement of PS nanocomposites through the POSS surfactants. *Journal of Nanomaterials* (2008), Article ID 739613, doi:10.1155/2008/739613.

33. H.-K. Fu, C.-F. Huang, J.-M. Huang, and F.-C. Chang, Studies on thermal properties of PS nanocomposites for the effect of intercalated agent with side groups. *Polymer*, 49 (2008), 1305–11.

34. J. Zhu, A. B. Morgan, F. J. Lamelas, and C. A. Wilkie, Fire properties of polystyrene–clay nanocomposites. *Chemistry of Materials*, 13 (2001), 3774–80.
35. J. Zhu, F. M. Uhl, A. B. Morgan, and C. A. Wilkie, Studies on the mechanism by which the formation of nanocomposites enhances thermal stability. *Chemistry of Materials*, 13 (2001), 4649–54.
36. H.-W. Wang, K.-C. Chang, H.-C.h Chu, S.-J. Liou, and J.-M. Yeh, Significant decreased dielectric constant and loss of polystyrene–clay nanocomposite materials by using long-chain intercalation agent. *Journal of Applied Polymer Science*, 92 (2004), 2402–10.
37. H.-W. Wang, K.-C. Chang, J.-M. Yeh, and S.-J. Liou, Synthesis and dielectric properties of polystyrene–clay nanocomposite materials. *Journal of Applied Polymer Science*, 91 (2004), 1368–73.
38. J.-M. Yeh, S.-J. Liou, H.-J. Lu, and H.-Y. Huang, Enhancement of corrosion protection effect of poly(styrene-co-acrylonitrile) by the incorporation of nanolayers of montmorillonite clay into copolymer matrix. *Journal of Applied Polymer Science*, 92 (2004), 2269–77.
39. A. B. Morgan, L.-L. Chu, and J. D. Harris, A flammability performance comparison between synthetic and natural clays in polystyrene nanocomposites. *Fire and Materials*, 29 (2005), 213–29.
40. M. H. Kim, C. I. Park, W. M. Choi, J. W. Lee, J. G. Lim, O. O. Park, and J. M. Kim, Synthesis and material properties of syndiotactic polystyrene/organophilic clay nanocomposites. *Journal of Applied Polymer Science*, 92 (2004), 2144–50.
41. S. Bruzaud, Y. Grohens, S. Ilinca, and J.-F. Carpentier, Syndiotactic polystyrene/organoclay nanocomposites: Synthesis via in situ coordination-insertion polymerization and preliminary characterization. *Macromolecular Materials and Engineering*, 290 (2005), 1106–14.
42. Y. Cai, Q. Wei, G. Chen, N. Wu, and W. Gao, Comparison between effects of two different cationic surfactants on structure and properties of HIPS/OMT nanocomposites. *Journal of Reinforced Plastics and Composites*, 28 (2009), 2161–72.
43. D.-R. Yei, H.-K. Fu, Y.-H. Chang, S.-W. Kuo, J.-M. Huang, and F.-C. Chang, Thermal properties of polystyrene nanocomposites formed from rigid intercalation agent-treated montmorillonite. *Journal of Polymer Science, Part B: Polymer Physics*, 45 (2007), 1781–7.
44. J. U. Calderon, B. Lennox, and M. R. Kamal, Thermally stable phosphonium–montmorillonite organoclays. *Applied Clay Science*, 40 (2008), 90–98.
45. M. R. Kamal, J. U. Calderon, and R. B. Lennox, Surface energy of modified nanoclays and its effect on polymer/clay nanocomposites. *Journal of Adhesion Science and Technology*, 23 (2009), 663–88.
46. J. W. Gilman, W. H. Awad, R. D. Davis, J. Shields, R. H. Harris, C. Davis, A. B. Morgan, T. E. Sutto, J. Callahan, P. C. Trulove, and H. C. DeLong, Polymer/layered silicate nanocomposites from thermally stable trialkylimidazolium-treated montmorillonite. *Chemistry of Materials*, 14 (2002), 3776–85.
47. W. H. Awad, High-throughput method for the synthesis of high performance polystyrene nanocomposites. *Polymer–Plastics Technology and Engineering*, 45 (2006), 1117–22.
48. S. Bourbigot, D. L. Vanderhart, J. W. Gilman, W. H. Awad, R. D. Davis, A. B. Morgan, and C. A. Wilkie, Investigation of nanodispersion in polystyrene–montmorillonite nanocomposites by solid-state NMR. *Journal of Polymer Science, Part B: Polymer Physics*, 41 (2003), 3188–3213.

49. F. A. Bottino, E. Fabbri, I. L. Fragala, G. Malandrino, A. Orestano, F. Pilati, and A. Pollicino, Polystyrene–clay nanocomposites prepared with polymerizable imidazolium surfactants. *Macromolecular Rapid Communication*, 24 (2003), 1079–84.

50. L. Abate, I. Blanco, F. A. Bottino, G. Di Pasquale, E. Fabbri, A. Orestano, and A. Pollicino, Kinetic study of the thermal degradation of PS/MMT nanocomposites prepared with imidazolium surfactants. *Journal of Thermal Analysis and Calorimetry*, 91 (2008), 681–6.

51. F. A. Bottino, G. Di Pasquale, E. Fabbri, A. Orestano, and A. Pollicino, Influence of montmorillonite nano-dispersion on polystyrene photo-oxidation. *Polymer Degradation and Stability*, 94 (2009), 369–74.

52. A. B. Morgan and J. D. Harris, Exfoliated polystyrene–clay nanocomposites synthesized by solvent blending with sonication. *Polymer*, 45 (2004), 8695–703.

53. J. Zhao, A. B. Morgan, and J. D. Harris, Rheological characterization of polystyrene–clay nanocomposites to compare the degree of exfoliation and dispersion. *Polymer*, 46 (2005), 8641–60.

54. Z. M. Wang, T. C. Chung, J. W. Gilman, and E. Manias, Melt-processable syndiotactic polystyrene/montmorillonite nanocomposites. *Journal of Polymer Science, Part B: Polymer Physics*, 41 (2003), 3173–87.

55. M. C. Costache, M. J. Heidecker, E. Manias, R. K. Gupta, and C. A. Wilkie, Benzimidazolium surfactants for modification of clays for use with styrenic polymers. *Polymer Degradation and Stability*, 92 (2007), 1753–62.

56. W. H. Awad, J. W. Gilman, M. Nydena, R. H. Harris, T. E. Sutto, J. Callahan, P. C. Trulove, H. C. DeLongc, and D. M. Fox, Thermal degradation studies of alkyl-imidazolium salts and their application in nanocomposites. *Thermochimica Acta*, 409 (2004), 3–11.

57. D. Wang and C. A. Wilkie, A stibonium-modified clay and its polystyrene nanocomposite. *Polymer Degradation and Stability*, 82 (2003), 309–15.

58. N. Greesh, P. C. Hartmann, and R. D. Sanderson, Preparation of polystyrene/clay nanocomposites by free-radical polymerization in dispersion. *Macromolecular Materials and Engineering*, 294 (2009), 787–94.

59. Z. Tang, D. Lu, J. Guo, and Z. Su, Thermal stabilities of vermiculites/polystyrene (VMTs/PS) nanocomposites via in-situ bulk polymerization. *Materials Letters*, 62 (2008), 4223–5.

60. J. Zhang and C. A. Wilkie, A carbocation substituted clay and its styrene nanocomposite. *Polymer Degradation and Stability*, 83 (2004), 301–7.

61. M. Zhao and P. Liu, Halloysite nanotubes/polystyrene (HNTs/PS) nanocomposites via in situ bulk polymerization. *Journal of Thermal Analysis and Calorimetry*, 94 (2008), 103–7.

62. X. Fan, C. Xia, and R. C. Advincula, Grafting of polymers from clay nanoparticles via in situ free radical surface-initiated polymerization: Monocationic versus bicationic initiators. *Langmuir*, 19 (2003), 4381–9.

63. A. Samakande, P. C. Hartmann, V. Cloete, and R. D. Sanderson, Use of acrylic based surfers for the preparation of exfoliated polystyreneeclay nanocomposites. *Polymer*, 48 (2007), 1490–99.

64. A. Samakande, J. J. Juodaityte, R. D. Sanderson, and P. C. Hartmann, Novel cationic raft-mediated polystyrene/clay nanocomposites: Synthesis, characterization, and thermal stability. *Macromolecular Materials and Engineering*, 293 (2008), 428–37.

65. A. Samakande, R. D. Sanderson, and P. C. Hartmann, RAFT-mediated polystyrene–clay nanocomposites prepared by making use of initiator-bound MMT clay. *European Polymer Journal*, 45 (2009), 649–57.
66. P. Liu, Polymer modified clay minerals: A review. *Applied Clay Science*, 38 (2007), 64–76.
67. Y. Zang, W. Xu, D. Qiu, D. Chen, R. Chen, and S. Su, Synthesis, characterization and thermal stability of different polystyryl quaternary ammonium surfactants and their montmorillonite complexes. *Thermochimica Acta*, 474 (2008), 1–7.
68. S. Su, D. D. Jiang, and C. A. Wilkie, Study on the thermal stability of polystyryl surfactants and their modified clay nanocomposites. *Polymer Degradation and Stability*, 84 (2004), 269–77.
69. S. Su, D. D. Jiang, and C. A. Wilkie, Novel polymerically-modified clays permit the preparation of intercalated and exfoliated nanocomposites of styrene and its copolymers by melt blending. *Polymer Degradation and Stability*, 83 (2004), 333–46.
70. B. N. Jang, M. Costache, and C. A. Wilkie, The relationship between thermal degradation behavior of polymer and the fire retardancy of polymer/clay nanocomposites. *Polymer*, 46 (2005), 10,678–87.
71. M. Sepehr, L. A. Utracki, X. Zheng, and C. A. Wilkie, Polystyrenes with macro-intercalated organoclay. Part I. Compounding and characterization. *Polymer*, 46 (2005), 11557–68.
72. S. S. Han, Y. S. Kim, S. G. Lee, J. H. Lee, K. Zhang, and H. J. Choi, Rheological properties of polystyrene–organophilic layered silicate nanocomposites. *Macromolecular Symposium*, 245–6 (2006), 199–207.
73. S. Su, D. D. Jiang, and C. A. Wilkie, Polybutadiene-modified clay and its nanocomposites. *Polymer Degradation and Stability*, 84 (2004), 279–88.
74. S. Su, D. D. Jiang, and C. A. Wilkie, Polybutadiene-modified clay and its polystyrene nanocomposites. *Journal of Vinyl and Additive Technology*, 10 (2004), 44–51.
75. S. Sen, N. Nugay, and T. Nugay, Effects of the nature and combinations of solvents in the intercalation of clay with block copolymers on the properties of polymer nanocomposites. *Journal of Applied Polymer Science*, 112 (2009), 52–63.
76. X. Zheng and C. A. Wilkie, Flame retardancy of polystyrene nanocomposites based on an oligomeric organically-modified clay containing phosphate. *Polymer Degradation and Stability*, 81 (2003), 539–50.
77. G. Chigwada, P. Jash, D. D. Jiang, and C. A. Wilkie, Synergy between nanocomposite formation and low levels of bromine on fire retardancy in polystyrenes. *Polymer Degradation and Stability*, 88 (2005), 382–93.
78. J. Zhang, D. D. Jiang, D. Wang, and C. A. Wilkie, Mechanical and fire properties of styrenic polymer nanocomposites based on an oligomerically-modified clay. *Polymers for Advanced Technologies*, 16 (2005), 800–806.
79. J. Zhang, D. D. Jiang, D. Wang, and C. A. Wilkie, Styrenic polymer nanocomposites based on an oligomerically-modified clay with high inorganic content. *Polymer Degradation and Stability*, 91 (2006), 2665–74.
80. J. Zhang, D. D. Jiang, and C. A. Wilkie, Polyethylene and polypropylene nanocomposites based on a three component oligomerically-modified clay. *Polymer Degradation and Stability*, 91 (2006), 641–8.
81. R. Chen, D. Chen, and S. Su, Synthesis and characterization of novel oligomeric poly(styrene-co-acrylonitrile)–clay complexes. *Journal of Applied Polymer Science*, 112 (2009), 3355–61.

82. X. Zheng, D. D. Jiang, D. Wang, and C. A. Wilkie, Flammability of styrenic polymer clay nanocomposites based on a methyl methacrylate oligomerically-modified clay. *Polymer Degradation and Stability*, 91 (2006), 289–97.

83. X. Zheng, D. D. Jiang, and C. A. Wilkie, Polystyrene nanocomposites based on an oligomerically-modified clay containing maleic anhydride. *Polymer Degradation and Stability*, 91 (2006), 108–13.

84. J. U. Calderon, B. Lennox, and M. R. Kamal, Polystyrene/phosphonium organoclay nanocomposites by melt compounding. *International Polymer Processing*, 23 (2008), 119–28.

4

Poly(ethylene terephthalate) nanocomposites using nanoclays modified with thermally stable surfactants

EVANGELOS MANIAS,[a] MATTHEW J. HEIDECKER,[a,c]
HIROYOSHI NAKAJIMA,[a,d] MARIUS C. COSTACHE,[b,e]
AND CHARLES A. WILKIE[b]

[a]*Polymer Nanostructures Lab–Center for the Study of Polymer Systems (CSPS)*
and The Pennsylvania State University
[b]*Marquette University*
[c]*Emerson Climate Technologies*
[d]*Sumitomo Chemical Co. Ltd.*
[e]*Rutgers University*

4.1 Introduction

The term "nanocomposite" is widely used to describe a very broad range of materials, where one of the phases has a submicrometer dimension [1–4].[1] In the case of polymer-based nanocomposites, this typically involves the incorporation of "nano" fillers with one (platelets), two (fibers, tubes), or all three dimensions at the submicrometer scale. However, strictly speaking, simply using nanometer-scaled fillers is not sufficient for obtaining genuine/true nanocomposites [5]: these fillers must also be well dispersed down to individual particles *and* give rise to intrinsically new properties, which are not present in the respective macroscopic composites or the pure components. In this chapter, we shall use a broader definition, encompassing also "nanofilled polymer composites" [5], where – even without complete dispersion or in the absence of any new/novel functionalities – there exist substantial concurrent enhancements of multiple properties (for example, mechanical, thermal, thermomechanical, barrier, and flammability). Further, we shall limit our discussion to one example, focusing on poly(ethylene terephthalate) (PET) with mica-type layered aluminosilicates.

The fact that nanometer-thin layered inorganics can be dispersed in macromolecular matrices is ubiquitous in nature, and it has also long been established in the laboratory; for example, synthetic polymers were shown to disperse appropriately-modified clay minerals before the 1960s [6, 7]. However, the field of polymer/layered-silicate

---

[1] This work is an overview of projects supported through various funding agencies and industrial sources. In particular, financial support by Bayer MaterialScience and by the Sumitomo Chemical Company is acknowledged. Work at Marquette University was supported by the Habermann Fund, MJH was supported through an ES&F ARL fellowship, HN by Sumitomo Chemical (Japan), and EM by the National Science Foundation (NSF DMR-0602877) and the "Virginia and Philip Walker" endowment. The authors would like to thank Dr. Cheng Fang Ou for providing us with his data, plotted in Figure 4.7d.

nanocomposites has gained substantial new momentum from the perspective of high-performance composite materials in the last decade. This renewed research interest was catalyzed by three main breakthroughs: first, pioneering work by the Toyota research group, who reported the preparation of a high-performance polyamide-6/layered-silicate nanocomposite [8, 9]; subsequently, the discovery by Giannelis and co-workers that it is possible to melt-process such composites [10], eliminating the need for organic solvents or new polymerization schemes; and finally, the work at NIST where it was first unveiled how nanosized clay fillers can impart a general flame-retardant character to polymers [11]. Since then, the field has been actively pursued, mostly because of the opportunities for concurrent enhancements in mechanical, thermal, barrier, and flammability properties [12–16] afforded by the addition of small amounts of clay (cf. nanofilled polymer composites).

Typically, marked property enhancements in nanocomposites originate from the nanometer-scale dispersion of highly anisotropic inorganic fillers in the polymer matrix and, thus, appropriate organic modification of the inherently hydrophilic clay fillers is a crucial step in the design and preparation of high-performing materials [16–19]. Among the modifiers used for nanoclays, quaternary ammonium surfactants are the most common because – besides their low cost and commercial availability – they can render these fillers miscible with a broad range of polymer matrices [18]. However, in the case of PET, alkylammonium surfactants are inadequate because, although they possess the proper favorable thermodynamics of mixing, they do not have the required thermal stability, neither for in situ PET polymerization nor for melt-processing of PET nanocomposites.

At this point, we should also mention that this chapter is not intended to provide an extensive review of the polymer nanocomposites field – the reader interested in such reviews can refer to a number of related books [1–4], numerous compilations of relevant symposia and conference proceedings, or recent review articles [12–15, 20, 21]. This chapter is rather an attempt to establish design principles toward the formation of PET nanocomposites with layered silicates bearing thermally stable surfactants, as well as linking these design principles to the relevant underlying fundamentals.

4.2 Melt-processable poly(ethylene terephthalate)/organosilicate nanocomposites

4.2.1 Challenges of PET nanocomposite formation

Because both melt-processing and polymerization of PET necessitate high temperatures (250–300 °C), it becomes obvious at the outset that any organically modified layered silicates that are intended as reinforcing fillers for PET should employ surfactants with appropriately high thermal stability. The typical alkylammoniums, for example, decompose below these temperatures. Two examples of higher-temperature surfactants that have been employed as modifiers for layered silicates in PET nanocomposites are pyridinium and phosphonium; specifically, cetylpyridinium, via solution dispersion [22], and dodecyltriphenylphosphonium, via in situ polymerization [23]. In these two cases, both the

pyridinium and phosphonium surfactants possess sufficiently high decomposition temperatures to survive, for example, the second stage of the polymerization reaction (typically
[23], 2 h at 280 °C). At the same time, the single alkane of the surfactant, dodecyl or cetyl,
satisfies the thermodynamic requirements for mixing [17], because in these two examples it
is only required that the organo-montmorillonites be miscible with small-molecule organics: for the solution mixing, the organofiller should disperse in a solvent (3:1 phenol:
chloroform) in which PET is dissolved [22], or for the in situ polycondensation, the
organofiller should disperse in the ethylene glycol, which is later reacted with dimethyl
terephthalate [23].

For melt-processing of PET nanocomposites, the requirements for the organically modified layered silicates are even more stringent, because, in addition to high temperatures,
melt-processing requires prolonged exposure to oxidizing environments. Thus, successful preparation of good melt-processed PET nanocomposites presents great challenges.
Successful nanocomposite preparation, in this case, implies thermodynamically favored
nanometer-scale dispersion of the layered silicates, with no marked decomposition of the
organofillers or of the PET matrix, processed under typical extrusion and injection-molding
conditions for PET. Specifically, all the following requirements should be satisfied by the
organic modification of the layered silicates:

1. *The organosilicate must possess sufficiently favorable thermodynamics of mixing with
   PET*: that is, the surfactant should promote mixing in PET nanocomposites with thermally stable organoclays within the typical residence time in an extruder (ca. 1–3 min).
   This requirement translates [16, 17] into the existence of one or two long-alkyl chains,
   for example, hexadecyl [24, 25] or octadecyl, or two hexadecyls [26]. Shorter alkyls
   are also possible, although they lead to poorer dispersion [24]. The incorporation of
   additional polar or polarizable groups into the surfactant, such as hydroxyls or phenyls,
   substantially *reduces* the thermodynamic free energy of mixing with PET.
2. *The surfactant must have high enough thermal stability to survive the typical melt-
   processing conditions of PET*: namely, it should have a decomposition temperature
   above 300 °C. This requirement is not met, for example, by alkylammoniums that have a
   decomposition temperature of about 250 °C. However, this requirement can be satisfied
   by large classes of surfactants, such as phosphonium-, imidazolium-, and pyridinium-
   based molecules.
3. *The organosilicates should be deprived of water as much as possible*, because PET is
   sensitive to water and will decompose under melt-processing conditions in the presence
   of water traces. This requirement is very important but, surprisingly, rather neglected.
   Most organically modified silicates contain measurable amounts of water that cannot
   be removed by drying: for example, alkylammonium and alkylphosphonium montmorillonites contain 5–8 wt% water that is hydrogen-bonded to the silicon oxide cleavage
   planes, which cannot be removed even after drying indefinitely under vacuum slightly
   above 100 °C; this water is often termed "structural water" and its removal requires
   drying temperatures of 350–450 °C under vacuum [27].

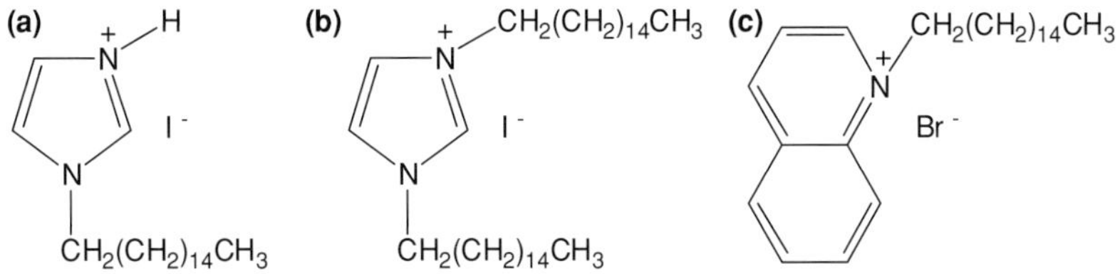

Figure 4.1 Chemical structures of (a) hexadecyl-imidazolium [26], (b) dihexadecyl-imidazolium [26], and (c) hexadecyl-quinolinium [25].

4.2.2 High–thermal stability surfactants for PET nanocomposites

In this chapter, we focus on *melt-processed* PET/nanoclay nanocomposites, where the nanoclay is alkylimidazolium montmorillonite or alkylquinolinium montmorillonite. The chemical structures of these thermally stable alkyl cations are shown in Figure 4.1, and both the surfactants and the respective organically modified clays can be synthesized easily (see Section 4.5). These imidazolium and quinolinium surfactants satisfy all three requirements for application to melt-processable PET nanocomposites, as enumerated in Section 4.2.1. Specifically,

1. These surfactants satisfy the requirement for *favorable thermodynamics of mixing with PET*, by virtue of their one or two long alkyl chains [17]. For example, Vaia and Giannelis [17, 18] employed a balance of entropic and enthalpic contributions – akin to those for polymer blends – to quantify the free energy change for dispersion of organically modified nanoclays in a polymer. In a first approximation, this mode defines the enthalpy change per interlayer-gallery area upon mixing by[2]

$$\Delta H \propto \varphi_p \varphi_a (\gamma_{ap} + \gamma_{sp} - \gamma_{sa}) \tag{4.1}$$

where the subscripts correspond to the various system components (layered silicate s, alkyl surfactant a, and polymer p); φ_p, φ_a are the interlayer volume fractions of polymer and surfactant; and γ_{ij} are the interfacial surface tensions describing the interactions between i and j components. Further, the gain in enthalpy upon filler dispersion, viz., the "competitive adsorption" interactions ($\Delta\gamma$, the difference of interfacial surface tensions in the parentheses in eq. (4.1)) can be calculated in a mean-field manner through pairwise interfacial surface tensions,

$$\gamma_{ij} = \gamma_{ij}^{LW} + \gamma_{ij}^{AB}, \text{ with } \begin{cases} \gamma_{ij}^{LW} \cong \left(\sqrt{\gamma_i^{LW}} - \sqrt{\gamma_j^{LW}} \right)^2 \\ \gamma_{ij}^{AB} \cong 2 \left(\sqrt{\gamma_i^+} - \sqrt{\gamma_j^+} \right) \left(\sqrt{\gamma_i^-} - \sqrt{\gamma_j^-} \right) \end{cases} \tag{4.2}$$

---

[2] In eq. (4.1) a number of proportionality constants [such as the monomeric volume fractions of polymer and surfactant in the interlayer, the surface area per surfactant, and the gallery height ($h = d_{001} - 0.97$ nm) of the intercalated structure] are omitted. Readers interested in the detailed calculation of ΔH are referred to the theoretical papers [17, 18].

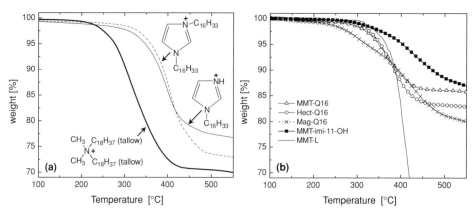

Figure 4.2 Thermal stability of organically modified clays: (a) TGA curves for hexadecylimida-zolium montmorillonite and dihexadecylimidazolium montmorillonite compared with di(tallow-alkyl)ammonium montmorillonite (heating rate 10 °C/min under nitrogen [26]). (b) TGA curves of MMT cationically exchanged with 1-(11-hydroxy-undecyl)-2,3-dimethyl-3H-imidazol-1-ium) [Imi-11OH] and three hexadecylquinolinium modified clays (montmorillonite [MMT-Q16], hectorite [Hect-Q16], and magadiite [Mag-Q16], all run at a heating rate of 20 °C/min under nitrogen [25]).

where the γ_{ij} interfacial surface tension are calculated based on the surface tension con-tributions from each component (γ_i and γ_j) following standard geometric combination rules and the van Oss–Chaudhury–Good formalization [28]; the LW, AB superscripts denote the nature of the interactions (apolar, Lifschitz–van der Waals LW; polar, Lewis-acid/Lewis-base AB), with γ_i^+ representing the electron-acceptor character of i, and γ_i^- representing the electron-donor character (thus, $\gamma_i^{AB} \cong 2\sqrt{\gamma_i^+ \gamma_i^-}$). The γ component values ($\gamma^{LW}, \gamma^+, \gamma^-$) are well established for all materials considered here, with $(66, 0.7, 36 \text{ mJ/m}^2)$ for montmorillonite [18, 29]; $(28, 0, 0 \text{ mJ/m}^2)$ for long-chain (C_{16}–C_{18}) alkyls [29]; and $(43.5, 0.01, 6.8 \text{ mJ/m}^2)$ for PET [30]. Substituting these values into eq. (4.2) gives excess enthalpy upon mixing PET with alkyl-modified montmorillonite that is approximately $\Delta\gamma \approx -8.5 \text{ mJ/m}^2$, indicating a rather high favorable energy of mixing (comparable to that of polyamide-6,6 in alkyl montmorillonite $\Delta\gamma \approx -9.9 \text{ mJ/m}^2$, and larger than that of polystyrene in alkyl montmorillonite $\Delta\gamma \approx -5.5 \text{ mJ/m}^2$).

Despite the approximations and the simplicity of this model, this line of thought clearly indicates that PET would show very good dispersion in hexadecyl montmorillonite (better than the intercalated PS structure, and comparable with the mostly exfoliated polyamide); this fact has been confirmed experimentally [24] and is also seen for our systems (see Section 4.3).

2. These surfactants also possess *sufficiently high thermal stability to survive the melt-processing conditions of any PET grade*. As shown in Figure 4.2, the decomposition temperatures of nanoclays organically modified with alkylimidazolium are above 300–320 °C, with a peak decomposition temperature around 400 °C, whereas the typical

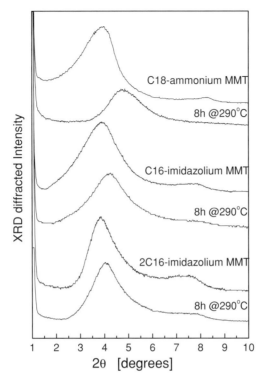

Figure 4.3 Thermal stability of organically modified MMT as reflected in XRD patterns (top) before and (bottom) after heat treatment at 290 °C for 8 h in vacuo. From top to bottom, three organo-MMT systems are shown: MMT modified by octadecylammonium (top two patterns), hexadecylimidazolium (middle two patterns), and dihexylimidazolium (bottom two patterns). Figure reproduced with permission from Wang *et al.* [26], © 2003 Wiley.

extrusion temperatures for PET are 240–280 °C and the typical injection molding temperature for PET is 260–300 °C (depending on application). In comparison, the same nanoclays modified by alkylammonium have a decomposition temperature of about 250 °C (Figure 4.2), which effectively renders them inapplicable under PET melt-processing conditions. Alternatively, instead of TGA, the thermal stability of the same organoclays can be quantified by X-ray diffraction after isothermal annealing at 290 °C in vacuo (Figure 4.3), which further demonstrates that alkylimidazolium montmorillonites can survive these temperatures, whereas alkylammonium montmorillonites cannot, as was indicated by the TGA behavior (Figure 4.2). In addition, substitution of methyl groups for the two hydrogens in the 2 and 3 positions of the imidazole ring offers even higher thermal stability [24, 31, 32] than the surfactants shown in Figures 4.1 (a, b), but may have lower efficacy for eliminating water from within the nanoclay interlayer galleries and for attaching two pending hydrophobic tails (cf. requirements 3 and 1). Along the same lines, even when PET/nanoclay composites are produced through an in situ polymerization scheme (typically polycondensation of ethylene glycol with dimethyl terephthalate in the presence of the nanofiller), the organic modification of the fillers still needs to survive a few hours at 250–280 °C in the second stage of the reaction. Although here the temperatures are slightly lower than for melt-processing, the alkylammonium surfactants still cannot meet the thermal stability requirements, whereas alkylimidazolium [33], -pyridinium [22], or -phosphonium [23] surfactants, or even acids [34], do.

3. Finally, a third important requirement is that *surfactant-treated fillers intended for PET nanocomposites should be deprived of water* as much as possible. This is particularly important in melt-processed PET nanocomposites, because water can be liberated from the nanofillers during extrusion and dramatically decompose the PET (hydrolysis). Removal of water becomes even more critical for clay nanofillers, because the pristine silicates are very hydrophilic and contain substantial amounts of structural water [6]. This structural water cannot be removed by standard practices of drying [27], but needs to be driven out of the interlayer gallery spaces by organic modification. This requirement implies that the surfactant should create a very hydrophobic environment in the interlayer gallery that would drive any structural water out and would prevent subsequent rehydration. Alternatively, water can be removed efficiently from alkylammonium or alkylphosphonium clays, if during the last stages of the organoclay preparation, washing is done with distilled organic solvents that form organoclay suspensions; however, these organoclays would readily rehydrate under ambient storage conditions. In contrast, alkylimidazolium surfactants do create a sufficiently hydrophobic environment in the interlayer gallery, which limits, or completely prevents, subsequent rehydration. Along these lines, dialkyl surfactants are more effective than monoalkyl equivalents – by virtue of their stronger hydrophobic character – and imidazole is probably more effective than quinoline or dimethyl-substituted imidazole – both of which tend to arrange flat on the silicate cleavage plane, when intercalated or surface end-tethered.

In summary, when surfactants are selected for nanofillers intended for PET nanocomposites, there are important requirements that should be met, substantially limiting the range of appropriate surfactants. As a first approach, alkylammoniums should be avoided, whereas alkylimidazoliums should be preferred. This last class of surfactants are applicable to high-performance nanocomposites beyond PET; for example, they were successfully used in polystyrene and polyamide matrices [32], epoxies [35], and ABS polymers [36, 37], as well as for fillers other than clays, such as polyhedral oligomeric silsesquioxane (POSS) [38] and carbon nanotubes [39].

4.3 Characterization and performance of poly(ethylene terephthalate)/clay nanocomposites

As examples of the utilization of thermally stable imidazolium-based surfactants with montmorillonite (MMT), allowing high-temperature melt-blending of PET nanocomposites,

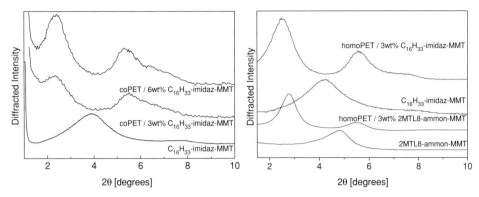

Figure 4.4 X-ray diffraction of alkylimidazolium montmorillonite (MMT) and the respective melt-processed nanocomposites with PET: (Left) XRD curves for hexadecylimidazolium MMT and its nanocomposites with copolymer-PET. (Right) XRD curves of MMT organically modified with dimethyl-tallow-2-ethylhexyl quaternary ammonium [2MTL8-ammon-MMT] and with hexadecylim-idazolium [$C_{16}H_{33}$-imidaz-MMT] and their respective nanocomposites with homopolymer-PET.

we compare two categories of poly(ethylene terephthalate): PET homopolymer [*homo-PET*, with high molecular weight ($M_w = 61$ kg/mol) and high intrinsic viscosity (IV = 0.95 dL/g), Voridian-12822] and PET copolymer (*co-PET*, poly(ethylene terephthalate-*co*-isophthalate), $M_w = 33$ kg/mol, IV = 0.64 dL/g, Kosa-1101). The focus is on elucidating the effect of the organo-MMT nature on the composite morphology of melt-processed PET nanocomposites and, further, compare the mechanical, thermal and fire properties between selected melt-processed PET nanocomposites.

4.3.1 Filler dispersion and composite morphology

The composite morphology was studied by Bragg-reflection powder X-ray diffraction (XRD, Figure 4.4) and by transmission electron microscopy (TEM, Figure 4.5). As a first approach, XRD can be used to assess the nanocomposite structure, because the d_{001} basal reflection is indicative of filler–filler separation (*intercalated* composite morphologies). In Figure 4.4 we compare the XRD patterns of organically modified montmorillonites before and after melt-processing them with PET (extrusion followed by injection molding). For both copolymer (blow-molding grade) PET and homopolymer (crystalliz-able) PET, a definitive shift of the d_{001} basal reflection to higher *d*-spacings (lower 2θ diffraction angles) denotes the insertion of PET within the interlayer gallery of the hexade-cylimidazolium montmorillonites; at the same time, the disappearance of the diffraction peak that corresponds to the organoclay (at $2\theta \sim 4°$ or $d_{001} \sim 2.2$ nm) denotes that all the organoclay was swollen or dispersed by PET. In addition, similar XRD studies of alkylammonium-modified MMT (Figure 4.4) also show gallery expansion of the organ-oclay upon melt-blending with PET, despite very considerable thermal degradation of the alkyl quaternary ammonium and some decomposition of the PET. However, the XRD can

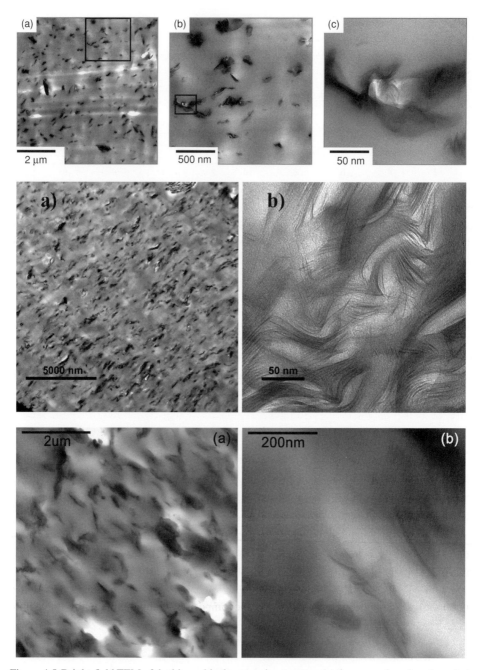

Figure 4.5 Bright-field TEM of the hierarchical composite structure (at the μm and nm length scales) of melt-processed PET/organo-MMT nanocomposites. (top) Melt-processed copolymer-PET/3 wt% $C_{16}H_{33}$–imidazolium MMT; boxes indicate the region of the subsequent higher-magnification image [44]. (middle) Melt-processed homopolymer-PET/3 wt% $C_{16}H_{33}$–imidazolium MMT [44]. (bottom) Melt-processed homopolymer-PET/3 wt% $C_{16}H_{33}$–quinolinium MMT [25]. © 2010, 2006 Wiley, reproduced with permission.

only detect the distance between periodically stacked layers; *disordered* (bunched together but not parallel stacked) or *exfoliated* layers are not detected by powder XRD. Although detailed quantitative analysis of XRD data in the low-2θ range, based on carefully prepared samples and the use of model reference samples, can yield substantially more information about the nanocomposite structure [40], powder XRD is still insufficient to capture and characterize the complete nanocomposite structure.[3] Thus, the XRD results need to be complemented by scattering techniques [41, 42], or by direct imaging of the composite structure, such as through transmission electron microscopy (TEM), to properly describe the diversity of nanostructures present in the nanocomposite [43].

In Figure 4.5, we show representative bright-field TEM pictures highlighting the hierarchical structures across length scales – from µm (agglomerates, tactoids) to nm (single montmorillonite layers). The structure of nanocomposites for co-PET and homo-PET with alkylimidazolium- and alkylquinolinium-modified MMT is shown. In all cases, organoclay dispersion in the polymer matrix is generally good, with no extended MMT agglomerates present. At the µm scale, filler tactoids are well dispersed throughout the polymer, most often separated by clay stacks of 2 to 4 layers and – to a lesser extent – single (exfoliated) MMT layers. At the nm scale, the higher-magnification images illustrate both intercalated and disordered structure, as individual layers of the fillers are expanded by the polymer penetration, often maintaining some of their parallel registry. It is tempting to note that as viscosity increases, so does the apparent tactoid dispersion, with homo-PET here showing better dispersion than co-PET; however, it must be mentioned that homo-PET is also more thermodynamically favored to disperse these organoclays than co-PET, given the aliphatic comonomer in the latter. These morphologies are in excellent agreement with the XRD observations, that is, shifted, broad, low-intensity d_{001} diffraction peaks and raised-intensity background (Figure 4.4). These composite structures are also consistent with a well-dispersed clay nanofiller composite morphology, showing the typical coexistence of intercalated, disordered, and exfoliated organoclay layers.[4]

4.3.2 Mechanical properties

The mechanical properties of the bulk PET and its nanocomposites were measured by tensile testing on injection-molded microtensile dogbones (ASTM-D638, Type IV) and by dynamic mechanical analysis (DMA) on injection-molded bars.

Tables 4.1 and 4.2 show the tensile properties of homo-PET and co-PET and their montmorillonite nanocomposites. For all nanocomposites, both with thermally stable imidazolium surfactants and with alkylammoniums, the tensile moduli increased compared to the respective unfilled PET (by 15% to 30%, for 3 wt% inorganic content) because of the montmorillonite addition. The magnitude of this modulus improvement is as expected

---

[3] In general, for the type of particles used here, that is, natural clays with medium (ca. 1 µm) lateral-size platelets, even with favorable thermodynamics for polymer nanocomposite formation, the composite structure is almost always characterized by the coexistence of exfoliated, intercalated, and disordered layers. Thus, a silent XRD may hide a large number of disordered tactoids, whereas an XRD with an intercalated peak does not reveal the extent of exfoliation.

[4] See note 3.

Table 4.1 *Tensile properties of homopolymer PET and its nanocomposites with alkylimidazolium (hexadecylimidazolium, im$_{16}$) and alkylammonium (dimethyl tallow 2ethylhexylammonium, 2MTL8, am$_{18.8}$) montmorillonite*

MMT content φ_{mmt} (wt%)	Tensile modulus E (MPa)	Yield strength σ_{yield} (MPa)	Elongation at break ε_{max} (%)
0	1550 ± 50	55 ± 2	230 ± 15
3 im$_{16}$	1830 ± 25	59 ± 1	202 ± 35
0	1550 ± 50	55 ± 2	230 ± 15
3 am$_{18.8}$	1860 ± 25	60 ± 1	115 ± 15
6 am$_{18.8}$	2080 ± 75	–	3 ± 0.4

Table 4.2 *Tensile properties of copolymer PET and its nanocomposites with hexadecylimidazolium, im$_{16}$, montmorillonite, as processed (extruded and injection-molded) and after annealing at 140 °C for 1.5 h under N$_2$*

MMT content φ_{mmt} (wt%)	Tensile modulus E (MPa)	Yield strength σ_{yield} (MPa)	Strengthk at break σ_{max} (MPa)	Elongation at break ε_{max} (%)
	co-PET/C$_{16}$-imidaz-MMT, as processed			
0	1780	62	31	82
3 im$_{16}$	2066	60	34	90
6 im$_{16}$	2308	59	46	5
	co-PET/C$_{16}$-imidaz-MMT, annealed			
0	2130	78	37	17
3 im$_{16}$	2512	–	73	4
6 im$_{16}$	2829	–	64	3

for the relatively stiff matrices considered here (unfilled polymer modulus 1.5–2 GPa); however, in most cases, increasing the filler loading resulted in concurrent decrease of the elongation at break, and eventually in dramatic embrittlement of the composite (ε_{max} < 10%). For example, addition of alkylammonium montmorillonite (2MTL8-MMT) to homo-PET at 3 wt% decreases ε_{max} from 230% to 115% (Table 4.1), and increasing the filler concentration to 6 wt% further reduces ε_{max} to 3% (the composite breaks under tensile deformation in a brittle manner, without showing a definitive yield point). Beyond any PET decomposition, this behavior was traced [44] in the changes of the PET crystallization upon addition of alkylammonium montmorillonite, which acts a heterogeneous nucleating agent for homo-PET, nucleating crystallites (spherulites, dendrites, etc.) with sizes comparable to the filler–filler separation distance [33, 45]; with increasing filler concentration the PET crystallites become increasingly small, unable to accommodate high tensile strains, causing the composite to catastrophically fail at very low strains. The tensile behavior of the PET/alkylammonium MMT systems is very similar to that of PET polymerized in the presence of surfactant-free/Na$^+$-MMT [45], hinting that the alkylammonium may have decomposed markedly during melt-processing. The tensile behavior of the PET/MMT nanocomposites can be improved by modifying the clay nanofillers with thermally stable surfactants; for example, a 3 wt% hexadecylimidazolium montmorillonite only reduces the elongation at break from 230% to 200%, but it is still necessary to keep the nanofiller concentration below the percolation threshold for these fillers (below ca. 4.5 wt% inorganic).

To further explore the embrittlement of PET upon addition of nanofillers, we investigated the tensile properties of the nanocomposites of PET copolymer with hexadecylimidazolium montmorillonite (Table 4.2). This blow-molding grade of co-PET already contains a noncrystallizable comonomer that reduces the PET crystallization. For practical purposes these polymers are considered to be mainly amorphous, and addition of small amounts of organoclay, for example, 3 wt% MMT, only increases the Young's modulus without affecting the yield point or the stress and strain at break (σ_{max}, ε_{max}). Increasing the organoclay concentration to 6 wt% dramatically reduces the elongation at break to 5%, because of nanofiller percolation. The tensile behavior upon crystallization in these systems can be revealed by annealing the nanocomposites (140 °C for 1.5 h under nitrogen), which allows crystallization to develop, increasing the crystal fraction and thus the Young's modulus and decreasing the ε_{max} (Table 4.2). As with the homo-PET, these annealed systems show very low ε_{max} (3–4%) and the absence of a definitive yield point. These results illustrate that in a mostly amorphous PET nanocomposite, the ε_{max} is also affected for high nanofiller loadings, where a percolated filler network associated with the dispersed silicate layers [46] causes decreased mobilities for the PET chains; whereas, for crystallizable PET nanocomposites, ε_{max} is dramatically decreased when the crystallite size reduction leads to inability of the crystallites to align properly during deformation, causing brittle failure. Thus, when the composites based on the two different PET matrices are compared, despite the apparent similar tensile behavior (E increases systematically with organoclay addition, and ε_{max} decreases dramatically beyond the percolation filler concentration) different mechanisms are responsible in each case. However, in both cases, the magnitude of improvement in E suggests that the interfacial adhesion between PET and imidazolium–MMT is sufficiently strong to transfer the external stress to the filler (up to $E \sim 2$–2.5 GPa), and is stronger than that of alkylammonium MMT; further, to maintain the PET ductility in MMT-based nanocomposites, the organofiller loading needs to be below the percolation filler concentration for both amorphous and crystallizable PET matrices.

Trends similar to those for the tensile properties can be observed in the thermomechanical properties of these nanocomposites. In Figure 4.6 we compare the dynamic mechanical response of a homo-PET with that of its nanocomposites at 3 wt% MMT (melt-processed with alkylammonium, 2MTL8, and alkylimidazolium, C$_{16}$-imidaz, montmorillonites). It is clear that the storage moduli of the nanocomposites are higher than that of the unfilled PET

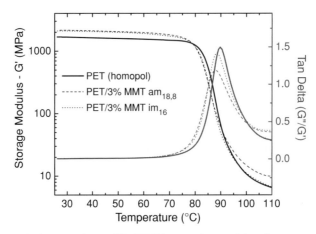

Figure 4.6 DMA analysis comparing unfilled PET homopolymer with melt-processed PET nanocomposites at 3 wt% inorganic loading, based on hexadecylimidazolium MMT (im$_{16}$) and on 2MTL8 alkylammonium MMT (am$_{18,8}$). The storage modulus (G') and the tanδ (G''/G') are plotted.

matrix (2.10 GPa at 30 °C, compared to 1.67 GPa for unfilled PET). This improvement in modulus persists up to the softening temperature of the matrix, around 80 °C, and is also present – albeit with a smaller relative increase – at high temperatures (G' plateau). Here it is of interest to note the difference in behavior around the softening temperature, for example, as manifested in the tanδ peak, where the composites based on alkylammonium MMT depart significantly more from the PET matrix response than those with the more thermally stable alkylimidazolium surfactants. This difference reflects the better thermal stability of the PET/alkylimidazolium MMT composites, because thermal decomposition of the PET in the presence of alkylammonium MMT leads to reduced polymer molecular weights, which, in turn, lead to a lowering of the apparent glass transition temperature (T_g).

4.3.3 Thermal and fire properties

The TGA results of co-PET and its melt-processed nanocomposites are given in Figure 4.7 (a, b). The addition of alkylimidazolium montmorillonite has no effect on the decomposition temperature at the maximum weight-loss rate ($T_{d\,max}$) of the co-PET under inert conditions (nitrogen), whereas this organoclay's addition caused an increase of the $T_{d\,max}$ of the co-PET in the oxidizing environment (air). These results indicate that the nanocomposites exhibit enhanced thermal stability, as would be expected. This behavior is identical for the homo-PET reinforced with the same alkylimidazolium montmorillonite (also melt-processed), as well as for the homo-PET reinforced by alkylammonium montmorillonite (Table 4.3); in this last case, TGA probably records the PET decomposition in the presence of "bare" MMT fillers/layers [45], because most of the alkylammoniums decomposed during the melt-processing. In all cases under air, the second decomposition temperature $T_{d\,max}^2$ probably includes contributions from the MMT dehydroxylation – resulting in a broader

Table 4.3 *Summary of TGA*

Materials	Under nitrogen[a]		In air[a]		
	$T_{d\,max}$ [b] (°C)	Char[c] (wt%)	$T^1_{d\,max}$ [b] (°C)	$T^2_{d\,max}$ [d] (°C)	Char[c] (wt%)
Co-PET (unfilled)	435	2.7	430	549	0.5
Co-PET/3wt% im$_{16}$	435	5.2	437	560	4.8
Co-PET/6wt% im$_{16}$	434	13.6	435	562	6.2
Homo-PET (unfilled)			439	550	0.0
Homo-PET/3wt% im$_{16}$			441	569	3.1
Homo-PET/3wt% am$_{18,8}$			438	553	4.4

[a] Gas flow rate 100 mL/mm; heating rate 10 °C/min.
[b] Decomposition temperature at the maximum weight-loss rate.
[c] Amount of the nonvolatile residue measured at the highest temperature (900 °C).
[d] Decomposition temperature at second maximum weight-loss rate.

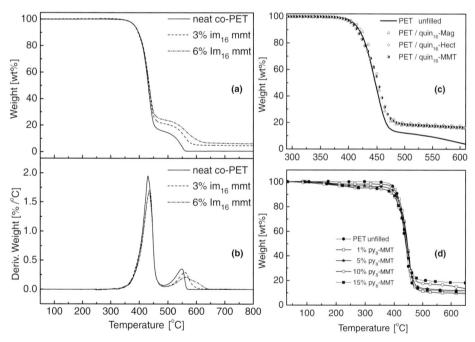

Figure 4.7 Thermogravitometric analysis (TGA) of various PET nanocomposites employing clays with thermally stable organic surfactants: (a,b) Melt-processed PET copolymer with hexadecylimidazolium MMT; (c) Melt-processed PET homopolymer with various clays bearing hexadecylquinoliniums [25]. (d) Solution-processed PET with cetylpyridinium MMT [22]. All TGA was done under N$_2$ flow.

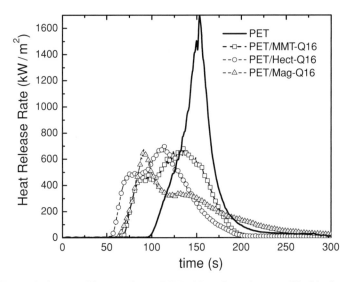

Figure 4.8 Cone calorimetry of homopolymer PET with various clays modified by hexadecylquino-linium (montmorillonite (MMT), hectorite (Hect), and magadiite (Mag)]. Figure adopted with permission from Costache *et al.* [25], © 2006 Wiley.

peak in the differential TGA, Figure 4.7 (b) – and any improvements in $T_{d\,\text{max}}^2$ cannot be definitively attributed to thermal improvements in the nanocomposites. This behavior also strongly resembles that of PET/organoclay melt-processed nanocomposites reinforced with alkylquinolinium MMT [Figure 4.7 (c)], and also of PET/alkylpyridinium-MMT nanocomposites formed by solution mixing [22] [Figure 4.7 (d)]; thus, this TGA response is characteristic of PET in the presence of MMT bearing thermally stable surfactants. It is interesting to note that the addition of imidazolium montmorillonite in co-PET greatly improved the amount of char, from 2.7% in the case of virgin PET to about 14% for the 6 wt% nanocomposite, in close agreement with what was previously observed for homo-PET/alkylquinolinium MMT nanocomposites [25].

Finally, because the TGA showed no deterioration in the PET thermal decomposition, it is of interest to characterize the fire performance of these nanocomposites as well. This was done by cone calorimetry for the alkylquinolinium MMT–reinforced homo-PET [25]. In Figure 4.8 we show the response of PET and its nanocomposites with three clays bearing thermally stable alkylquinolinium surfactants, measured at an incident flux of 35 kW/mm^2 on thermally thin (3-mm) specimens. These cone results were typical for polymer/clay nanocomposites, with the total heat release remaining unchanged, similarly to that of the unfilled PET, whereas the peak heat release rate (PHRR) is significantly reduced compared to the respective unfilled PET. At the same time, the mass loss rates are approximately constant for the three montmorillonite nanocomposites, and there is a more prolonged burning time for all the nanocomposites (evidenced by the shape of the HRR curves) accompanied by a decrease in the time-to-ignition (Figure 4.8). This previous study [25] concluded that this fire performance is probably due to filler-induced increased char formation, because the reductions in PHRR are roughly comparable for all four composites, despite their morphological variations (ranging from very good nanoscale and mesoscale dispersions for the MMT-based nanocomposites to rather poor dispersions and conventional composite structures for the magadiite-based composites [25]). In addition, when the fire behavior of these PET nanocomposites [25] was compared with that of PS nanocomposites [47] (based on both alkylquinolinium- and alkylammonium-modified MMT), it was suggested that the thermally stable quinolinium surfactants are more effective in fire-resistance improvement than the alkylammonium surfactants [25].

4.4 Conclusions

Nanoclays with thermally stable surfactant modifications were explored for use in melt-processed PET nanocomposites. Both alkylimidazolium and alkylquinolinium surfactants offer viable alternatives to the common alkylammoniums, and both resulted in well-performing PET nanocomposites. The melt-processing of PET composites with clays bearing these thermally stable surfactants does not require any changes in the standard practices of PET processing. The composite morphology of the melt-processed PET nanocomposites showed good nanofiller dispersion, which, in turn, resulted in good mechanical and thermal properties for the nanocomposites. For optimum mechanical improvement, the filler loading needs to be smaller than the filler percolation threshold concentration, for both amorphous and crystalline PET; for filler concentrations higher than this, marked embrittlement of the PET nanocomposite is observed.

4.5 Experimental

4.5.1 Synthesis of alkylimidazoles

As an example of a monoalkylimidazole surfactant, we describe the synthesis of 1-hexadecylimidazolium iodide [Figure 4.1 (a)] based on prior work [26]. An amount of 0.500 g (7.35 mmol) of imidazole was completely dissolved in 50 mL of tetrahydrofuran (THF) in a 100-mL flask with a reflux condenser. Subsequently, 2.590 g (7.35 mmol) of 1-iodohexadecane was added dropwise to the flask while it was stirred at 55–60 °C.

The reactants were heated to reflux with stirring for approximately 12 h. At this point, complete consumption of imidazole can be verified by ^{1}H NMR, through the disappearance of the peak at the 11.8-ppm chemical shift (assigned to the proton connected to the imidazole ring nitrogen: cf. Figure 1 in Ref. [26]). The solution was dried by removal of the THF, and a yellow solid was obtained; this solid was washed three times in 20 mL of hexane. The complete removal of unreacted 1-iodohexadecane can be verified by the absence of the ^{1}H peak at 3.28 ppm attributed to the CH_2 adjacent to the iodine in the 1-iodohexadecane. The resulting solid was fully protonated in 30 mL of 1 wt% hydrochloric acid solution in methanol for 1 h, and subsequently dried under vacuum to get a yellow solid, yielding 2.3 g of product (74% yield). To obtain a dialkylimidazole [Figure 4.1 (b)], a similar synthesis can be carried out, using a higher reaction temperature and an excess of 1-iodohexadecane [26]. An amount of 0.500 g (7.35 mmol) of imidazole was dissolved in 50 mL of THF and 7.770 g (three times the moles of imidazole) of 1-iodohexadecane was added dropwise to the flask while it was stirred. The solution was heated to reflux for approximately 48 h at 60–75 °C. After the removal of THF, the resulting purple-yellow solid was washed with large quantities of pentane to remove the unreacted 1-iodohexadecane, protonated with 1 wt% hydrochloric acid solution in methanol for 1 h, and subsequently dried under vacuum to yield 2.8 g of product (34% yield).

4.5.2 Synthesis of alkylquinoline

Hexadecylquinolinium [Figure 4.1 (c)] was prepared by the reaction of quinoline with 1-bromohexadecane. In a 250-mL flask, 10.0 g (77.4 mmol) of quinoline was dissolved in 150 mL of acetone by stirring for a few minutes. To this solution, 23.6 g (77.4 mmol) of bromohexadecane was added gradually, and then the mixture was refluxed for 48 h. Most of the solvent was removed under vacuum, followed by cooling to room temperature, upon which crystallization occurred. The sample was then washed with ether and filtered (5% yield).

4.5.3 Preparation of organically modified montmorillonites

For all cationic surfactants, the preparation of the organically modified nanoclays is carried out via a common cation exchange reaction with Na^+ montmorillonite. For example, in the case of alkylimidazolium [26], a large excess of the surfactant (twice the CEC of montmorillonite), was dissolved in ethanol at 50 °C and was added to a 1 wt% aqueous suspension of the montmorillonite under vigorous stirring. The mixture was stirred for 8 h at 50 °C, before the imidazolium-exchanged silicates were collected by filtration. The solids were subsequently washed with hot ethanol and paper-filtered 3–4 times, until an $AgNO_3$ test indicated the absence of halide anions. The filtrate was dried at room temperature, ground, and further dried at 80 °C under vacuum for 24 h, before being hermetically stored in a desiccator at nearly 0% RH.

4.5.4 Preparation of PET nanocomposites

Nanocomposites were prepared by melt-blending PET polymer with organically modified MMT. A variety of processing equipment was used, typically operated at 280 °C: a Braben-der plasticorder twin-head kneader (7 min at 280 °C and 60 rpm) for PET/quinolinium–MMT, a Prism TSE 16TC extruder with an L/D ratio of 16 (280 °C at a screw speed of 280–330 rpm) for large-scale PET/imidazolium–MMT (10–30 lb/h), and a laboratory-scale Haake counter-rotating twin screw extruder with an L/D ratio of 20 (280 °C at a screw speed of 50 rpm) for smaller-scale PET/imidazolium–MMT (5 lb/h). Prior to extrusion, all materials were dried overnight under vacuum at 100 °C and were tumbled-mixed for 20 min. Subsequent preparation of tensile dogbone specimens was done in a Boy 22D 24-ton injection molding machine (operated at 295 °C, with the mold at ambient temperature).

4.5.5 Mechanical characterization

The mechanical properties of the bulk PET and its nanocomposites were measured by tensile testing on injection-molded tensile bars (dogbones) and by DMA on injection-molded bars. The dogbones are ASTM D638 type IV specimens with a molded thickness of approximately 3.18 mm. An Instron 5866 tensile tester was operated with a cross-head speed of 50.8 mm/min. The Young's modulus, yield strength, and elongation at break are reported per the calculations from stress–strain curves done with the Instron software. The elongation at break is reported from the cross-head travel, as a strain extensometer with sufficient travel was unavailable. It is also important to note that the tensile behavior was typically measured on the as-molded tensile bars with no postmolding annealing, unless otherwise noted. DMA was also utilized to examine the thermomechanical behavior on a TA Instruments Q800 instrument with a 35-mm dual-cantilever setup. Such tests probe the response of the material to oscillatory deformation (1 Hz, at a constant strain of 0.01%, ramping temperature from 25 to 170 °C at a rate of 4 °C/min) and determine the storage (G') and loss (G'') modulus, and through $\tan\delta = G''/G'$ the material's character for energy dissipation (the temperature of the $\tan\delta$ peak is also a good indicator of the apparent glass transition temperature, T_g).

References

1. T. J. Pinnavaia and G. W. Beall, eds., *Polymer–Clay Nanocomposites* (West Sussex, UK: Wiley, 2000).
2. L. Utracki, *Clay-Containing Polymeric Nanocomposites* (Shropshire, UK: Rapra Tech, 2004).
3. Y. Mai and Z. Yu, eds., *Polymer Nanocomposites* (Cambridge, UK: Woodhead, 2006).
4. A. B. Morgan and C. A. Wilkie, eds., *Polymer Nanocomposite Flammability* (Hoboken, NJ: Wiley, 2006).
5. E. Manias, Nanocomposites – Stiffer by design. *Nature Materials*, 6 (2007), 9–11.
6. B. K. G. Theng, *Formation and Properties of Clay–Polymer Complexes* (Amsterdam: Elsevier, 1979).
7. B. K. G. Theng, *Chemistry of Clay–Organic Reactions* (New York: Wiley, 1974).
8. Y. Kojima, A. Usuki, M. Kawasumi, A. Okada, Y. Fukushima, T. T. Kurauchi, and O. Kamigaito, Synthesis and mechanical properties of nylon-6/clay hybrid. *Journal of Materials Research*, 8 (1993), 1179–84 and 1185–9.
9. Y. Kojima, A. Usuki, M. Kawasumi, A. Okada, T. T. Kurauchi, and O. Kamigaito, Synthesis of nylon-6/clay hybrid by montmorillonite intercalated with ε-caprolactam. *Journal of Polymer Science, Part A: Polymer Chemistry*, 31 (1993), 983–6.

10. R. A. Vaia, H. Ishii, and E. P. Giannelis, Synthesis and properties of 2-dimensional nanostructures by direct intercalation of polymer melts in layered silicates. *Chemistry of Materials*, 5 (1993), 1694–6.
11. J. Gilman, C. Jackson, A. Morgan, R. Harris, E. Manias, E. Giannelis, M. Wuthenow, D. Hilton, and S. Phillips, Flammability properties of polymer/layered-silicate nanocomposites: Polypropylene and polystyrene nanocomposites. *Chemistry of Materials*, 12 (2000), 1866–73.
12. P. C. LeBaron, Z. Wang, and T. J. Pinnavaia, Polymer-layered silicate nanocomposites: An overview. *Applied Clay Science*, 15 (1999), 11–29.
13. E. P. Giannelis, R. Krishnamoorti, and E. Manias, Polymer-silicate nanocomposites: Model systems for confined polymers and polymer brushes. *Advances in Polymer Science*, 138 (1999), 107–47.
14. M. Alexandre and P. Dubois, Polymer-layered silicate nanocomposites: Preparation, properties and uses of a new class of materials. *Materials Science and Engineering R: Reports*, 28 (2000), 1–63.
15. S. S. Ray and M. Okamoto, Polymer/596 layered silicate nanocomposites: A review from preparation to processing. *Progress in Polymer Science*, 28 (2003), 1539–1641.
16. E. Manias, A. Touny, L. Wu, K. Strawhecker, B. Lu, and T. C. Chung, Polypropylene/montmorillonite nanocomposites. Review of the synthetic routes and materials properties. *Chemistry of Materials*, 13 (2001), 3516–23.
17. R. A. Vaia and E. P. Giannelis, Lattice model of polymer melt intercalation in organically-modified layered silicates. *Macromolecules*, 30 (1997), 7990–99.
18. R. A. Vaia and E. P. Giannelis, Polymer melt intercalation in organically modified layered silicates: Model predictions and experiment. *Macromolecules*, 30 (1997), 8000–8009.
19. A. C. Balazs, C. Singh, and E. Zhulina, Modeling the interactions between polymers and clay surfaces through self-consistent field theory. *Macromolecules*, 31 (1998), 8370–81.
20. F. Leroux and J. Besse, Polymer interleaved layered double hydroxide: A new emerging class of nanocomposites. *Chemistry of Materials*, 13 (2001), 3507–15.
21. X. L. Xie, Y.-W. Mai, and X. P. Zhou, Dispersion and alignment of carbon nanotubes in polymer matrix: A review. *Materials Science and Engineering R: Reports*, 49 (2005), 89–112.
22. C. Ou, M. Ho, and J. Lin, Synthesis and characterization of poly(ethylene terephthalate) nanocomposites with organoclay. *Journal of Applied Polymer Science*, 91 (2004), 140–45.
23. J. Chang, S. Kim, Y. Joo, and S. Im, Poly(ethylene terephthalate) nanocomposites by in situ interlayer polymerization: The thermo-mechanical properties and morphology of the hybrid fibers. *Polymer*, 45 (2004), 919–26.
24. C. H. Davis, L. J. Mathias, J. W. Gilman, D. A. Schiraldi, J. R. Shields, P. C. Trulove, T. E. Sutto, and H. C. de Long, Effects of melt-processing conditions on the quality of poly(ethylene terephthalate) montmorillonite clay nanocomposites. *Journal of Polymer Science, Part B: Polymer Physics*, 40 (2002), 2661–6.
25. M. C. Costache, M. J. Heidecker, E. Manias, and C. A. Wilkie, Preparation and characterization of poly(ethylene terephthalate)/clay nanocomposites by melt blending using thermally stable surfactants. *Polymers for Advanced Technologies*, 17 (2006), 764–71.

26. Z. M. Wang, T. C. Chung, J. W. Gilman, and E. Manias, Melt-processable syndiotactic polystyrene/montmorillonite nanocomposites. *Journal of Polymer Science, Part B: Polymer Physics*, 41 (2003), 3173–87.

27. R. C. MacKenzie, ed., *The Differential Thermal Investigation of Clays* (London: Mineralogical Society, 1957).

28. C. J. van Oss, M. K. Chaudhury, and R. J. Good, Interfacial Lifschitz–van der Waals and polar interactions in macroscopic systems. *Chemical Reviews*, 88 (1988), 927–41.

29. C. J. van Oss, *Interfacial Forces in Aqueous Media* (New York: Dekker, 1994).

30. W. Wu, R. F. Giese, and C. J. van Oss, Evaluation of the Lifshitz–van der Waals/acid–base approach to determine surface tension components. *Langmuir*, 11 (1995), 379–82.

31. D. M. Fox, W. H. Awad, J. W. Gilman, P. H. Maupin, H. C. de Long, and P. C. Trulove, Flammability, thermal stability, and phase change characteristics of several trialkylimidazolium salts. *Green Chemistry*, 5 (2003), 724–7.

32. J. W. Gilman, W. H. Awad, R. D. Davis, J. Shields, R. H. H. Jr., C. Davis, A. B. Morgan, T. E. Sutto, J. Callahan, P. C. Trulove, and H. C. DeLong, Polymer/layered-silicate nanocomposites from thermally stable trialkylimidazolium-treated montmorillonite. *Chemistry of Materials*, 14 (2002), 3776–85.

33. S. Monemian, V. Goodarzi, P. Zahedi, and M. Angaji, PET/imidazolium based OMMT nanocomposites via in situ polymerization: Morphological, thermal, and nonisothermal crystallization studies. *Advances in Polymer Technology*, 26 (2007), 247–57.

34. A. Vassiliou, K. Chrissafis, and D. Bikiaris, Thermal degradation kinetics of in situ pre-pared PET nanocomposites with acid-treated multi-walled carbon nanotubes. *Journal of Thermal Analysis and Calorimetry*, 100 (2010), 1063–71.

35. J. Langat, S. Bellayer, P. Hudrlik, A. Hudrlik, P. H. Maupin, J. W. Gilman, Sr., and D. Raghavan, Synthesis of imidazolium salts and their application in epoxy montmoril-lonite nanocomposites. *Polymer*, 47 (2006), 6698–709.

36. M. Modesti, S. Besco, A. Lorenzetti, M. Zammarano, V. Causin, C. Marega, J. W. Gilman, D. M. Fox, P. C. Trulove, H. C. de Long, and P. H. Maupin, Imidazolium-modified clay-based ABS nanocomposites: A comparison between melt-blending and solution-sonication processes. *Polymers for Advanced Technologies*, 19 (2008), 1576–83.

37. M. Modesti, S. Besco, A. Lorenzetti, V. Causin, C. Marega, J. W. Gilman, D. M. Fox, P. C. Trulove, H. C. De Long, and M. Zammarano, ABS/clay nanocomposites obtained by a solution technique: Influence of clay organic modifiers. *Polymer Degradation and Stability*, 92 (2007), 2206–13.

38. D. M. Fox, P. H. Maupin, R. H. Harris, Jr., J. W. Gilman, D. V. Eldred, D. Katsoulis, P. C. Trulove, and H. C. De Long, Use of a polyhedral oligomeric silsesquioxane (POSS)–imidazolium cation as an organic modifier for montmorillonite. *Langmuir*, 23 (2007), 7707–14.

39. S. Bellayer, J. Gilman, N. Eidelman, S. Bourbigot, X. Flambard, D. Fox, H. De Long, and P. Trulove, Preparation of homogeneously dispersed multiwalled carbon nanotube/polystyrene nanocomposites via melt extrusion using trialkyl imidazolium compatibilizer. *Advanced Functional Materials*, 15 (2005), 910–16.

40. R. A. Vaia and W. D. Liu, X-ray powder diffraction of polymer/layered silicate nanocomposites: Model and practice. *Journal of Polymer Science, Part B: Polymer Physics*, 40 (2002), 1590–1600.

41. R. A. Vaia, W. D. Liu, and H. Koerner, Analysis of small-angle scattering of suspensions of organically modified montmorillonite: Implications to phase behavior of polymer

nanocomposites. *Journal of Polymer Science, Part B: Polymer Physics*, 41 (2003), 3214–36.

42. H. J. M. Hanley, C. D. Muzny, D. L. Ho, C. J. Glinka, and E. Manias, A SANS study of organoclay dispersions. *International Journal of Thermophysics*, 22 (2001), 1435–48.

43. A. Morgan and J. Gilman, Characterization of polymer-layered silicate (clay) nanocomposites by transmission electron microscopy and X-ray diffraction: A comparative study. *Journal of Applied Polymer Science*, 87 (2003), 1329–38.

44. M. J. Heidecker, H. Nakajima, and E. Manias, Structure and properties of PET/montmorillonite nanocomposites prepared by melt-blending. *Polymers for Advanced Technologies* (in press).

45. Z. Chen, P. Luo, and Q. Fu, Preparation and properties of organo-modifier free PET/MMT nanocomposites via monomer intercalation and in situ polymerization. *Polymers for Advanced Technologies*, 20 (2009), 916–25.

46. L. Xu, H. Nakajima, E. Manias, and R. Krishnamoorti, Tailored nanocomposites of polypropylene with layered silicates. *Macromolecules*, 42 (2009), 3795–803.

47. G. Chigwada, D. Wang, and C. Wilkie, Polystyrene nanocomposites based on quinolinium and pyridinium surfactants. *Polymer Degradation and Stability*, 91 (2006), 848–55.

5

Thermally stable polyimide/4,4′-bis(4-aminophenoxy)phenylsulfone-modified clay nanocomposites

VLADIMIR E. YUDIN[a] AND JOSHUA U. OTAIGBE[b]

[a]*Institute of Macromolecular Compounds, Russian Academy of Sciences, Russia*
[b]*School of Polymers and High Performance Materials, The University of Southern Mississippi, USA*

5.1 Polyimide/clay nanocomposites

5.1.1 Introduction

One of the most effective ways to develop new types of polymer materials is via hybrid organic–inorganic nanocomposites, where the size of the inorganic phase does not exceed 100 nm [1]. The formation of hybrid organic–inorganic nanocomposites with improved mechanical and gas/liquid barrier properties from hybridization of organic (polymer) and inorganic (nanoparticle) constituents at the molecular and supramolecular levels yields completely new polymer structures and properties that are not just a simple sum of those of the constituents [1]. In this context, polymer/clay nanocomposites have attracted significant academic and industrial interest in recent years because they exhibit unique microstructures with enhanced mechanical, thermal, and barrier properties compared with those of conventional microcomposites at identical filler concentrations [1–9]. It is noteworthy that chemical modification of nanoparticles such as clay prior to their incorporation into polymer nanocomposites provides a convenient route to improving dispersion and modifying interfacial properties that may in turn improve the properties of nanocomposites.

Polyimides (PI) are widely used in microelectronics and photonics because of their outstanding electrical properties, heat resistance, and chemical stability [10–12]. PI/clay nanocomposites have been reported to reduce the coefficient of thermal expansion, amount of moisture absorption, and dielectric constant for improved performance in these application areas already mentioned [13–22]. For example, Yano *et al.* [13] prepared a PI (pyromellitic anhydride–4,4′-oxydianiline)/clay composite film [(PMDA–ODA)/clay] by solution-mixing of polyamic acid (PAA) and a dimethylacetamide (DMA) dispersion of clay. They used dodecylamine as the clay modifier, and the film showed reduced thermal

Financial support of this work by the U.S. National Science Foundation under Contract Grant Numbers DMR 0652350, CBET 0317646 and CBET 0752150, and by the Russian Foundation of Basic Research under Contract Grant Number 07–03-00846-a is gratefully acknowledged. We are indebted to our collaborators, with whom we had the privilege of working on projects cited in this chapter. The research work of J.U.O.'s former graduate students and postdocs is gratefully acknowledged.

expansion and lowered gas permeability. However, they did not report the thermal and mechanical properties of the nanocomposites films. Yang *et al.* [14] fabricated and characterized a PI (pyromellitic anhydride–4,4′-diamino-3,3′-dimethyldiphenylmethane)/clay composite film [(PMDA-MMDA)/clay]. They used hexadecylamine as the clay modifier, and prepared a mixture of PAA with the clay dispersed in DMA followed by subsequent in situ polymerization of PAA in the mixture to yield the (PMDA–MMDA)/clay composite film. The mechanical and thermal properties of the film were improved even in the presence of agglomerated clay particles, as confirmed by a TEM image of the film. Delozier *et al.* [15] reported the morphology of a PI/clay nanocomposite film containing clay modified with a long-chain aliphatic quaternary ammonium salt. They showed that the modified clay particles were dispersed as stacked silicate layers structures in the polymer matrix. Delozier *et al.* further suggested that the poor exfoliation of the clay particles in the polymer matrix might be due to decomposition of the organic modifier, leading to stacking of the clay particles during thermal imidization. These reported studies indicate that the thermal stability of organic modifiers is important when modified clay is utilized with high-performance polymers such as PI that must be processed at relatively high temperatures.

5.1.2 Practical limitations of thermally stable modification of clay

Compared with the chemical modifiers already mentioned, aromatic ammonium salts are expected to provide improved thermal stability and compatibility between the PI matrix and modified clay. To our knowledge, only a few research groups have investigated PI nanocomposites containing clay modified with aromatic modifiers [16–22]. For example, Tyan *et al.* modified natural clay with selected aromatic ammonium salts having multiamine functional groups [16–18]. They prepared PAA and the modified clay in DMA dispersion by solution-mixing. The solution-mixing involved chemical reaction between amine groups of the modified clay and anhydride end groups of the PAA. The resulting PI/clay nanocomposite film displayed the intercalated and exfoliated structure of modified clay and showed enhanced thermal and mechanical properties. Vora *et al.* [19] and Liang *et al.* [20, 21] reported studies on a PAA/modified clay dispersion, where the clay was modified with aromatic ammonium salts, followed by in situ polymerization of PAA to yield the PI/clay nanocomposite film. Campbell and Scheiman [22] prepared nanocomposites of PMR-15 PI and a diamine-modified clay, where the clay was modified with four types of aromatic ammonium salts having diamine functional groups. Few studies on PI/clay nanocomposites have been reported in the literature, and the effects of the modification method on dispersions of clay in thermoplastic and thermosetting PI matrices are relatively poorly understood.

This chapter describes recent progress in studies aimed at understanding effects of various chemical modification methods on the state of dispersion of clay particles in a high-temperature PI matrix and on the physical properties of polymer/clay nanocomposite films. Three different modification methods and their effects on the PI/clay nanocomposite film properties are described. 4,4′-Bis(4-aminophenoxy)phenylsulfone (BAPS) with the chemical structure

was used as an aromatic modifier to improve the thermal stability of montmorillonite (MMT)-type clay. Special attention is devoted to the method of MMT modification by BAPS and the different methods (i.e., melt- or solution-blending) of incorporation of MMT–BAPS particles into the PI matrix. It is anticipated that the work described in this chapter will stimulate better understanding of the effects of modification methods on the quality of clay dispersions in PI matrices, enhancing our ability to prepare PI/clay nanocomposites with improved properties.

5.2 Polyimide/4,4′-bis(4-aminophenoxy)phenylsulfone modified clay nanocomposites: Synthesis and characterization

5.2.1 4,4′-bis(4-aminophenoxy)phenylsulfone-modified clay

Natural montmorillonite clay, Na-MMT [Cloisite Na$^+$, cation exchange capacity (CEC) = 92.6 meq/100 g, Southern Clay Products, Inc.] was organically modified with BAPS ammonium salts (Wakayama Seika Kogyo Co. in Japan) following the steps shown in Figure 5.1. A mixture of BAPS and hydrochloric acid (36.5% concentration) in deionized water (DI water) was prepared. Subsequently, either Na-MMT or an aqueous dispersion of Na-MMT that had been agitated for 3 h was added to the previously prepared mixture. The resulting mixture was agitated simultaneously with a mechanical stirrer and ultrasound at 60 °C for 6 h, and the mixture was then left at room temperature for 12 h. The resulting white precipitate was filtered and washed repeatedly with DI water at 60 °C to remove superfluous ammonium salts and Cl$^-$ ions. Removal of Cl$^-$ ions was monitored by titration with addition of 0.1N AgNO$_3$ into the filtered liquid (based on whether or not a white precipitation of AgCl was obtained). The filtered cake was freeze-dried for 12–18 h to yield BAPS-modified clay (MMT-BAPS).

The three different modification methods used are summarized in Table 5.1. The Na-MMT powder was added to the solution containing DI water + HCl + BAPS to give the modified clay denoted as Clay A. Note that the pH of the resulting Clay A mixture was adjusted to 1.6 after agitation of the mixture. For the modified clays denoted as Clay B and Clay C, Na-MMT was predispersed in DI water with ultrasound for 2 h prior to being mixed with solution A, containing DI water + HCl + BAPS, and the pH of the resulting Clay B and Clay C solutions was adjusted to 9.7. Note that Clay B and Clay C differed only in the concentration of BAPS used as depicted in Table 5.1. The ratio of the amount of BAPS to the CEC of Na-MMT was maintained at 1:1 for Clay A and Clay B and at 2.5:1 for Clay C.

Table 5.1 *Material compositions used in chemical modification of natural montmorillonite clay (Na-MMT) [46]*

	Solution A			Solution B	
Modified clay	Water (g)	HCl (g)	BAPS (g)	Water (g)	Na-MMT (g)
Clay A	150	1	0.8		2
Clay B	50	1	0.8	100	2
Clay C	50	1	2	100	2

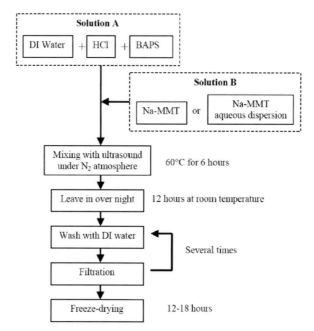

Figure 5.1 Elementary steps for modification of natural montmorillonite Na-MMT. Reproduced with permission from [46].

The XRD patterns of the MMT-BAPS, together with that of Na-MMT, are shown in Figure 5.2. A diffraction peak at around $2\theta = 7.3°$ (corresponding to an interlayer spacing of 1.2 nm) was observed in the Na-MMT. Diffraction peaks at $2\theta = 5.8°$ (equivalent to interlayer spacing of 1.5 nm) were displayed by the MMT-BAPS regardless of the modification methods used. These results indicate that modification of the Na-MMT with BAPS increases its interlayer spacing and are consistent with those reported in the literature on clay particles modified with aromatic ammonium salts [19, 22]. For example, Vora *et al.* [19] and Campbell and Scheiman [22] reported a diffraction peak located at $2\theta \sim 6°$ in the XRD patterns of MMT-BAPS and of clay modified with various diamines such as

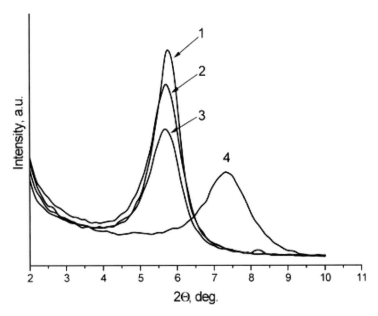

Figure 5.2 XRD patterns of modified clay particles and natural montmorillonite clay particles (Na-MMT). See text for sample nomenclature. Reproduced with permission from [46].

p-phenylenediamine, methylenedianiline, 4,4′(1,4-phenylenebismethylene)bisaniline, and 4,4′-bis(4-aminobenzyl)diphenylmethane.

TGA data for the respective clays are shown in Figure 5.3, together with those of the BAPS monomer for comparison. Onset of significant weight loss of the MMT-BAPS is evident at about 350 °C and can be ascribed to the loss of BAPS monomers, which are known to decompose at these temperatures (Figure 5.3). The weight loss of MMT-BAPS at temperatures ranging from 350 to 800 °C was found to be 14, 13, and 32 wt% in Clay A, Clay B, and Clay C, respectively. If all of the BAPS monomers used were chemically (ionic) bonded with the Na-MMT, the weight loss of the modified MMT should be 28.6 wt% for Clay A and Clay B, and 50 wt% for Clay C, according to the amount of BAPS used during the modification (see Table 5.1). The appearance of two thermal degradation steps in the TGA data is clearly evident for Clay C and not so clearly evident for Clay A and Clay B (Figure 5.3). This thermal degradation is thought to be connected with the presence of BAPS monomers in the MMT-BAPS powder that are chemically unreacted with the platelet surface of MMT. An additional indication of the presence of BAPS monomers that are not ionically bonded with MMT is the X-ray data for Clay C in the range of $2\theta = 17\text{–}22°$, where the crystalline form of BAPS has strong reflections at $2\theta = 17.8°$ and $18.2°$. It is worth noting that these reflections are completely absent for Na-MMT, suggesting that a portion of the BAPS monomer in the MMT-BAPS powder exists in the crystalline form and is not ionically bonded with MMT platelets.

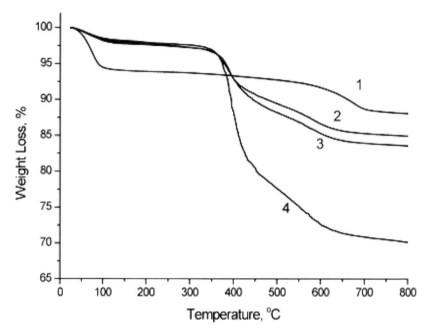

Figure 5.3 TGA scans for Na-MMT, modified clay particles, and BAPS monomers at a heating rate of 10 °C/min under nitrogen (see text for sample nomenclature). Reproduced with permission from [46].

5.2.2 Dispersion of MMT-BAPS particles in N-methyl-2-pyrrolidone solvent

For the processing of PI/MMT-BAPS nanocomposite films by solvent casting, it is imperative to determine optimal conditions for efficient dispersion of the aprotic *N*-methyl-2-pyrrolidone (NMP) solvent that is commonly used for dissolving R-BAPS type PI. To investigate the feasibility of MMT-BAPS forming a homogeneous nanodispersion in NMP, a rheology method was used that was shown previously to be an effective method of characterizing the exfoliation of clay in a suitable solvent [23, 24]. Homogeneous dispersion of nanoparticles in a suitable solvent leads to gelation of the solution, which exhibits thixotropic behavior that is ascribed to formation of a percolation-type network at some critical concentration of the nanoparticles used. Therefore, it is possible to estimate the efficiency of MMT chemical treatment with BAPS monomers in dispersing these MMT-BAPS particles in NMP and to compare the particle concentration dependencies of the gelation process in NMP.

Dispersions of different MMT-BAPS concentrations (from 1 to 20 wt%) in NMP solvent were typically prepared in separate beakers that were subjected to ultrasonic mixing for 1 h in an ultrasound bath (VWR ultrasonic cleaner, Model 50HT, frequency 40 kHz, average sonic power 45 W). The resulting translucent MMT-BAPS dispersions in NMP were used to perform rheological measurements. Note that all the solutions showed typical gel-like

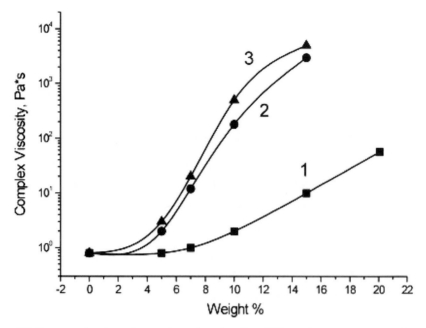

Figure 5.4 Concentration dependence of complex viscosity of Clay A (1), Clay B (2), and Clay C (3) suspensions in NMP. $T = 25\,°C$, frequency $\omega = 1$ rad/s, strain $\varepsilon = 1\%$. Reproduced with permission from [46].

behavior at appropriate concentrations of the MMT-BAPS particles depending on type of treatment (i.e., Clay A, B, or C). The resulting gels were all observed to be very stable for a long period of time (i.e., ≥ 1 week) in a closed bottle.

Dynamic frequency sweeps with a linear strain of 1% at room temperature in the cone/plate geometry were used to characterize the equilibrium state of the nanodispersions of MMT-BAPS in NMP. Figure 5.4 shows the dependence of complex viscosity on the weight concentration of the MMT-BAPS particles in NMP. Clearly, dramatic increases in the viscosities of the ClayB/NMP and ClayC/NMP nanodispersions are evident at approximately the same weight fraction of the MMT type particles (i.e., $\sim$5–7 wt%), which is remarkably close to that obtained from Cloisite-15A (or 20A)/xylene nanodispersions [23, 24]. In contrast, the viscosity increase of the ClayA/NMP nanodispersion occurred at relatively high concentrations of the MMT-BAPS particles ($\sim$12–15 wt%), indicating low dispersibility of Clay A in NMP compared to that of the Clay B and Clay C nanodispersions.

In an effort to find the reason for the obvious differences in the quality of dispersion of Clay A and Clay B in NMP solvent, the role of pH was explored by measuring the XRD pattern of an aqueous solution of Na-MMT with varying pH and constant concentration of Na-MMT (i.e., 2 wt% Na-MMT was added to DI water of pH 1.5 and 7.0, respectively). After the dispersions were agitated under ultrasonic mixing for 3 h, the pH increased from 1.5 to 1.7 and from 7.0 to 9.7, respectively, because of the well-known basic properties of

Table 5.2 *Properties and molecular weights of R-BAPS
type PI and oligoimides S1 and S2 [46]*

Sample	Stoich. offset, %	T_g °C	M_n g/mole	M_w g/mole	PDI
OI-1	50	153	3609	11553	3.2
OI-2	20	201	8845	23603	2.7
OI-3	5	220	18558	50133	2.7

Note: T_g is glass transition temperature; M_n is number average
molecular weight; M_w is weight average molecular weight; PDI
is polydispersity.

Na-MMT in water [25, 26]. The aqueous dispersion of Na-MMT at pH 9.7 did not show any diffraction peak in the XRD pattern, suggesting that the Na-MMT particles are well exfoliated in DI water. On the other hand, the aqueous dispersion of Na-MMT at pH 1.7 exhibited a significant diffraction peak at around $2\theta = 4.1°$, indicating that the Na-MMT particles exist as stacked silicate layers structures in the DI water at pH 1.7. During the preparation of Clay A, Na-MMT powder was added into solution A (see Table 5.1 and Figure 5.1) having pH 1.5. Under this low-pH condition, the Na-MMT does not exfoliate. Therefore, it can be concluded that the MMT is modified by the BAPS salts, whereas the stacked silicate layers structures are preserved. In contrast, Clay B was prepared by dispersing Na-MMT in DI water at pH 7 to yield a mixture with a pH of 9.7 after agitation with ultrasonic mixing for 3 h. Thus, when the aqueous dispersion of Na-MMT (solution B) and solution A, which includes BAPS, are combined, exfoliated Na-MMT particles are homogeneously modified with the BAPS salt. This experimental fact supports the conclusion that the finely dispersed structure of MMT-BAPS in Clay B and Clay C is preserved even after washing, filtering, and freeze-drying. The pH effect just described explains the difference in the possibility of obtaining homogeneous dispersion in NMP for Clay A, Clay B, and Clay C. Therefore, Clay B was selected for processing nanocomposites with PI by a solvent casting method because Clay B showed good dispersibility in NMP compared to Clay A; and the content of BAPS monomers in Clay B is approximately twice lower than that in Clay C.

5.2.3 Dispersion of MMT-BAPS particles in R-BAPS type polyimides and oligoimides

A number of oligoimides (OI) with different molecular weights was synthesized by poly-condensation of 1,3-bis(3',4,-dicarboxyphenoxy)benzene (R) and BAPS and by controlling the ratio of monomers to phthalic anhydride (PA) endcapper in accordance with the procedure described in [23]. Table 5.2 summarizes the properties and molecular weights of the R-BAPS-PA type OIs obtained from gel permeation chromatography. The lowest molecular weight of the oligoimide that was synthesized (OI-0) corresponds to the diamine BAPS endcapped by PA ($M_n \sim 700$ g/mole):

As previously reported in [23], the application of a strong shear flow field to R-BAPS-PA type OI melt-blended with MMT-15A clay particles (Southern Clay Products) can lead to relatively good dispersion of MMT-15A particles in the polymer volume. In addition, the significant viscosity increase exhibited by the OI/MMT nanocomposites after application of strong shear flow fields may be attributed to the partial exfoliation of the organo-MMT particles and formation of the percolating (network) structure at ~ 10 wt% MMT-15A particles in OI melt [23]. This finding suggests that OI/MMT clay nanocomposites are excellent model systems for investigating exfoliation of MMT-BAPS particles in PI, as follows.

A simple solution-mixing method was used to prepare a mixture of NMP solution of OIs with modified MMT. In this method, the modified clay was first dispersed in NMP with the aid of an ultrasonic mixer for 1 h, and various amounts of the modified clay were used to obtain the final OI/MMT mixtures containing 3–20 wt% of the modified MMT clay. Subsequently, an NMP solution of OI (20 wt% OI concentration) was added to the modified MMT/NMP dispersion and the combined mixture was stirred with a magnetic stirrer for 5 h, followed by ultrasonic mixing for 1 h. The resulting NMP dispersion of OI with modified MMT was poured on a metal substrate coated with Teflon and was subsequently dried at 100 °C for 1 h, 200 °C for 1 h, 280 °C for 0.5 h to remove NMP solvent completely (controlled with the aid of TGA). Note that the Teflon coating facilitates release (or collection) of the OI/MMT-BAPS blends in the form of film or powder after drying.

Figure 5.5 shows the dependence of complex viscosity (frequency $\omega = 1$ rad/s, strain $\varepsilon = 1\%$) for the OI blends just described. The viscosities of the blends were estimated at temperatures corresponding to that of the plastic (fluid) state of OIs. A significant increase of viscosity of at least four decades is typical for the OI-1/Clay B or OI-2/Clay B blends at concentrations of 5–10 wt% of the nanoparticles. This result is in relatively good agreement with that obtained in [23]. It is worth noting that relative to the results reported in [23], the gelation (or solidlike behavior) of the OI/MMT samples occurs not after melt-blending with application of a strong shear flow field but after simple solution-mixing of OI and MMT-BAPS particles and subsequent evaporation of NMP. The same result but with less (i.e., two decades) increase of viscosity is typical for OI-3/Clay B blends over approximately the same range of clay concentration. This typical gelation for OI/Clay B blends at clay concentrations of 5–10 wt% is not observed for the OI/Na-MMT blend, and its viscosity is changed linearly with varying concentrations of Na-MMT particles (see

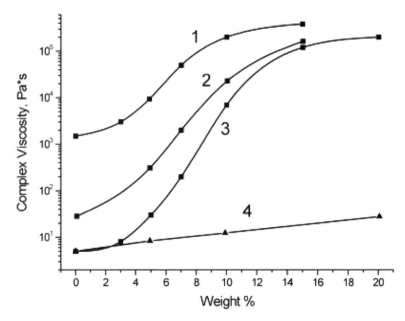

Figure 5.5 Complex viscosity as a function of concentration of the following blends prepared from NMP: 1 – OI-3/Clay B, $T = 360\,°C$; 2 – OI-2/Clay B, $T = 260\,°C$; 3 – OI-1/Clay B, $T = 260\,°C$; 4 – OI-1/Na-MMT, $T = 260\,°C$. Frequency $\omega = 1$ rad/s, strain $\varepsilon = 1\%$. Reproduced with permission from [46].

Figure 5.5). This last result clearly shows good dispersibility in OI of pure Na-MMT type particles, whereas appropriate surface treatment of Clay B particles is required to homogeneously disperse the clay in the polymer volume, even without applying a strong shear flow field.

It is important to note that the X-ray data shown in Figure 5.6 do not reveal complete exfoliation of the OI/Clay B nanocomposite at a particle concentration of 10 wt%. Therefore, the significant increase in viscosity at concentrations of the platelet-type particles near the percolation threshold may be ascribed to partial exfoliation of the Clay B particles in the polymer volume. This suggests that some of the clay platelets in the samples with relatively high concentrations of MMT particles (i.e., >5 wt%) may aggregate after NMP solvent evaporation during the processing of OI/MMT blends. This hypothesis is consistent with the observed intensity of the peak at $2\theta = 6.5°$ ($d = 1.35$ nm), which indirectly suggests the aggregation of clay platelets. The intensity of this peak decreases slightly with decreasing molecular weight of the polymer in the nanocomposites (i.e., from sample OI-3 to OI-1). Note that for the nanocomposite samples containing the lowest–molecular weight polymer (i.e., sample OI-0), the intensity of the peak at $2\theta = 6.5°$ is negligibly small for the OI-0/10 wt% polymer/Clay B nanocomposite, implying the absence of clay aggregation for the oligoimide OI-0 with chemical structure very similar to that of the BAPS monomer (see Figure 5.6). These results suggest that a decrease in the molecular weight of OI is

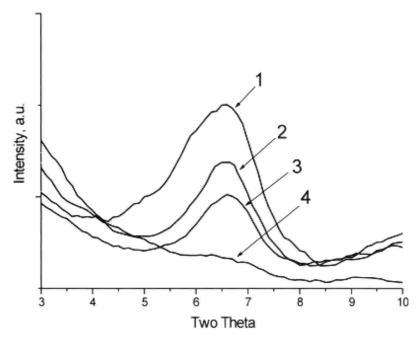

Figure 5.6 XRD patterns of the blends containing 10 wt% Clay B prepared from NMP: 1 – OI-3/Clay B; 2 – OI-2/Clay B; 3 – OI-3/Clay B; 4 – OI-0/ Clay B. See text for sample nomenclature. Reproduced with permission from [46].

preferable to achieving improved clay dispersion in the polymer volume. This improved clay dispersion is facilitated by the compatibility of the modified clay with the polymer matrix already described.

5.3 Novel polyimide nanocomposites based on silicate nanoparticles with different morphology

MMT nanoparticles are well known to be a mixture of several natural compounds with nonuniform composition and particle size [27]. For example, MMT minerals from different deposits might differ considerably in composition. This variation in composition of MMT significantly complicates the task of making functional nanocomposites with prescribed properties for targeted applications. Therefore, there is a need to develop synthetic nanofillers with prescribed particle composition, shape, and size for use as fillers in polymer nanocomposites with well-defined properties. In this area, synthetic nanodimensional silicates may provide a number of opportunities in polymer nanocomposites that is relatively little studied and poorly understood relative to the well-studied polymer nanocomposites filled with natural layered MMTs [28–32]. In contrast to the commonly used layered MMT compounds, it is envisaged that use of nanoparticles with different morphology (e.g.,

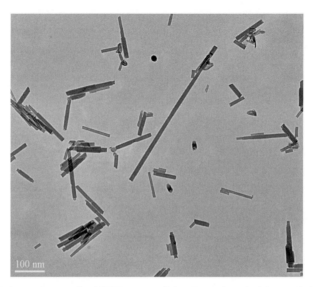

Figure 5.7 TEM photomicrograph of SNT-type particles. Reproduced with permission from [47].

nanotubes and nanoparticles of isometric shape) might provide additional benefits in polymer nanocomposites not possible with MMT.

In this section, we describe the feasibility of using as nanofillers for PI matrices new silicate nanotubes (SNT) and zirconium dioxide (ZrO_2) nanoparticles (with morphology different from that of layered MMT platelets) that were custom synthesized specifically for nanocomposite application [30–32]. The rheology and mechanical behavior of the PI nanocomposites as a function of the nanofiller shapes (i.e., platelets, nanotubes, and isometric form) is presented and discussed to provide a starting point for further experiments that will improve our understanding of optimal methods of incorporating nanoparticles with varying shapes and sizes into PI matrices and of the effects of the nanoparticles' morphology on the structure and properties of PI nanocomposites in general.

5.3.1 Silicate nanotubes and ZrO_2 nanoparticles

SNT with a chrysotile-type structure of $Mg_3Si_2O_5(OH)_4$ were synthesized using a hydrothermal method that is facilitated by high-pressure autoclaves, as described elsewhere [30]. The SNT density is 2.5 g/cm³. TEM showed that the SNT particles have average outer diameter 15 nm and average interior diameter 3 nm, as illustrated in Figure 5.7. The length of the SNT ranges from 50 to 600 nm and their average aspect ratio is 10.

Zirconium dioxide (ZrO_2) particles were synthesized using a hydrothermal method [31]. These particles have spherical shapes with an average diameter of 20 nm, and a density of ~6 g/cm³.

To improve compatibility of the nanoparticles with the PI matrix, the SNT and ZrO_2 particles were treated with *m*-aminophenyltrimethoxysilane (Gelest, Inc.) following the procedure described in [33]. The amount of silane needed to obtain minimum uniform multilayer coverage was estimated from the known values of the specific wetting surface of the silane (i.e., ~350 m^2/g) and the surface area of the filler (i.e., ~100 m^2/g) and found to be about 3.5 times less than the amount of filler used. Specifically, one gram of nanoparticles (SNT or ZrO$_2$) was dispersed in 50 mL of ethanol with the aid of an ultrasonic mixer (40 kHz, average sonic power 45 W) for 1 h. A quantity of 0.3 mL of silane was added to the sonicated suspension, followed by additional sonication for 10 min. The resulting suspension of the nanoparticles was centrifuged and the supernatant ethanol was decanted to yield the silane-treated nanoparticles. The silane-treated nanoparticles were rinsed twice with ethanol, followed by curing of the silane layer and subsequent drying until a constant weight was achieved in a vacuum oven maintained at 60 °C for 5 h.

5.3.2 Rheology of oligoimide nanocomposites filled with silicate nanoparticles with different morphology

To evaluate the feasibility of homogeneous dispersion of the nanoparticles in a PI matrix, a rheological method was used that was previously reported to be an effective method for characterizing the dispersion of nanoparticles in a variety of polymers [23, 24, 34–36]. Homogeneous dispersion of nanoparticles in the polymer should lead to time-dependent rheological behavior (i.e., thixotropy) that is characteristic of the formation of a percolation-type network at some concentration of the nanoparticles. Generally, it is possible to estimate this percolation threshold P_c theoretically [37] by using the following equations for a cylinder:

$$P_c = 0.6/r, \tag{5.1}$$

and for an ellipsoid:

$$P_c = 1.27/r, \tag{5.2}$$

where the aspect ratio $r = L/d$, L is the length of the cylinder or the diameter of the ellipsoid, and d is the diameter of the cylinder or the thickness of the ellipsoid. The preceding equations and the average aspect ratio of ~10 for SNT were used to estimate the percolation threshold (P_c) and the P_c (SNT) was found to be ~6 vol%. Assuming an average diameter (lateral dimension) of ~200 nm for the MMT particles [3, 38] and a thickness of ~1 nm, the percolation threshold for the MMT particles was estimated as P_c (MMT) = 0.64 vol%.

Figure 5.8 shows the dependence of complex viscosity (frequency $\omega = 1$ rad/s, strain $\varepsilon = 1\%$) on the concentration of the nanofillers (i.e., MMT-BAPS, SNT, and ZrO$_2$) for the current OI-1 nanocomposites prepared as already described. The figure shows significant increases (~3 decades) for the OI/MMT-BAPS and OI/SNT nanocomposites occurring at 2–3 vol% MMT–BAPS and 8–12 vol% SNT nanofiller concentrations, respectively. For the two types of nanocomposites just mentioned, the critical nanofiller concentrations corresponding to the dramatic rise in viscosity were found to be both higher than the percolation threshold values estimated theoretically as already described [i.e., P_c

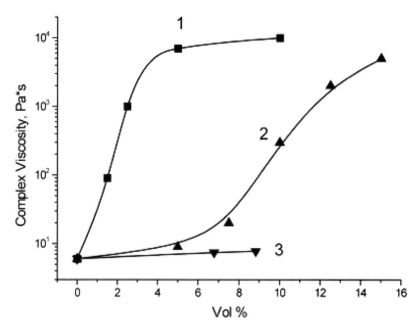

Figure 5.8 Dependence of complex viscosity on volume concentration of nanoparticles in oligoimide OI-1. Temperature $T = 260\,°C$, frequency $\omega = 1$ rad/s, strain $\varepsilon = 1\%$. Reproduced with permission from [47].

(MMT) $= 0.64$ vol% and P_c (SNT) $= 6$ vol%]. This trend in the viscosity data is consistent with that previously reported in the literature for similar polymer nanocomposites and has been attributed to different degrees of exfoliation of the nanofillers in the polymer matrix [39].

These results are consistent with our expectations of strong influence of the particle morphology and aspect ratio of the nanofillers (i.e., MMT platelets or SNT cylinders). For example, the relatively low–aspect ratio SNT was found to exhibit a dramatic viscosity rise that is known to be consistent with the formation of a percolation network structure at relatively higher concentrations than that exhibited by the relatively higher–aspect ratio MMT (Figure 5.8).

As can be seen in Figure 5.8, the concentration dependence of viscosity for the OI/ZrO$_2$ nanocomposite was found to be consistent with that predicted by the classical Thomas equation [40]

$$\eta = 1 + 2.5\phi + 10.05\phi^2 + A\,\exp(B\phi) \qquad (5.3)$$

where ϕ is volume fraction of particles, $A = 0.0273$, and $B = 16.6$. In this case, this relatively modest increase in viscosity with increasing volume fraction (up to 9 vol% as depicted in Figure 5.8) is somewhat similar to that reported in the literature for polymers filled with spherical particles of various sizes [41]. The observed behavior of the OI/SNT and OI/MMT nanocomposites suggests that anisotropic particles such as MMT or SNT are more effective in inducing formation of percolating network structure (i.e., gelation) than spherical particles such as ZrO_2 [42].

5.3.3 Processing of PI/MMT–BAPS nanocomposite thin films and sheets by solvent casting and compression molding respectively

Nanocomposite films based on poly(pyromellitic dianhydride-co-4,4'-oxydianiline) (PM) were prepared with different concentrations of nanoparticles (SNT, ZrO_2, or MMT-BAPS) by adding the desired amount of nanoparticles to NMP. The resulting suspension of particles in NMP was homogenized with the aid of an ultrasonic mixer (40 kHz, average sonic power 45 W) for 1 h. The sonicated suspension was transferred into a three-necked round-bottom flask equipped with a mechanical stirrer, a nitrogen gas inlet, and a drying tube outlet filled with calcium sulfate. After the nanoparticle solution was stirred for 10 min, PAA of PM was added to the nanoparticle suspension and the stirring of the mixture was continued for an additional 60 min until a constant viscosity was obtained. The solid content of the nanoparticles/PAA-PM was 10 wt% in NMP. Thin PI-PM nanocomposite films (30–40 μm thick) with varying nanoparticle weight concentrations were prepared from the nanoparticles/PAA-PM solution. Imidization was achieved by placing the films on soda lime glass plates in an oven for curing in air at 100 °C/1 h, 200 °C/1 h, 300 °C/1 h, and 350 °C/30 min. The cast films were removed from the glass plates after complete imidization by soaking in water. Using the material densities (1.42 g/cm^3 for PI-PM, 2.5 g/cm^3 for SNT, 6 g/m^3 for ZrO_2, and ~2 g/cm^3 for MMT-BAPS), the corresponding volume concentrations of nanoparticles in the PI nanocomposites were estimated.

A HAAKE MiniLab micro compounder was used for melt-mixing the nanoparticles with PI type R-BAPS [23, 43, 44, 46] based on diamine BAPS and dianhydride 1, 3-bis(3′,4,-dicarboxyphenoxy)benzene (R):

The materials were dried before mixing for a minimum of 6 h at 80 °C. Five grams of mixtures of R-BAPS and nanoparticles were extruded to obtain blend concentrations of 2, 6, 10, and 15 wt% nanoparticles in the PI. The melt-compounding was carried out at a barrel temperature of 340 °C and a screw speed of 100 rpm for a period of 20 min. Extruded nanocomposite pellets were injection molded into standard specimens using a DACA Instruments MicroInjector. The test specimens had length 20 mm, width 5 mm, and thickness 1 mm. These specimens were prepared using a barrel temperature of 380 °C, a mold temperature of 90 °C, and an injection pressure of 100 bar.

Table 5.3 *Mechanical properties of nanocomposite films based on PI-PM type PI matrix [47]*

Particles	vol %	E_t GPa	σ_t MPa	ε_t %
MMT-BAPS	0	2.85 ± 0.05	131 ± 3	30 ± 2
	1.5	3.13 ± 0.04	120 ± 2	22 ± 6
	2.5	3.34 ± 0.11	117 ± 5	16 ± 3
	3.4	3.65 ± 0.07	110 ± 4	13 ± 3
	4.8	3.96 ± 0.10	108 ± 6	5 ± 2
SNT	2.8	3.07 ± 0.07	123 ± 3	24 ± 2
	5.4	3.5 ± 0.10	97 ± 2	12 ± 3
	7.9	3.7 ± 0.10	86 ± 3	6 ± 2
ZrO_2	2.3	3.00 ± 0.05	105 ± 2	15 ± 2
	3.4	3.06 ± 0.07	95 ± 4	10 ± 5
	4.5	3.13 ± 0.10	90 ± 3	7 ± 3

Notes: E_t – tensile modulus; σ_t – tensile strength; ε_t – elongation at break.

After molding or preparation of the films by solvent casting, the specimens were sealed in a polyethylene bag and placed in a vacuum desiccator for a minimum of 24 h prior to measurements of the properties described in the following sections.

5.3.4 *Properties of PI nanocomposites containing silicate nanoparticles*

The mechanical properties (tensile modulus E_t, tensile strength σ_t, and elongation at break ε_t) of the PI-PM based nanocomposite films are summarized in Table 5.3. Clearly, this table shows that incorporation of SNT and MMT-BAPS into a PI matrix improves its modulus considerably. Figure 5.9 compares the nanofiller concentration dependencies of the relative modulus (i.e., modulus E_t divided by the modulus of the pure PI-PM matrix [$E_0 =$ 2.8 GPa]) for the three types of nanocomposites studied. It is evident from this figure that the PI-PM/MMT-BAPS showed the largest increase in modulus with increasing nanofiller concentration. This finding is consistent with the nanofiller concentration dependency of the viscosity of the nanocomposites already discussed (see Figure 5.8). As before, the large increase in modulus just mentioned is thought to be due to the relatively high aspect ratios and enhanced dispersion of the MMT-BAPS nanofiller compared with the isometric ZrO_2 nanofillers. The results confirm our expectation that the morphology of the nanoparticles (platelets, tubes, or isometric form), and especially their aspect ratios, strongly influences the viscoelastic properties of the nanocomposites in both their solid and liquid states. Therefore, understanding the effect of the nanofiller variables such as morphology, aspect ratio, and composition on the properties of polymer nanocomposites may lead to development of new materials with optimal properties for targeted application areas.

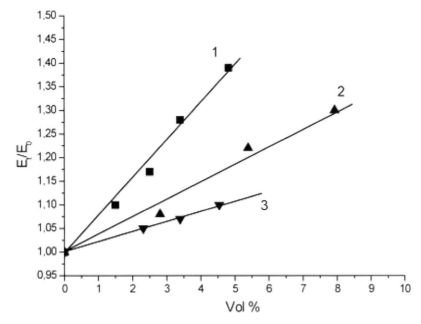

Figure 5.9 Dependence of tensile modulus E_t of nanocomposite films PI-PM on volume concentration of the nanoparticles indicated (E_0 is the modulus of unfilled PI-PM film as described in the text). Reproduced with permission from [47].

As shown in Table 5.3, the tensile strength and deformation at break of the PI-PM based nanocomposite films decreases slightly with increasing concentration of the nanofiller used, to an extent that depends on the specific characteristics of the nanofiller. This decrease in tensile strength may be attributed to the less than optimum adhesion between the nanofillers and the matrix and the possible formation of an inhomogeneous network structure density in the nanocomposite, such as others have reported for other polymer nanocomposites [1, 38]. It is worth noting that the formation of an inhomogeneous network structure is exacerbated by the strong rise in viscosity near the percolation threshold, making it crucial to determine the optimal preparation conditions that may reduce or eliminate any potential microheterogeneity in the resultant nanocomposite films. Despite the reduction of tensile strength caused by incorporation of the nanofillers into the PI nanocomposites, it is interesting to note that incorporation of relatively high concentrations of SNT and ZrO_2 (up to 10 vol%) does not lead to the catastrophic decrease of elongation at break (ε_t) that is widely reported for PI nanocomposites containing organoclay nanofillers [1–9, 38]. This last point suggests the possibility of using the current SNT and ZrO_2 to develop useful PI nanocomposite films with improved thermal and barrier properties, and adequate mechanical properties for coating applications where coating film flexibility is an important performance requirement. This important benefit offered by SNT and ZrO_2 is believed to be due to the desirable intrinsic properties of these ceramic particles, which can be prepared

Table 5.4 *Mechanical properties of nanocomposite samples based on R-BAPS type PI matrix*

Particles	Vol %	E_b GPa	σ_b MPa	ε_b %	G GPa
MMT-BAPS	0	2.61 ± 0.07	170 ± 5	>12	1.30
	2.5	3.42 ± 0.09	182 ± 7	>12	1.61
	4.8	4.05 ± 0.11	204 ± 4	8	1.82
SNT	1.0	2.95 ± 0.07	188 ± 5	>12	1.46
	3.0	3.41 ± 0.10	193 ± 4	>12	1.70
	4.9	3.65 ± 0.09	200 ± 3	10	1.81
	7.1	3.98 ± 0.05	187 ± 7	7	2.0

Notes: E_b – bending modulus; *σ_b* – bending strength; *ε_b* – elongation at break; *G* – shear modulus.

with prescribed properties for specific applications, including polymer nanocomposites, as already mentioned. The studies described in this chapter may stimulate better understanding of the effects of methods of incorporation of nanoparticles with varying shapes and sizes into PI matrices and the effects of nanoparticle morphology on the structure and properties of PI nanocomposites, enhancing our ability to prepare useful polymer nanocomposites with improved properties for targeted applications where common polymer nanocomposite systems are not usable.

Otaigbe and co-workers have reported that application of high shear stress facilitates the homogeneous dispersion of nanoparticles in a PI matrix [32, 45]. They found that increasing MMT-BAPS concentration to ~5 vol% or SNT concentration to ~7 vol% in PI increases bending modulus and strength by ~50% and ~20%, respectively (Table 5.4). The relatively smaller increase in strength as compared with modulus for both types of particles is thought to be due to partial aggregation of the nanoparticles. Note that aggregation of MMT-BAPS particles up to micrometer size defects is more probable than aggregation of SNT particles, because SNT particles can be relatively well dispersed in a suitable or compatible polymer environment [32, 45]. This last point explains why it is possible to increase the concentration of SNT particles significantly relative to that of MMT-BAPS particles in a PI matrix without catastrophic decrease in elongation at break and associated embrittlement of the resulting nanocomposite materials. Therefore, as other researchers have reported for other polymer nanocomposites, the mechanical properties of the current PI nanocomposites discussed in this chapter strongly depend on dispersion of the nanoparticles as well on their structure, morphology, and aspect ratio.

5.4 Conclusions and outlook

This chapter surveys the strong influence of the quality of dispersion, structure, morphology, and aspect ratio on the properties of PI nanocomposites containing various nanoparticles with different shapes and sizes. X-ray diffraction and TGA measurements confirm that natural montmorillonite clay (Na-MMT) can be organically modified with 4,4′ bis(4″-aminophenoxy)diphenylsulfone (BAPS) to improve its compatibility and dispersion in PI nanocomposites with enhanced benefits. The TGA data show weight loss by organically modified clay, starting at 350 °C, which corresponds to the degradation temperature of the BAPS monomer. This suggests that the organically modified clay (MMT-BAPS) can be incorporated into a PI matrix or other high-performance polymers that must be processed at elevated temperatures without exhibiting the detrimental effects of thermal degradation of commercial clays at elevated temperatures on the performance properties of the nanocomposites. Homogeneous dispersion of MMT-BAPS nanoparticles in NMP (a common solvent for dissolving R-BAPS type PI) leads to gelation of the solution at some critical concentration of the nanoparticles used, to an extent that depends on the quality of the clay pretreatment or chemical modification. As shown here, predispersing the Na-MMT in DI water before modification with aromatic ammonium salts is the most effective method of improving the dispersion of MMT-BAPS particles in NMP and subsequently in R-BAPS type PI. This is thought to be an important result that will enhance the performance of PI nanocomposites containing MMT-BAPS nanoparticles. On the other hand, relatively low–molecular weight oligoimides are preferable for achieving optimal dispersion of clay in the polymer volume using the solution-mixing method only. This important result can be exploited with enhanced benefits in the processing of PI nanocomposites by first mixing the MMT-BAPS particles with relatively low–molecular weight prepolymers and then subsequently transforming the oligoimide/MMT-BAPS mixture into the corresponding PI nanocomposites via a chemical reaction in the melt state.

Novel PI nanocomposites based on synthetic silicate nanoparticles such as chrysotile-type nanotubes and zirconium dioxide have recently been reported by Otaigbe and co-workers. To analyze the possibility of homogeneously dispersing these nanoparticles in a PI matrix, these researchers used rheological methods to investigate model systems of oligoimide (OI)/nanoparticles mixtures. As expected, thixotropic behavior of this OI/nanoparticles system was observed at some critical concentration of the nanoparticles that depended strongly on the morphology and aspect ratio of the nanoparticles used. Specifically, MMT particles with the highest aspect ratio relative to the silicate nanotubes (SNT) and ZrO_2 particles showed a huge increase in the viscosity of the OI/MMT mixture at low (i.e., ~3 vol%) particle concentrations. This viscosity increase exhibited by the OI/SNT mixture occurred at an SNT concentration in the range 8–12 vol% [a concentration range that is higher than the theoretically estimated percolation threshold (P_c) value of ~6 vol%]. This disagreement between the experimental and theoretical estimates of P_c is thought to be due to agglomeration of the nanoparticles, which was indirectly confirmed by XRD data for the OI/MMT nanocomposites. In contrast, a relatively modest increase in the viscosity of OI/ZrO_2 nanocomposites is typically observed at 9 vol% ZrO_2 particle concentration.

The tensile modulus of PI-PM films filled with nanoparticles is consistent with the rheological data, showing a more significant increase in the tensile modulus of the PI-PM films containing MMT-BAPS particles than in that of the films containing SNT or ZrO_2 particles. In contrast to conventional PI/clay nanocomposites reported in the literature, relatively high concentrations of SNT and ZrO_2 (up to 10 vol%) in the PI-PM films do not lead to catastrophic decrease of the elongation of these films at break, as reported for PI films filled with organoclay.

In general, it is worth noting that the SNT and ZrO_2 nanoparticles described in this chapter can be used to develop useful PI nanocomposite films with adequate thermomechanical properties for a number of coating applications where coating film flexibility and thermooxidative stability are important performance requirements. Overall, current results confirm our expectation that the morphology of the nanoparticles (platelets, tubes, or isometric form) and especially their aspect ratios strongly influence the viscoelastic properties of the nanocomposites in both their solid and liquid states. It is hoped that the work described in this chapter may stimulate better understanding of the effects of functional chemical modification methods on the quality of clay dispersion in PI matrices, enhancing our ability to prepare useful PI/clay nanocomposites with improved properties for targeted high-temperature applications where common polymer nanocomposite systems are not useable.

References

1. S. Ray and M. Okamoto, Polymer/layered silicate nanocomposites: A review from preparation to processing. *Progress in Polymer Science*, 28 (2003), 1539–1641.
2. E. P. Giannelis, Polymer layered silicate nanocomposites. *Advanced Materials*, 8 (1996), 29–35.
3. E. P. Giannelis, R. Krishnamoorti, and E. Manias, Polymer-silicate nanocomposites: Model systems for confined polymers and polymer brushes. *Advanced Polymer Science*, 138 (1999), 107–47.
4. P. C. LeBaron, Z. Wang, and T. J. Pinnavaia, Polymer-layered silicate nanocomposites: An overview. *Applied Clay Science*, 15 (1999), 11–29.
5. R. A. Vaia, G. Price, P. N. Ruth, H. T. Nguyen, and J. Lichtenhan, Polymer/layered silicate nanocomposites as high performance ablative materials. *Applied Clay Science*, 15 (1999), 67–92.
6. M. Biswas and S. S. Ray, Recent progress in synthesis and evaluation of polymer–montmorillonite nanocomposites. *Advanced Polymer Science*, 155 (2001), 167–221.
7. R. Xu, E. Manias, A. J. Snyder, and J. Runt, New biomedical poly(urethane urea)–layered silicate nanocomposites. *Macromolecules*, 34 (2001), 337–9.
8. R. K. Bharadwaj, Modeling the barrier properties of polymer layered silicate nanocomposites. *Macromolecules*, 34 (2001), 1989–92.
9. Y. Kojima, A. Usuki, M. Kawasumi, Y. Fukushima, A. Okada, T. Kurauch, and O. Kamigaito, Synthesis of nylon 6–clay hybrid. *Journal of Materials Research*, 8 (1993), 1179–84.
10. M. I. Bessonov, M. M. Koton, V. V. Kudryavtsev, and L. A. Laius, *Polyimides – Thermally Stable Polymers* (New York: Plenum, 1987).
11. D. Wilson, H. D. Stengenberger, and P. M. Hergenrother, *Polyimides* (New York: Chapman & Hall, 1990).
12. M. K. Ghose and K. L. Mittal, eds., *Polyimides – Fundamentals and Applications* (New York: Dekker, 1996).
13. K. Yano, A. Usuki, A. Okada, T. Kurauchi, and O. Kamigaito, Synthesis and properties of polyimide clay hybrid. *Journal of Polymer Science, Part A: Polymer Chemistry*, 31 (1993), 2493–8.

14. Y. Yang, Z.-K. Zhu, J. Yin, X.-Y. Wang, and Z.-E. Qi, Preparation and properties of hybrids of organo-soluble polyimide and montmorillonite with various chemical surface modification. *Polymer*, 40 (1999), 4407–14.

15. D. M. Delozier, R. A. Orwoll, J. F. Cahoon, N. J. Johnston, J. G. Smith, Jr., and J. W. Connell, Preparation and characterization of polyimide/organoclay nanocomposite. *Polymer*, 43 (2002), 813–22.

16. H.-L. Tyan, Y.-C. Liu, and K.-H. Wei, Thermally and mechanically enhanced clay/polyimide nanocomposite via reactive organoclay. *Chemistry of Materials*, 11 (1999), 1942–7.

17. H.-L. Tyan, C.-M. Liu, and K.-H. Wei, Effect of reactivity of organics-modified montmorillonite on the thermal and mechanical properties of montmorillonite/polyimide nanocomposites. *Chemistry of Materials*, 13 (2001), 222–6.

18. H.-L. Tyan, K.-H. Wei, and T.-E. Hsieh, Mechanical properties of clay–polyimide (BTDA–ODA) nanocomposites via ODA-modified organoclay. *Journal of Polymer Science, Part B: Polymer Physics*, 38 (2000), 2873–8.

19. R. H. Vora, P. K. Pallathadka, S. H. Goh, T. S. Chung, Y. X. Lim, and T. K. Bang, Preparation and characterization of 4,4-bis(4-aminophenoxy)diphenyl sulfone based fluoropoly(ether-imide)/organo-modified clay nanocomposites. *Macromolecular Materials and Engineering*, 288 (2003), 337–56.

20. Z.-M. Liang, J. Yin, and H.-J. Xu, Polyimide/montmorillonite nanocomposites based on thermally stable, rigid-rod aromatic amine modifiers. *Polymer*, 44 (2003), 1391–9.

21. Z.-M. Liang and J. Yin, Poly(etherimide)/montmorillonite nanocomposites prepared by melt intercalation. *Journal of Applied Polymer Science*, 90 (2003), 1857–63.

22. S. Campbell and D. Scheiman, Orientation of aromatic ion exchange diamines and the effect on melt viscosity and thermal stability of PMR-15/silicate nanocomposites. *High Performance Polymers*, 14 (2002), 17–30.

23. V. E. Yudin, G. M. Divoux, J. U. Otaigbe, and V. M. Svetlichnyi, Synthesis and rheological properties of oligoimide/montmorillonite nanocomposites. *Polymer*, 46 (2005), 10866–72.

24. Y. Zhong and S.-Q. Wang, Exfoliation and yield behavior in nanodispersions of organically modified montmorillonite clay. *Journal of Rheology*, 47 (2003), 483–95.

25. S. Laribi, J.-M. Fleureau, J.-L. Grossiord, and N. Kbir-Ariguib, Effect of pH on the rheological behaviour of pure and interstratified smectite clays. *Clays and Clay Minerals*, 54 (2006), 29–37.

26. J. Tarchitzky and Y. Chen, Rheology of humic substances/ Na- and Ca-montmorillonite suspensions. *Soil Science Society of America Journal*, 66 (2002), 406–12.

27. A. Cadene, S. Durand-Vidal, P. Turqa, and J. Brendle, Study of individual Na-montmorillonite particles size, morphology, and apparent charge. *Journal of Colloid and Interface Science*, 285 (2005), 719–30.

28. G. Falini, E. Foresti, G. Lesci, and N. Roveri, Structural and morphological characterization of synthetic chrysotile single crystals. *Chemical Communication*, (2002), 1512–13.

29. Y. Zhang, S. Lu, Y. Li, Z. Dang, J. Xin, S. Fu, G. Li, R. Guo, and L. Li, Novel silica tube/polyimide composite films with variable low dielectric constant. *Advanced Materials*, 17 (2005), 1056–9.

30. E. Korytkova, A. Maslov, L. Pivovarova, I. Drozdova, and V. Gusarov, Formation of $Mg_3Si_2O_5(OH)_4$ nanotubes under hydrothermal conditions. *Glass Physics and Chemistry*, 30 (2004), 51–5.

31. O. Pozhidaeva, E. Korytkova, D. Romanov, and V. Gusarov, Formation of ZrO2 nanocrystals in hydrothermal media of various chemical compositions. *Russian Journal of General Chemistry*, 72 (2002), 849–53.

32. V. Yudin, J. Otaigbe, S. Gladchenko, B. Olson, S. Nazarenko, E. Korytkova, and V. Gusarov, New polyimide nanocomposites based on silicate type nanotubes: Dispersion, processing and properties. *Polymer*, 48 (2007), 1306–15.

33. Gelest, Inc., Applying a silane coupling agent. In: *Silane Coupling Agents: Connecting across Boundaries*, ed. B. Arkles, 2nd ed. (Morrisville, PA: Gelest, 1998), pp. 88–9.

34. P. Potschke, T. D. Fornes, and D. R. Paul, Rheological behaviour of multiwalled carbon nanotube/polycarbonate composites. *Polymer*, 43 (2002), 3247–55.

35. G. Wu, J. Lin, Q. Zheng, and M. Zhang, Correlation between percolation behavior of electricity and viscoelasticity for graphite filled high density polyethylene. *Polymer*, 47 (2006), 2442–7.

36. C. Liu, J. Zhang, J. He, and G. Hu, Gelation in carbon nanotube/polymer composites. *Polymer*, 44 (2003), 7529–32.

37. E. Garboczi, K. Snyder, J. Douglas, and M. Thorpe, Geometrical percolation threshold of overlapping ellipsoids. *Physical Review E*, 52 (1995), 819–28.

38. Q. Zeng, A. Yu, G. Lu, and D. Paul, Clay-based polymer nanocomposites: Research and commercial development. *Journal of Nanoscience and Nanotechnology*, 5 (2005), 1574–92.

39. J.-N. Paquiena, J. Galya, J.-F. Gerarda, and A. Pouchelon, Rheological studies of fumed silica–polydimethylsiloxane suspensions. *Colloids and Surfaces A: Physicochemical and Engineering Aspects*, 260 (2005), 165–72.

40. D. J. Thomas, Transport characteristics of suspension. VIII. A note on the viscosity of Newtonian suspensions of uniform spherical particles. *Journal of Colloid and Interface Science*, 20 (1965), 267–77.

41. L. Nielsen, *Mechanical Properties of Polymers and Composites* (New York: Dekker, 1994).

42. H. Barnes, Thixotropy – A review. *Journal of Non-Newtonian Fluid Mechanics*, 70 (1997), 1–33.

43. V. M. Svetlichnyi, T. I. Zhukova, V. V. Kudriavtsev, V. E. Yudin, G. N. Gubanova, and A. M. Leksovskii, Aromatic polyetherimides as promising fusible film binders. *Polymer Engineering and Science*, 35 (1995), 1321–4.

44. V. V. Nesterov, V. V. Kudryavtsev, V. M. Svetlichnyi, N. V. Gazdina, N. G. Belnikevich, O. I. Kurenbin, and T. I. Zhukova, Study of soluble poly(amic acid)s and poly(ester imide)s by methods of exclusion liquid chromatography. *Polymer Science, Series A*, 39 (1997), 953–7.

45. B. G. Olson, J. J. Decker, S. I. Nazarenko, V. E. Yudin, J. U. Otaigbe, E. N. Korytkova, and V. V. Gusarov, Aggregation of synthetic chrysotile nanotubes in the bulk and in solution probed by nitrogen adsorption and viscosity measurements. *Journal of Physical Chemistry: Part C*, 112 (2008), 12943–50.

46. T. Kurose, V. E. Yudin, J. U. Otaigbe, V. M. Svetlichnyi, Compatibilized polyimide (R-BAPS)/BAPS-modified clay nanocomposites with improved dispersion and properties. *Polymer*, 48 (2007), 7130–38.

47. V. E. Yudin, J. U. Otaigbe, V. M. Svetlichnyi, E. N. Korytkova, O. V. Almjasheva, V. V. Gusarov, Effects of nanofiller morphology and aspect ratio on the rheo-mechanical properties of polyimide nanocomposites, *eXPRESS Polymer Letters*, 2 (2008), 485–93.

6

Clays modified with thermally stable ionic liquids with applications in polyolefin and polylactic acid nanocomposites

JIN UK HA AND MARINO XANTHOS

New Jersey Institute of Technology

6.1 Introduction

A variety of discontinuous (short) functional fillers may be combined with thermoplastic or thermoset matrices to produce composites. The fillers may differ in shape (fibers, platelets, flakes, spheres, or irregulars), aspect ratio, and size. When the fully dispersed (exfoliated or deagglomerated) fillers are of nanoscale dimensions, the materials are known as nanocomposites. They differ from conventional microcomposites in that they contain a significant number of interfaces available for interactions between the intermixed phases. As a result of their unique properties, nanocomposites have great potential for applications involving polymer property modification utilizing low filler concentrations for minimum weight increase; examples include mechanical, electrical, optical, and barrier properties improvement and enhanced flame retardancy.

In addition to the inherent properties of the nanofiller and its size/shape, the properties of nanocomposites depend to a significant extent on the filler surface chemistry and the interaction of the composite components at the phase boundaries; these parameters control the filler dispersion in the polar or nonpolar matrix and the adhesion between the components. The thermal stability of the filler itself, and that of the interface, are important considerations if the composites are to be melt-compounded and shaped at elevated temperatures, as, for example, by extrusion or injection molding.

At the present time, important nanofillers include certain nanoclays (montmorillonite, hydrotalcite in platelet form), nanofibers (single- and multiple-wall carbon nanotubes), and nanosized particulate metal oxides. Several technological advances will undoubtedly contribute to additional growth in the usage of these fillers. Examples of such advances include [45]

- Clay modification with thermally stable additives that would promote deagglomeration, dispersion, and exfoliation of agglomerated nanoclays, particularly in high-temperature thermoplastic matrices.
- For carbon nanotube composites, appropriate interfacial modification to improve dispersion and adhesion and minimize reagglomeration.
- In both these cases, equipment/process modification to ensure the desired orientation and maintain the fiber's or platelet's high aspect ratio.

143

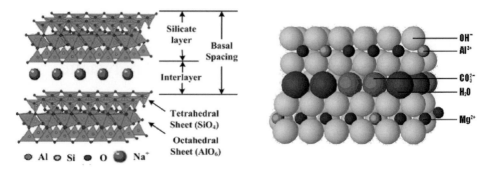

Figure 6.1 Structure of montmorillonite MMT [30] (left) and hydrotalcite HT [27] (right).

The present chapter presents examples of ionic liquids as a new class of thermally stable nanofiller modifiers that are also expected to address the issues that have been raised in regard to improved dispersion and adhesion. The examples are mostly drawn from polyolefin composites (polypropylene, PP, and polyethylene, PE) and to a lesser extent from polylactic acid (PLA) composites containing nanoclays.

6.2 Clays and clay nanocomposites

6.2.1 Types of clays

Nanoclays can be categorized into cationic and anionic types. Cationic nanoclays are based on smectite clays. An example is montmorillonite (MMT), a hydrated Al, Mg silicate that may contain cations such as Na^+ and Ca^{++} between the anionic layers. In contrast, anionic clays contain cationic layers and anions such as Cl^- and CO_3^{2-} in the interlayer space. Typical examples include layered double hydroxides (LDH) and hydrotalcite (HT), a mostly synthetic hydrated magnesium and aluminum carbonate salt. Whereas MMT is commonly used as a nanofiller to improve thermal, mechanical, and barrier properties, LDHs have many attractive properties that lead to application as surfactant adsorbents, biohybrid materials, antacid food formulations, acid neutralizers, and active pharmaceutical ingredients' excipients [37, 13, 28, 14, 35].

Figure 6.1 shows the lamellar structures of montmorillonite and hydrotalcite clays. Because the ions in the interlayer space can easily be replaced by foreign ions (cation/anion), these clays can be modified under appropriate conditions by intercalation with suitable ions to satisfy various practical applications.

6.2.2 Fabrication and properties of clay nanocomposites

Three different methods have been used to prepare polymer–clay nanocomposites. In the in situ intercalative polymerization, the first method, a precursor solution (monomer liquid or monomer solution) is inserted into the basal spaces of expanding clay layers, followed by dispersion of the layers into the matrix during polymerization. The polymerization can be initiated either by heat or radiation, by diffusion of a suitable initiator, or by an organic initiator or catalyst fixed inside the interlayer through cationic exchange before the swelling step by the monomer. This method was initially investigated by the Toyota research team, which produced clay/nylon-6 nanocomposites [32], and has been applied to a wide range of polymer systems, because it is able to achieve fully exfoliated nanocomposites.

The second preparation method uses polymer solutions and involves a strategy similar to the in situ polymerization method. However, this method is only suitable for polymers such as polyethylene oxide (PEO) and polyacrylic acid (PAA) that are soluble in polar solvents, or for soluble polymeric precursors in the case of polymers insoluble in organic solvents, such as polyimide (PI) [33, 4, 46].

In the third processing method, nanoclays are mixed with the polymer in the molten state. If the clay surfaces are compatible with the selected polymer, the polymer can increase the basal space, resulting in either intercalated or exfoliated nanocomposites; however, if the clay surface is incompatible with the selected polymer, microsized composites will be formed. Although the efficiency of melt-processing may not be as high as that of in situ polymerization, this method is most commonly applied by the polymer processing industry to produce nanocomposites by conventional techniques, such as extrusion and injection molding [43, 18, 49]. The nanoclays, commonly referred to as "organoclays" after modification with suitable organic modifiers, are blended with the polymer matrix above its melting temperature in order to optimize interactions between the components. The polymer chains significantly lose their conformation during intercalation, and this serves as a force for disordered or delaminated exfoliation of the clays in the polymer matrix [19].

Depending on the types of clay and polymer, their physiochemical interactions, and the melt-mixing conditions, clays may not be fully dispersed, forming an agglomerated tactoid. In this case, clays are dispersed as a phase segregated and separated from the polymer matrix, and the properties of the composites are in the same range as in traditional microcomposites [1].

In that state, the mechanical properties of each individual layer in the clay particles are not fully realized, as a result of the low aspect ratio and the weak interlayer van der Waals forces, which may act as fracture initiation sites; higher clay loadings are needed to achieve adequate improvement of the modulus, whereas strength and toughness of the polymer are often reduced [18]. Therefore, to fully utilize the property improvements resulting from the high aspect ratios at low loadings, including flammability performance, exfoliation of clay layers is necessary. Figure 6.2 shows various stages of clay dispersion in a polymeric matrix.

Exfoliation and dispersion of agglomerated MMT clay platelets (Figure 6.3) in a given polymer may be facilitated by the proper screw design and feeding sequence in the compounding equipment and the addition of compatibilizing additives (Figure 6.4), and/or by premodification of the clay surface with suitable organic modifiers (usually quaternary ammonium salts).

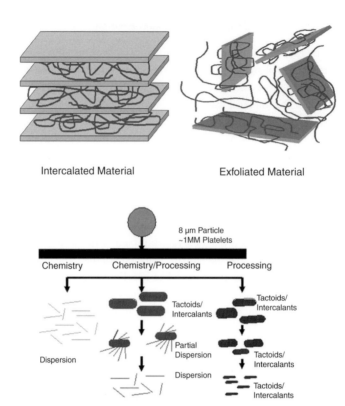

Intercalated Material   Exfoliated Material

8 μm Particle
~1MM Platelets

Chemistry   Chemistry/Processing   Processing

Tactoids/
Intercalants

Dispersion

Partial
Dispersion

Dispersion

Tactoids/
Intercalants

Tactoids/
Intercalants

Tactoids/
Intercalants

Figure 6.2 Schematics of various clay dispersion mechanisms [26]. With permission from Wiley–VCH.

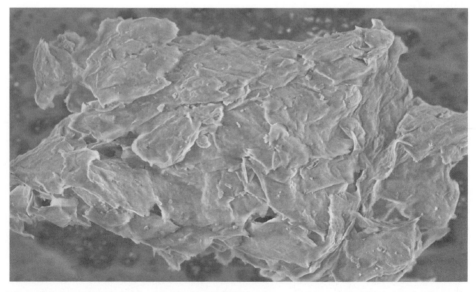

Figure 6.3 Scanning electron micrograph of MMT agglomerate (a natural cationic clay) prior to its dispersion into high–aspect ratio nanoplatelets at ×7,000 [45]. With permission from Wiley–VCH.

Table 6.1 *Organic modifiers for commercially available montmorillonite nanoclays*

	Cloisite 15A	Cloisite 20A	Cloisite 30B
Organic modifier, concentration	Dimethyl, dehydrogenated tallow, quaternary ammonium, 125 meq/100 g clay	Dimethyl, dehydrogenated tallow, quaternary ammonium, 95 meq/100 g clay	Methyl, tallow, bis-2-hydroxylethyl, quaternary ammonium, 90 meq/100 g clay
Structures	$\mathrm{CH_3}$ | $\mathrm{CH_3 - N^+ - HT}$ | HT	$\mathrm{CH_3}$ | $\mathrm{CH_3 - N^+ - HT}$ | HT	$\mathrm{CH_2CH_2OH}$ | $\mathrm{CH_3 - N^+ - T}$ | $\mathrm{CH_2CH_2OH}$
	HT is hydrogenated tallow (~65% C18; ~30% C16; ~ 5% C14)	HT is hydrogenated tallow (~65% C18; ~30% C16; ~ 5% C14)	T is tallow (~65% C18; ~30% C16; ~5% C14)
D-space	3.15 nm	2.42 nm	1.85 nm

Source: Southern Clay Products, Inc. (Product Bulletins). Available at http://www.scprod.com/product_bulletins.asp, accessed Feb. 19, 2010.

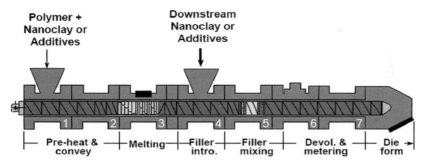

Figure 6.4 Typical setup for nanoclay incorporation in a twin screw extruder [2].

Table 6.1 includes typical quaternary ammonium salts used as organic modifiers in commercial montmorillonite nanoclays, which are to be primarily dispersed in nonpolar matrices. Figure 6.5 shows that hydrotalcite platelets of nanodimensions present a much smaller degree of agglomeration than the MMT particles.

Products that take advantages of the attributes of MMT nanocomposites (stiffening, barrier properties, lighter weight) are mostly low–melt temperature resins, because the existing alkylammonium modifiers have limited thermal stability during the processing of

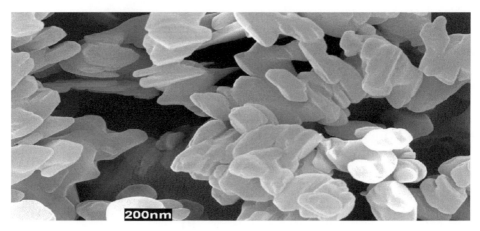

Figure 6.5 Scanning electron micrograph of hydrotalcite particle at ×30,000 [12].

high–temperature thermoplastics such as polyethylene terephthalate, PET, polycarbonate, PC, and polyamides, PA. Furthermore, for packaging applications, the existing modifiers may adversely affect cost, sealability, color under heat aging, or UV stability. Significant R&D efforts have recently been initiated toward alternative modifiers. In particular, studies on ionic liquids as novel modifiers have been increased [20, 41, 21, 25].

6.3 Ionic liquids

Ionic liquids (ILs) are organic molten salts with low melting temperatures, typically below 100 °C. They usually consist of organic cations containing nitrogen (imidazolium or pyridinium) or phosphorus; corresponding anions are halides, tetrafluoroborate, hexafluorophosphates, and so forth. ILs have attractive solvent properties and are considered as potentially environmentally benign solvents featuring low volatility, thermal stability, a broad liquid-temperature range, low flammability, and compatibility with organic and inorganic materials [23]. They have been investigated for organic synthesis [44], bioprocessing operations, catalysis, gas separation [47], and corrosion inhibitors for magnesium alloys [24]. Recently, ILs have been evaluated as nonvolatile plasticizers and as external or internal lubricants in several polymers, including polyvinyl chloride [38], polymethyl methacrylate [39], and PLA [34].

   ILs can also be used to advantage to overcome the lower thermal stability of conventional organic modifiers that are used to enhance dispersion, wetting, and compatibility of inorganic particles in organic media; the high thermal stability of ILs and the ability to control their hydrophilic/lipophilic balance through proper selection of the constituent ions are some of their major attributes. Thus, it is not surprising that the treatment of functional fillers with ILs and, in particular, nanosized fillers such as cationic or anionic clays to impart organophilic properties has been the topic of recent publications on polymer composites [20, 3, 42, 11, 25].

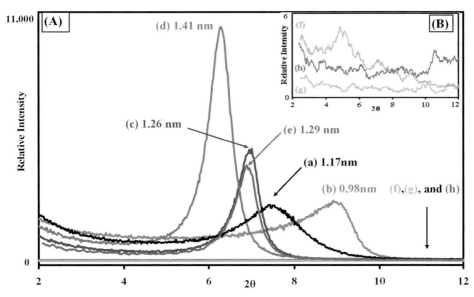

Figure 6.6 XRD results of (a) pristine MMT, (b) pristine MMT washed with water, (c) MMT/IL-1, (d) MMT/IL-2, (e) MMT-IL-3, (f) IL-1, (h) IL-2, and (g) IL-3 [25]. (With permission from Elsevier.)

6.4 Clay modification with thermally stable ionic liquids

Kim *et al.* [25] modified MMT-Na$^+$ by ion exchange with three different low–molecular weight ionic liquids, namely, 1-ethyl-3-methylimidazolium bromide (MW 191.07), IL-1; 1-hexyl-3-methylimidazolium chloride (MW 202.72), IL-2; and *N*-ethyl pyridinium tetrafluoroborate (MW 194.8), IL-3. Thermal stability of the three ILs was ranked as IL-3 > IL-1 > IL-2 according to TGA results. MMT modified with all three ILs showed enhanced thermal stability above 200 °C, as compared to commercial nanoclays, which used quaternary tallow-based ammonium salts as their modifiers. XRD data (Figure 6.6) show that MMT modified by intercalation with ILs had increased basal spacing, in particular IL-2, which has a relatively bulkier cation than the other two ILs. However, basal spacing increases were not as large as for the commercial nanoclays containing higher-*M*W cations at much higher concentrations (Table 6.1).

Although great attention has been given to quaternary nitrogen-based ionic liquids in recent years (e.g., ammonium, imidazolium, pyridinium), specific accounts of ionic liquids containing quaternary phosphorus cations have been less common. It has been shown that certain phosphonium-based ionic liquids have better properties than nitrogen-based ionic liquids (6, 42). For example, phosphonium-based ILs are thermally stable above 300 °C for many species, even though their decomposition point upon heating varies, depending on the anion (5). Hence, research efforts on nanoclay composites containing cationic clays modified with phosphonium-based ILs have recently increased as a result of the higher thermal stability of the modified clays and the anticipated higher thermal stability of the corresponding composites (41, 8, 9, 21).

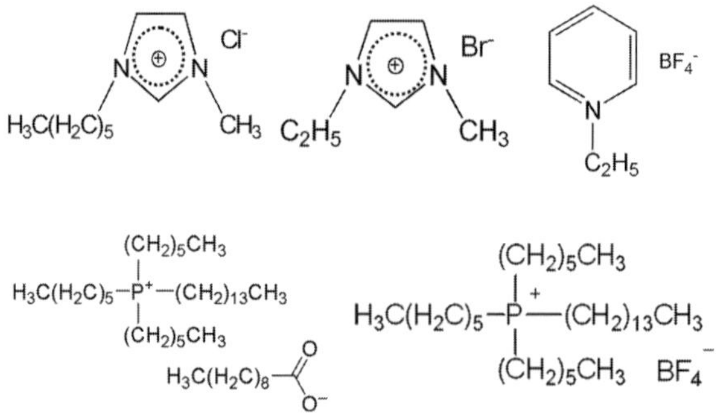

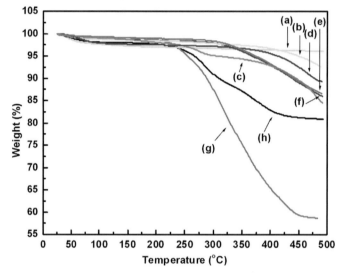

Figure 6.7 Structure of ionic liquids [21]. (With permission from Wiley.)

Figure 6.8 TGA results of (a) pristine MMT, (b) MMT/IL-1, (c) MMT/IL-2, (d) MMT/IL-3, (e) MMT/IL-4, (f) MMT/IL-5, (g) Cloisite 15A, and (h) Cloisite 30B (in N$_2$, 15 °C/min) [21]. (With permission from Wiley.)

Ha and Xanthos [21] used two different phosphonium-based ionic liquids that have the same cation but different anions, trihexyltetradecylphosphonium decanoate (IL-4) and trihexyltetradecylphosphonium tetrafluoroborate (IL-5) (Figure 6.7), and compared them with MMT previously modified with low-molecular weight imidazolium- and pyridinium-based ILs [25]. MMTs modified with the phosphonium-based ILs also showed significantly improved thermal stability compared with commercial nanoclays (Figure 6.8). XRD results showed decreased 2θ angles, which indicates increased basal spacings. These basal spacings are larger than those obtained with the lower–molecular weight ILs [25] because of the bulkier/larger intercalated phosphonium cations.

Stoeffler *et al.* [41] used different ILs to evaluate the effect of intercalated ionic liquids on the clays and properties of clays dispersed in a nonpolar polyolefin. ILs containing hexadecylpyridinium, 1-vinylhexadecylimidazolium, 1-vinyloctadecylimida-zolium, dihexadecylimidazolium, dioctadecylimidazolium, and tributylhexadecylphospho-nium as cations were used for clay modification. All modified clays showed higher thermal stability than a commercial nanoclay, Cloisite 20A (Southern Clay Products, Inc.), used as a control. The MMT modified with the phosphonium-based IL showed the highest thermal stability among the modified and commercial clays. However, the commercial nanoclay, Cloisite 20A, showed the largest basal spacing.

6.5 Polyolefins containing nanoclays modified with thermally stable ionic liquids

Ding *et al.* [15] prepared PP composites containing MMT modified with 1-methyl-3-tetradecylimidazolium chloride. MMT was modified with the IL in water (aq-MMT) and xylene (xo-MMT). The PP composites were prepared by solution polymer intercalation. Predissolved PP was mixed with pristine MMT, aq-MMT, and xo-MMT in anhydrous xylene at about 5 wt% filler. PP composites containing xo-MMT showed a fully exfoliated X-ray diffraction pattern. TGA results of PP composites containing xo-MMT tested under N_2 showed the highest thermal stability, which was attributed to the high degree of exfoliation. The well-dispersed xo-MMT in the PP matrix functions effectively as an insulator and mass transport barrier to the volatile products generated during decomposition and contributes to the great enhancement of the thermal stability of the composite. The temperature for 50% weight loss increased from 344 °C for neat PP, 421 °C for PP/Na$^+$-MMT, and 447 °C for PP/aq-MMT to 514 °C for PP/xo-MMT composites.

MMT modified with 1-octadecyl-3-methylimidazolium chloride had an increased inter-layer spacing. This was subsequently melt-mixed with PP [16, 17] without showing fully exfoliated clay dispersion; however, PP mixed with the modified MMT in the presence of maleated PP showed a much more fully exfoliated clay dispersion. XRD data showed that different IL loadings affected the IL adsorption in the basal spacing and also the degree of the clay dispersion in the PP matrix. The authors reported that the twice–equivalent IL loading into MMT was better than mono or triple IL loadings. PP composites containing the modified MMT showed enhanced thermal stability. The onset degradation temperature was increased from 270 to 380 °C.

Kim *et al.* [25] evaluated the thermal stability of PP/PP-g-MA mixtures containing 5 wt% unmodified MMT and MMT modified with different ILs by following torque versus time in a batch mixer. The sample containing MMT treated with thermally stable *N*-ethylpyridinium tetrafluoroborate was shown to be as stable after 25 min mixing time as the sample that contained the clay without modifiers.

Ha and Xanthos [21] compounded PP with unmodified MMT nanoclay, commer-cially modified nanoclay, and nanoclay modified with phosphonium-based ILs. After melt-mixing, the interspacing of the nanoclays in the composites increased somewhat as compared to that for the original clays. This was because PP molecular chains penetrated into the clay interlayer spacing during melt-mixing. Even though melt-processing helped

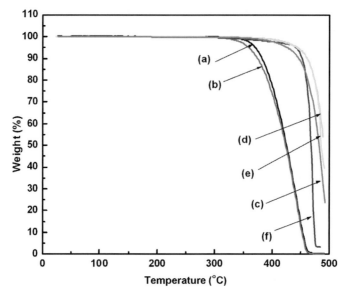

Figure 6.9 TGA results of (a) pure PP, (b) PP/PP-g-MA, (c) PP/MMT, (d) PP/MMT/IL-4, and (e) PP/MMT/IL-5, and (f) PP/Cloisite 15A (in N_2 15 °C/min) [21]. (With permission from Wiley.)

to increase the interlayer spacing of the nanoclays, none of the composites achieved exfoliated clay dispersion in the PP matrix, except the commercial nanoclay. Thus, nanoclays modified with phosphonium-based ILs were less efficient than the commercial nanoclay in producing well-dispersed structures in the PP matrix. The thermal stability of the composites was increased after adding, not only modified clays, but also the unmodified clay and the commercial nanoclay (Figure 6.9). Although all clay composites showed improved thermal stability, there were slight differences. The PP composite containing unmodified nanoclay showed a slightly earlier onset degradation temperature and the ones containing the clay modified with phosphonium-based ILs showed the highest thermal stability among the nanoclay composites. Burnside and Giannelis [7] attributed the increased thermal stability of the polymer composites to the hindered diffusion of the volatile decomposition products.

Figure 6.10 shows clay composites pressed into 1-mm thin disks. Clays modified with the large cations of the phosphonium-based ILs showed better dispersion in the PP/PP-g-MA matrix [Figure 6.10 (a) and (b)] than clays modified with the lower–molecular weight ILs and their corresponding smaller cations [Figure 6.10 (c) and (d)]. Figure 6.11 shows SEM images of fracture surfaces of the PP nanoclay composites. Unmodified MMT and MMT modified with low-MW ILs showed poor clay dispersion and agglomeration in the PP matrix, whereas clays modified with the phosphonium-based ILs and the commercial clay (Cloisite 15A) showed micro- and nanodispersions, respectively.

Similar studies were carried out with polyethylene, which is also a nonpolar polymer with limited affinity toward unmodified clays. Stoeffler *et al.* [41] used clays modified with

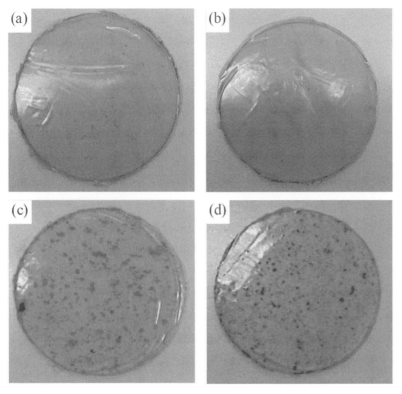

Figure 6.10 Thin disks of nanocomposites: (a) PP/MMT/IL-4, (b) PP/MMT/IL-5, (c) PP/MMT/IL-1, and (d) PP/MMT/IL-2 [21]. (With permission from Wiley.)

various ionic liquids and compounded with a linear low-density polyethylene (LLDPE) along with LLDPE-g-MA. The thermal stabilities of clays modified with various ionic liquids are higher than those of commercial organoclays, as measured by TGA in inert atmospheres. TGA in oxidative atmospheres proved to be very sensitive to the dispersion state of the organoclays, as shown by the drastic enhancement of thermal stability for partially exfoliated and intercalated composites. However, the inherent thermal stability of the organoclay did not appear to influence the overall thermal stability of the composite significantly in the range of temperatures investigated (160–230 °C). It appears, however, that in the case of matrices requiring higher processing temperatures, the thermal degradation of the intercalant would be critical, rendering the use of highly thermally stable organoclays necessary.

6.6 Polylactic acid composites

PLA is one of the best-known biodegradable polymers, whose mechanical and thermal properties are comparable with those of commodity polymers. PLA is a promising polymer

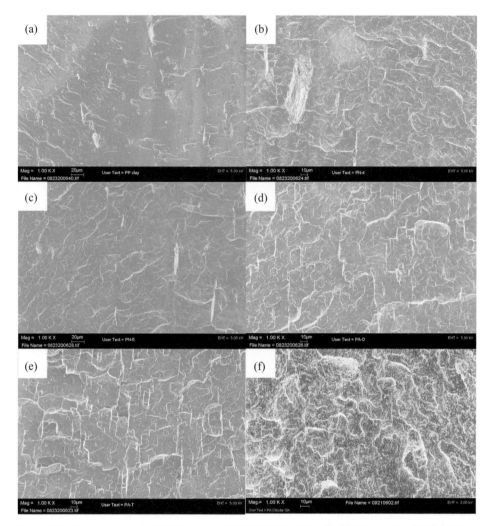

Figure 6.11 Fracture surface SEM images of (a) PP/MMT, (b) PP/MMT/IL-2, (c) PP/MMT/IL-3, (d) PP/MMT/IL-4, (e) PP/MMT/IL-5, and (f) PP/Cloisite 15A [21]. (With permission from Wiley.)

for various new applications, and it is not surprising that studies on PLA–clay nanocomposites have been increasing recently [1, 40]. Current studies on PLA nanocomposites are mainly focusing on conventional commercial MMT organoclays, whereas studies on PLA composites containing nanoclays modified with thermally stable ionic liquids are very rare. MMT increases the thermal stability of PLA, although not significantly, as in commodity polymer nanocomposites [36, 10]. Note that adding nanosized clays does not always increase the thermal stability of the PLA; the effects also depend on the type of PLA [48].

Ha and Xanthos [22] compounded PLA with anionic clays (LDH) and cationic clays (MMT). LDH was also calcined, and the calcined product was also treated with ionic liquids

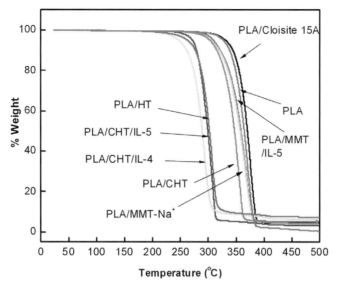

Figure 6.12 TGA results of PLA LDHs/MMTs composites (in air, 10 °C/min) [22]. (With permission from Elsevier.)

having the same cation but different anions. In general, PLA compounded with LDHs showed decreased thermal stability versus the unfilled polymer and a commercial MMT organoclay. Among the LDH/PLA composites, LDH modified with the decanoate anion of a phosphonium-based ionic liquid showed the lowest thermal stability, whereas a calcined LDH/PLA compound showed less of a decrease in thermal stability (Figure 6.12). The authors ascribed the reduced thermal stability of PLA to moisture present in the interlayer spacing of LDH. Because calcined LDH (at 500 °C for 5 h) contains less moisture, it did not significantly affect the polymer matrix. On the other hand, the reduced thermal stability of PLA containing LDH modified with decanoate or phosphinate anions of a phosphonium cation–based ionic liquid was mainly because of the catalytic effects of these anions (particularly the decanoate) and/or their inherent low thermal stability.

6.7 Conclusions

Cationic montmorillonite clays can be modified by ion exchange with ionic liquids. Literature data on their modification by a variety of cations of different molecular weights, including those present in thermally stable phosphonium–based ionic liquids, are presented and compared with data from commercial organoclays. Examples of polyolefin nanocomposites containing IL-modified montmorillonite are also presented. It appears, overall, that the relatively low melt-processing temperatures of the polyolefin composites do not permit full demonstration of the beneficial effects of the thermally stable phosphonium ionic liquid intercalants versus commercial modifiers, which are more prone to thermal degradation at processing temperatures. The few literature examples available on the modification of polylactic acid with anionic hydrotalcite nanoclays suggest no improvement in the thermal stability of the corresponding polymer composites through the addition of ionic liquid–modified clays.

References

1. M. Alexandre and P. Dubois, Polymer-layered silicate nanocomposites, properties and uses of a new class of materials. *Materials Science and Engineering R: Reports*, 28 (2000), 1–63.
2. P. G. Andersen, Compounding polymer nanocomposites. In: *NPE Educational Conference: Nanotechnology Applications for Polymers* (Chicago: NPE, June 21, 2009). [CD]
3. W. H. Awad, W. Gilman, M. Nyden, R. H. Harris, T. E. Sutto, J. Callahan, P. C. Trulove, H. C. DeLong, and D. M. Fox, Thermal degradation studies of alkyl-imidazolium salts and their application in nanocomposites. *Thermochimica Acta*, 409 (2004), 3–11.
4. J. Billingham, C. Breen, and J. Yarwood, Adsorption of polyamine, polyacrylic acid and polyethylene glycol on montmorillonite: An in situ study using ATR-FTIR. *Vibrational Spectroscopy*, 14 (1997), 19–34.
5. C. J. Bradaric, A. Downard, C. Kennedy, A. J. Robertson, and Y. Zhou, Industrial preparation of phosphonium ionic liquids. *Green Chemistry*, 5 (2003), 143–52.
6. L. G. Bonnet and B. M. Kariuki, Ionic liquids: Synthesis and characterization of triphenylphosphonium tosylates. *European Journal of Inorganic Chemistry*, 2 (2006), 437–46.
7. S. D. Burnside and E. P. Giannelis, Synthesis and properties of new poly(dimethylsiloxane) nanocomposites. *Chemistry of Materials*, 7 (1995), 1597–1600.
8. C. Byrne, T. McNally, and C. G. Armstrong, Thermally stable modified layered silicates for PET nanocomposites. Presented at: Polymer Processing Society Americas Regional Meeting, Quebec, Canada, 2005.
9. J. U. Calderon, B. Lennox, and M. R. Kamal, Polystyrene/phosphonium organoclay nanocomposites by melt compounding. *International Polymer Processing*, 1 (2008), 119–28.
10. J. H. Chang, Y. U. An, and G. S. Sur, Poly(lactic acid) nanocomposites with various organoclays. 1. Thermomechanical properties, morphology, and gas permeability. *Journal of Polymer Science, Part B: Polymer Physics*, 41 (2003), 94–103.
11. G. Chigwada, D. Wang, and C. A. Wilkie, Polystyrene nanocomposites based on quinolinium and pyridinium surfactants. *Polymer Degradation and Stability*, 91 (2006), 848–55.
12. G. Chouzouri, Modification of biodegradable polyesters with inorganic fillers. Master's Thesis, New Jersey Institute of Technology, Newark, NJ, 2003.
13. J. H. Choy, S. Y. Kwak, J. S. Park, Y. J. Jeong, and J. Portier, Intercalative nanohybrids of nucleoside monophosphates and DNA in layered metal hydroxide. *Journal of American Chemical Society*, 121 (1999), 1399–1400.
14. U. Costantino, V. Ambrogi, M. Nocchetti, and L. Perioli, Hydrotalcite-like compounds: Versatile layered hosts of molecular anions with biological activity. *Microporous Mesoporous Materials*, 107 (2008), 149–60.
15. Y. Ding, C. Guo, J.-Y. Dong, and Z. Wang, Novel organic modification of montmorillonite in hydrocarbon solvent using ionic liquids – Type surfactant for the preparation

of polyolefin–clay nanocomposites. *Journal of Applied Polymer Science*, 102 (2006), 4314–20.

16. Y. Ding, R. Xiong, S. Wang, and X. Zhang, Aggregative structure of the 1-octadecyl-3-methylimidazolium cation in the interlayer of montmorillonite and its effect of polypropylene melt intercalation. *Journal of Polymer Science, Part B: Polymer Physics*, 45 (2007), 1252–9.

17. Y. Ding, X. Zhang, R. Xiong, S. Wu, M. Zha, and H. Tang, Effects of montmorillonite interlayer micro-circumstance on the PP melting intercalation. *European Polymer Journal*, 44 (2008), 24–31.

18. F. Gao, Clay/polymer composites: The story. *Materials Today*, 20 (2004), 50–55.

19. E. P. Giannelis, Polymer-layered silicate nanocomposites: Synthesis, properties and applications. *Applied Organometallic Chemistry*, 12 (1998), 675–80.

20. J. W. Gilman, W. H. Awad, R. D. Davis, J. Shields, R. H. Haris, C. David, A. B. Morgan, T. E. Sutto, J. Callahan, P. C. Trulove, and H. C. DeLong, Polymer/layered silicate nanocomposites from thermally stable trialkylimidazolium-treated montmorillonite. *Chemistry Materials*, 14 (2002), 3776–85.

21. J. U. Ha and M. Xanthos, Functionalization of nanoclays with ionic liquids for polypropylene composites. *Polymer Composites*, 30 (2008), 534–42.

22. J. U. Ha and M. Xanthos, Novel modifiers for layered double hydroxides and their effects on the properties of polylactic acid composites. *Applied Clay Science*, 47 (2010), 303–10.

23. J. G. Huddleston, A. E. Visser, M. W. Reichert, H. D. Willauer, G. A. Broker, and R. D. Rogers, Characterization and comparison of hydrophilic and hydrophobic room temperature ionic liquids incorporating the imidazolium cation. *Green Chemistry*, 3 (2001), 156–64.

24. P. C. Howlett, S. Zhang, D. R. Macfarlane, and M. Forsyth, An investigation of a phosphinate-based ionic liquid for corrosion protection of magnesium alloy AZ31. *Australian Journal of Chemistry*, 60 (2007), 43–6.

25. N. H. Kim, S. V. Malhotra, and M. Xanthos, Modification of cationic nanoclays with ionic liquids. *Microporous Mesoporous Materials*, 96 (2006), 29–35.

26. K. Kamena, Nanoclays and their emerging markets. In *Functional Fillers for Plastics*, ed. M. Xanthos (Weinheim: Wiley-VCH, 2010), pp. 177–88.

27. Kyowa Chemical Industry. Available at: http://www.kyowa-chem.co.jp/indexe.html (accessed 2008).

28. C. F. Linares and M. Brikgi, Interaction between antimicrobial drugs and antacid based on cancrinite-type zeolite, *Microporous Mesoporous Materials*, 96 (2006), 141–8.

29. M. Modesti, A. Lorenzetti, D. Bon, and S. Besco, *Polymer Degradation and Stability*, 91 (2006), 672–80.

30. A. Mohanty, The future of nanomaterials. In: *Proceedings of Pira International Conference* (Miami, FL: Pira International, February 22–24, 2005). [CD]

31. J. Morawiec, A. Pawlak, M. Slouf, A. Galeski, E. Piorkowska, and K. Krasnikowa, Preparation and properties of compatibilized LDPE/organo-modified montmorillonite nanocomposites, *European Polymer Journal*, 41 (2005), 1115–22.

32. A. Okada, Y. Fukushima, M. Kawasumi, S. Inagaki, A. Usuki, S. Sugiyama, T. Kurauchi, and O. Kamigaito, *Composite Material and Process for Manufacturing Same*, United States Patent 4,739,007 (1988).

33. N. Ogata, S. Kawakage, and T. Ogihara, Poly(vinyl alcohol) clay and poly(ethylene oxide) clay blend prepared using water as solvent. *Journal of Applied Polymer Science*, 66 (1997), 573–81.

34. K. Park and M. Xanthos, A study on the degradation of polylactic acid in the presence of phosphonium ionic liquids. *Polymer Degradation and Stability*, 92 (2009), 1350–58.
35. S. H. Patel, Processing aids. In: *Functional Fillers for Plastics*, ed. M. Xanthos (Weinheim: Wiley–VCH, 2010), pp. 407–23.
36. M. A. Paul, M. Alexandre, P. Degee, C. Henrist, A. Rulmont, and P. Dubois, New nanocomposites materials based on plasticized poly(l-lactide) and organo-modified montmorillonites: Thermal and morphological study. *Polymer*, 44 (2003), 443–50.
37. P. C. Pavan, E. L. Crepaldi, and J. B. Valim, Sorption of anionic surfactants on layered double hydroxides. *Journal of Colloid and Interface Science*, 229 (2000), 346–52.
38. M. Rahman and C. S. Brazel, Effectiveness of phosphonium, ammonium and imidazolium based ionic liquids as plasticizers for poly(vinyl chloride): Thermal and ultraviolet stability. *Polymeric Preprints (American Chemical Society Division of Polymer Chemistry)*, 45 (2004), 301–2.
39. M. P. Scott, M. G. Benton, M. Rahman, and C. S. Brazel, *Ionic Liquids as Green Solvents: Progress and Prospects*, ACS Symposium Series (Washington, DC: American Chemical Society, 2003).
40. S. Sinha Ray and M. Okamoto, Polymer/layered silicate nanocomposites: A review from preparation to processing. *Progress in Polymer Science*, 28 (2003), 1539–1641.
41. K. Stoeffler, P. G. Lafleur, and J. Denault, Effect of intercalating agents on clay dispersion and thermal properties in polyethylene/montmorillonite nanocomposites. *Polymer Engineering and Science*, 48 (2008), 1449–66.
42. K. Stoeffler, P. G. Lafleur, and J. Denault, Thermal properties of polyethylene nanocomposites based on different organoclays. Presented at: Annual Technical Conference, Society of Plastics Engineers, Charlotte, NC, 2006, pp. 263–7.
43. R. A. Vaia and E. P. Giannelis, Lattice model of polymer melt intercalation in organically modified layered silicates. *Macromolecules*, 30 (1997), 7990–99.
44. B. Weyershausen and K. Lehmann, Industrial application of ionic liquids as performance additives. *Green Chemistry*, 7 (2005), 15–19.
45. M. Xanthos, Polymers and polymer composites. In: *Functional Fillers for Plastics*, ed. M. Xanthos (Weinheim: Wiley–VCH, 2010), 3–18.
46. K. Yano, A. Usuki, A. Okada, T. Kurauchi, and O. Kamigaito, Synthesis and properties of polyimide–clay hybrid, *Journal of Polymer Science, Part A: Polymer Chemistry*, 31 (1993), 2493–8.
47. H. Zhao and S. V. Malhotra, Applications of ionic liquids in organic synthesis. *Aldrichimica Acta*, 35 (2007), 75–83.
48. Q. Zhou and M. Xanthos, Nanosize and microsize clay effects on the kinetics of the thermal degradation of polylactides. *Polymer Degradation and Stability*, 94 (2009), 327–38.
49. L. Zhu and M. Xanthos, Effects of process conditions and mixing protocols on structure of extruded polypropylene nanocomposites. *Journal of Applied Polymer Science*, 93 (2004), 1891–9.

Part II
Flame retardancy

7

Introduction to flame retardancy of polymer–clay nanocomposites

TIE LAN[a] AND GÜNTER BEYER[b]

[b]Kabelwerk Eupen AG

7.1 Introduction

The use of plastic materials, from construction materials to consumer electronics, has been increasing substantially in the past few decades. Easy processing, low density, and possible recycling make plastic the first choice of materials for many applications, such as automobile parts and food packaging structures. Comparing with traditional materials such as metal and concrete, plastic materials are combustible in a fire. Enhancing the flame retardation of plastic materials has been a priority in material development for many researchers. Many fire retardation standards have been established for relevant industries. Several trade organizations have also created industrial standards for fire safety standards and testing procedures. Underwriters Laboratories (UL) is an independent product safety certification organization that has been testing products and writing standards for safety for more than a century. UL has extensive standards and testing protocols for building materials, energy, lighting, power, and control, as well as wire and cables. The International Electrotechnical Commission (IEC) also publishes extensive standards for materials used in the electrical and electronic industries.

Many researchers have dedicated their efforts to understanding the fire risks and hazards of plastic materials. The most important fire risks and fire hazards are rate of heat release, rate of smoke production, and rate of toxic gas release [1]. Other factors include ignitability, ease of extinction, flammability of generated volatiles, and smoke obscuration. An early and high rate of heat release causes fast ignition and flame spread; furthermore, it controls fire intensity, which is much more important than ignitability, smoke toxicity, or flame spread. The time for people to escape in a fire is also controlled by the heat release rate [2].

A lot of chemistries have been developed to reduce the flammability of plastics, including addition of flame retardation elements into the polymer backbone and additive technologies. Different technologies are available to control fire from various aspects. Halogen elements such as fluorine, chlorine, and bromine are commonly used as radical captors in a fire to reduce the fire spread in the gas phase. Metal hydrate compounds such as aluminum trihydrate (ATH) and magnesium hydroxide (MDH) were developed as flame retardants because they release water under heat to cool the burning materials and eventually put out the flame. Challenges still remain to plastic polymers, despite various flame retardation technologies, in property balancing, processing capability, and the match between resins and flame retardant compounds. For instance, MDH and ATH are very effective flame retardants in polyolefins such as PP and PE. However, it requires a very high loading level of these materials for the applied polymers to achieve a suitable flame retardancy rating. It is obvious that the high filling levels result in high density and lack of flexibility of the end products, poor mechanical properties, and problematic compounding and extrusion processes. Other flame retardant–based brominated compounds emit toxic fumes in the burning process. They also have a serious impact on the environment. Intumescent systems are relatively expensive and also lack wet electrical stability.

Plastic burning includes plastic melting, polymer decomposition, and generation of combustible gases. Decomposition is critical to the generation of combustible gases. The breakdown of the bulk polymer phases may change the decomposition path during the burning process. Addition of 1 nm–thick, high–aspect ratio (200–300) layered aluminum silicate layers makes phase separation possible.

Plastic additives based on layered aluminum silicate, montmorillonite clay, have drawn significant attention in the polymer and plastic community in the past 10 years. This article will focus on the general chemistry of polymer–clay nanocomposites and their applications as flame retardant additives.

7.2 Clay chemistry

Naturally occurring montmorillonite clays are one of the ideal candidates for fundamental building blocks of nanodimension fillers. The aluminum silicate layers are separated by exchangeable cations (Na^+ or Ca^{2+}) and the inner-layer regions can be modified from hydrophilic to hydrophobic to allow insertion of organic polymers. A single layer of montmorillonite has a thickness of 0.96 nm and lateral dimensions in the range of 200–300 nm. Modified montmorillonite clays (organoclays) have been used as additives in various plastics to enhance performance properties such as stiffness, barrier properties, and heat stability.

Montmorillonite clay is the major component of bentonite. Bentonite has been used as a thickener, sealant, binder, lubricant, or absorption agent for many years in a broad range of industries. The development of the petrochemical industries was dependent on the use of bentonite as a drilling agent and cracking catalyst. Deposits of bentonite on earth are quite abundant: more than 10 billion metric tons, based on a marketing report by Vicente Flynn International in 2006. The current usage level could be sustained for more than 700 years. World bentonite production in 2007 is estimated at 14.1 million tons, of which U.S. production is estimated at 4.5 million tons. However, not all bentonite deposits are suitable as plastic additives, because the requirements of the plastic industry for bentonite compositions and purity levels are much higher than those of traditional industries such as the paint and coating industry. Proper choice of clay mines and purification processes is needed prior to organophilic modification by an exchange reaction of the clay. Traditional bentonite refinement processes use gravity separation and water washing after soda ash treatment of the crude bentonite. Nanocor, a subsidiary of Amcol, has developed a patented processing technology (United States Patent 6,050,509) to extract montmorillonite clay from bentonite [3]. The core of this technology is using an ion exchange process to convert natural Ca-bentonite to Na-exchanged bentonite. This process is a revolution from the traditional soda ash clay treatment process. The ion exchange process does not introduce any electrolytes into the clay system. The Na-exchanged clay can be purified further by gravimetric methods such as centrifugation.

X-ray diffraction has been widely used to study the crystallinity and morphology of clay minerals. Figure 7.1 (A) is a typical diffraction pattern for a bentonite ore. Major components are listed in the figure, including montmorillonite, kaolin, limestone, quartz, and amorphous silica. Because of the hydration or swelling of Na^+-modified montmorillonite, the nonswollen materials will be removed by the centrifugation. Figure 7.1 (B) shows the X-ray pattern of a montmorillonite purified from bentonite clay ore. Major identified non-montmorillonite components have been removed. It is noted that some amorphous silica has a very fine nature and is very difficult to remove by gravimetric processes. It is also extremely difficult to quantify the amount of amorphous silica in the bentonite. Therefore, bentonite ores having visible amorphous silica diffraction should not be used as starting clay materials.

Organic modification is the key to modify the montmorillonite from hydrophilic to hydrophobic. Figure 7.2 shows a schematic ion exchange diagram. It is desired to carry out the exchange reaction in a medium such as water or a mixture of water and organic solvents. Dimethyl dehydrogenated tallow ammonium chloride (DMDHT) is one of the most commonly used surface modifiers. The ionic bonding of DMDHT to the negatively charged clay surface is quite strong. The DMDHT exchange reaction to replace Na is nearly 1:1 quantitative. After surface modification, the inner layer of the clay is hydrophobic. A large amount of organic substance can be adsorbed into the region. Interestingly, nonexchanged DMDHT chloride can also be adsorbed into the clay inner layer as ion pairs. Therefore, it is necessary to limit the amount of organic modifier to the minimum needed quantity.

DMDHT-modified montmorillonite clay is supplied in an agglomerate form. The basic process involves drying and grinding the modified organoclay. Air classification is recommended to remove large particles in the milled products. A typical particle size distribution curve is shown in Figure 7.3. The mean particle size is normally in the range of 15–25 μm. Top size (99%) should be less than 60 μm.

7.3 Polymer–clay nanocomposites

Preparation of polymer–clay nanocomposites has been a hot research topic for chemists and plastic engineers over the past two decades. Basically, the process involves deagglomerating micrometer-sized organoclay and introducing nanoscale interactions between a polymer matrix and the organoclay. It is possible to have two preparation routes: from monomer or

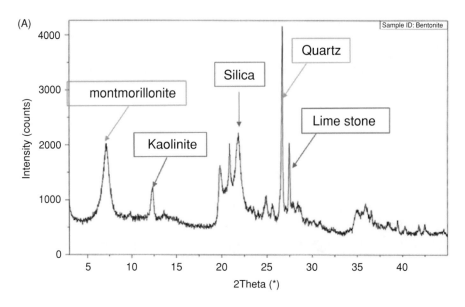

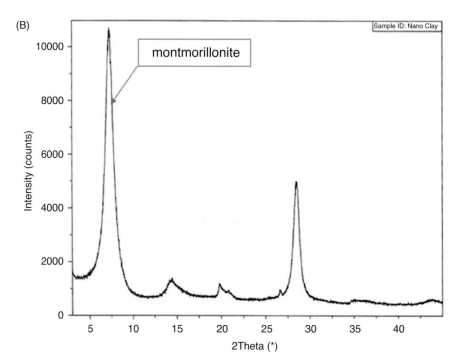

Figure 7.1 (A) Typical X-ray diffraction of bentonite clay and (B) Diffraction of purified montmo-rillonite clay.

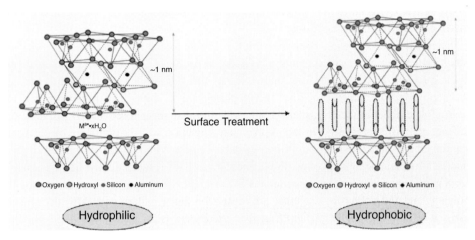

Figure 7.2 Schematic diagram of montmorillonite modification.

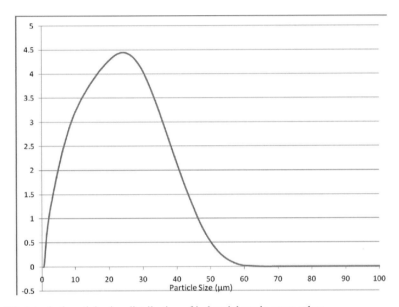

Figure 7.3 A typical particle size distribution of industrial grade organoclay.

oligomer and from polymer melts. In fact, the first polymer–clay nanocomposite technology developed by researchers used monomer in situ polymerization in organoclays [4]. It has been proven that the monomers of polyamide 6, caprolactam molecules, can enter the inner gallery resins before the polymerization. In the polymerization process, because of the polymer catalyst in the clay gallery, continuous migration of monomers into the gallery resin will expand the organoclay gallery further to a degree of exfoliation. Several attempts to use in situ polymerization of polymer–clay nanocomposites have been reported for both thermoset and thermoplastic resins.

For most thermoplastics, melt-processing is the most common and practical way to incorporate fillers or reinforcements. The hydrophobicity of the organoclay provides an organic moiety to allow nanoscale interaction of resin molecules with silicate layers of the organoclay. In a melt-processed thermoplastic nanocomposite, it is not realistic to desire full exfoliation of organoclay because the needed shear force tends to destroy or degrade the polymer matrix. In fact, organoclays function well in providing mechanical reinforcement, flame retardation, or barrier improvement with partially exfoliated or just intercalated organoclay structures. Polymers such as EVA (ethylene vinyl acetate) easily form nanocomposites by regular compounding processes. Low-polarity polymers such as polyethylene and polypropylene enter the organoclay inner gallery with difficulty unless assisted by modified polyethylene (PE) and polypropylene (PP). Maleic anhydride–grafted PP or PE resins are commonly used as compatibilizers to facilitate organoclay dispersion and polymer insertion into the organoclay inner region. The selection of PP-g-MA and processing conditions has significant impact on the performance of final PP nanocomposites [5]. The masterbatch process has been developed to minimize and eliminate these concerns [6]. Polyolefin–organoclay masterbatch products are produced by Nanocor for commercial application under the trademark nanoMax. A typical nanoclay masterbatch product contains 40–50% nanoclay. These nanoclay masterbatch products can be used with ease in commercial processing equipment. For flame retardation compounds, masterbatches can be used as regular additives in the compounding and injection molding process, in combination with other additives and fillers.

7.4 Flame retardation by polymer–clay nanocomposites

The comprehensive flame retardation of polymer–clay nanocomposite materials was reported by Dr. Jeff Gilman and others at NIST [7]. They disclosed that both delaminated and intercalated nanoclays improve the flammability properties of polymer–layered silicate (clay) nanocomposites. In the study of the flame retardant effect of the nanodispersed clays, XRD and TEM analysis identified a nanoreinforced protective silicate/carbon-like high-performance char from the combustion residue that provided a physical mechanism of flammability control. The report also disclosed that "The nanocomposite structure of the char appears to enhance the performance of the char layer. This char may act as an insulation and mass transport barrier showing the escape of the volatile products generated as the polymer decomposes." Cone calorimetry was used to study the flame retardation. The HRRs (heat release rates) of thermoplastic and thermoset polymer–layered silicate nanocomposites are reduced by 40% to 60% in delaminated or intercalated nanocomposites containing a silicate mass fraction of only 2% to 6%. On the basis of their expertise and experience in plastic flammability, they concluded that polymer–clay nanocomposites are very promising new flame-retarding polymers. In addition, they predict that the addition of organoclay into plastics has few or none of the drawbacks associated with other additives. The physical properties are not degraded by the additive (silicate); instead they are improved. Regarding the mechanism of flame retardation by organoclay, they concluded that the nanocomposite structure of the char appears to enhance the performance of the char layer. This layer may act as an insulator and a mass transport barrier showing the escape of the volatile products generated as the polymer decomposes.

Commercial use of polymer–clay nanocomposites started with cable jacket compounds developed by Dr. Beyer at Kabelwerk Eupen, Belgium [8]. Several research groups reported on the preparation and properties of EVA-based nanocomposites. EVA nanocomposites were prepared in a Brabender mixer by Camino *et al.* [9] and their thermal degradation was investigated. Hu *et al.* [10] prepared intercalated EVA nanocomposites. Filler content of only 5% improved the flame retardancy of the nanocomposites. Camino *et al.* [11] described the synthesis and thermal behavior of layered EVA nanocomposites. The organoclay was a synthetic modified fluorohectorite, which is a layered silicate. Protection against thermal oxidation and mass loss was observed in air. The modified silicates accelerated the deacetylation of the polymer but slowed the thermal degradation of the deacetylated polymer because of the formation of a barrier at the surface of the polymer. Zanetti, Camino, and Mulhaupt [12] mixed modified fluorohectorite with EVA in an internal mixer. They indicated that the accumulation of the filler on the surface of a burning specimen created a protective barrier to heat and mass loss during combustion. Antidripping was observed in vertical combustion in the case of nanocomposites, reducing the hazard of flame spread to surrounding materials. Melt-intercalated and additionally gamma-irradiated HDPE/EVA nanocomposites based on modified montmorillonite organoclays were prepared by Hu *et al.* [13, 14]. Increasing the clay content from 2% to 10% improved the flammability properties. Thermogravimetric analysis (TGA) data showed that the nanodispersion of the modified montmorillonite within the polymer inhibited the irradiation degradation of the HDPE/EVA blend, which led to nanocomposites with irradiation-resistant properties improved in comparison to those of the nonfilled blend. Other authors described the preparation of EVA-based nanocomposites in more detail. Sundararaji and Zhang [15] used a twin-screw extruder and found intercalation of modified montmorillonites with EVAs, differing in melt flow index and vinyl acetate content. The use of maleated EVA obviously improved the exfoliation, probably because of chemical interaction between the maleated EVA and the filler. Camino *et al.* [16] studied the effect of the compounding apparatus on the properties of EVA nanocomposites. A discontinuous batch mixer, a single-screw extruder, and a counterrotating and corotating intermeshing twin-screw extruder were used. Hu *et al.* [17] prepared EVA nanocomposites on a twin-screw extruder and a twin-roll mill. Morgan, Chu, and Harris [18] compared natural and synthetic clays with regards to polymer flammability. The natural clay was a U.S. mined and refined montmorillonite, whereas the synthetic clay was a fluorinated synthetic mica. Both clays were converted into organoclays by ion exchange with an alkylammononium salt and were then used to synthesize PS-based nanocomposites by melt-blending. Both nanocomposites reduced the peak heat release rate very similarly. The major differences between the natural and synthetic clay were improved color and better batch-to-batch consistency, but also higher costs, for the synthetic clay. Concerning the reaction mechanism of degradation and FR-behavior of EVA nanocomposites, Wilkie, Costache, and Jiang [19, 20] found that in early EVA degradation the loss of acetic acid seemed to be catalyzed by the hydroxyl groups that were present on the edges of the montmorillonite. The thermal degradation of EVA in the presence and in the absence of the modified clay showed that the formation of reaction products differed in quantity and identity. He found that the products were formed as a result of radical recombination reactions that could occur because the degrading polymer was contained within the layers for a long enough time. The formation of these new products explained the variation of heat release rates. In cases with multiple degradation pathways, the presence of the modified montmorillonite could promote one of these at the expense of another and thus lead to different products and hence a different rate of volatilization. Because the clay layers acted as a barrier to mass transport and led to superheated conditions in the condensed phase, extensive random scission of the products formed by radical recombination was an additional degradation pathway of polymers in the presence of clay. The polymers that showed good fire retardancy upon nanocomposite formation exhibited significant intermolecular reactions, such as interchain aminolysis/acidolysis, radical recombination, and hydrogen abstraction. In the case of polymers that degraded through a radical pathway, the relative stability of the radicals was the most important factor for the prediction of the effect of nanocomposite formation on reduction in the peak heat release rate. The more stable the radical produced by the polymer, the better was the fire retardancy, as measured by the reduction of the heat release rates of the polymer/clay nanocomposites.

Other nanostructured fillers were also described as flame retardants. Frache *et al.* [21] investigated the thermal and combustion behavior of PE–hydrotalcite nanocomposites. Hydrotalcites were synthesized and then intercalated with stearate anions, because of the compatibility of the long alkyl chain with the PE chains. The presence of the inorganic filler shielded PE from thermal oxidation and a reduction of the peak heat release rate by 55% was observed. Nelson, Yngard, and Yang [22] generated different kinds of nanocomposites using modified silica. PMMA–silica and PS–silica nanocomposites were obtained by single-screw extrusion. Although these nanocomposites exhibited higher thermal stability and oxygen indices, they burned faster than virgin polymers, according to horizontal burning tests, suggesting that nanocomposites themselves cannot be considered sufficiently flame-retardant materials. In combination with traditional flame-retardant additives, flame retardancy and better mechanical properties could be achieved using less flame-retardant additives in the presence of nanofillers. Zammarano *et al.* [23] studied the flame-retardant properties of modified layered double hydroxide (LDH) nanocomposites, which can be more effective than modified montmorillonites in the reduction of heat release rates. This may be related to the layered structure of LDHs and their hydroxyl groups and water molecules. Zammarano *et al.* [24] reported on synergistic effects of LDHs with ammonium polyphosphate in particular.

Recently Jho *et al.* [25] pointed out that modified montmorillonites alone are not sufficient as flame retardants in cable applications. Also, Wilkie [26] gave a clear statement: "It is apparent that nanocomposite formation alone is not the solution to the fire problem, but it may be a component of the solution. We, and others, have been investigating combinations of nanocomposites with conventional fire retardants."

Charring polymers such as PA-6 and PA-6 nanocomposites were used by Bourbigot *et al.* [27] to improve the flame retardancy of EVA. The organoclay increased the efficiency of the char as a protective barrier by thermal stabilization of a phosphorcarbonaceous structure in the intumescent char and additionally the formation of a "ceramic" layer. Hu *et al.* [28] used a blend of PA-6 and EVA nanocomposite to improve the flame retardancy of PP.

With greatly reduced burning rate, heat release, and char formation, it is possible to combine nanofiller with traditional flame retardants to achieve satisfactory regulatory approvals such as various UL ratings. In fact, several studies have revealed synergistic effects of combinations of nanofillers with traditional microsized flame retardants.

A halogen-free flame-retardant nanocomposite using PA-6, modified montmorillonite, MDH, and red phosphorus was reported by Hu *et al.* [29]. This system showed better mechanical and flame-retardant properties than a classical flame-retarded PA-6 and therefore a synergistic effect for all the three fillers. Leroy, Ferry, and Cuesta [30] partially substituted organoclays for MDH in flame-retardant EVA. Improvements in self-extinguishability were reported and the main mechanism was linked to intumescence leading to the formation of a foamlike structure during the preignition period. Horrocks, Kondola, and Padbury [31] demonstrated that the combination of organoclays with ammonium polyphosphate or polyphosphine oxide showed synergistic effects in flame retardancy for PA-6. Cogen and co-workers [32] investigated blends of EVA and ethylene-co-octene copolymers with MDH and modified montmorillonites for cable compounds. The time dependence of the char layer formation in such systems suggested that optimal loadings for the montmorillonite could be different for applications with different cable jacketing thicknesses and therefore the true performance of nanocomposite-based jacketing compounds needs to be assessed in actual cable construction.

Shen and Olsen [33, 34] reported improvements in flame retardancy from using a filler combination of modified montmorillonites, MDH, and zinc borate in EVA. A modified montmorillonite with a smaller particle size, formed in the presence of zinc borate that had very fine particle size, gave a better UL 94 performance than one with a larger size and a stronger char. London and co-workers [35] used modified montmorillonite as a partial substitute for ATH in PP/ATH composites and observed enhanced flame retardancy with composites containing both fillers. Wilkie and Zhang [36] studied the fire behavior of PE combined with ATH and a modified montmorillonite. The combination of PE with 2.5% modified montmorillonite and 20% ATH gave a 73% reduction in the peak heat release rate, which was the same as that obtained when 40% ATH was used alone. A further increase in the montmorillonite loading did not improve the fire properties. Mechanical properties, such as elongation at break, could be improved when compounds with or without montmorillonite were compared at the same reduction in peak heat release rate.

Cusak, Cross, and Hornsby [37] found that zinc hydroxystannate greatly enhanced the performance of an ATH/organoclay synergistic fire-retardant system in an EVA formulation that allowed reductions in the overall filler level with no or little compromise in terms of flame-retardant or smoke-suppressant properties.

This report reviews results of nanocomposites based on organoclays and the synergistic effects of these fillers with microsized aluminum trihydrate as a traditional flame retardant for cables. A new organoclay chemistry with greatly enhanced thermal stability will be disclosed. This may have flame retardation applications in many engineering plastics.

7.5 Ethylene vinyl acetate and polyethylene nanocomposites for wire and cable applications

Poly(ethylene vinyl acetate) copolymers are commonly used in the wire and cable jacket compounds because of easy processing and good compatibility with many traditional flame retardants such as ATH and MDH. Recently, EVA also demonstrated their ability to promote nanocomposite formation by melt-blending with organoclays [38, 39].

Low-density polyethylene (LDPE) BPD 8063 from Innovene (formerly BP Petrochemicals) was used as a nonpolar polymer matrix for organoclay. Maleic anhydride–grafted PE, Polybond 3109 from Chemtura, was used as compatibilizer for organoclay formulations.

Aluminum trihydrate (ATH) Martinal OL 104 LE from Albemarle was used as a conventional flame retardant.

A commercially available organoclay based on a layered silicate (montmorillonite modified by dimethyldistearylammonium cation exchange) was used. The content of the quaternary ammonium compound was 38 wt%. Octadecylamine-modified montmorillonite is used to study the thermal stability of organoclay.

Compound mixing was performed on different compounding equipment. For EVA organoclay-based nanocomposites, a laboratory twin-roll mill and an internal mixer heated to 145 °C were used. A corotating twin screw extruder from Leistritz, Germany, with a 27-mm screw diameter and an aspect ration of 40 L/D was used to generate polyethylene nanocomposites. The mass temperature was 190 °C at the extruder die.

X-ray diffraction (XRD) was used to measure the organoclay layer separation after compounding. An X-ray unit from PanAnalytical was used with CuKα as the X-ray source.

A cone calorimeter was used to characterize the burning of flame-retardant compounds. The tests were carried out for 35 kW/m^2 or 50 kW/m^2 heat fluxes with horizontal orientation of the samples (plates [100 mm $\times$ 100 mm $\times$ 3 mm] or cut cables) in accord with ASTM E 1354. The reported data were averages of three measurements for each sample with a standard uncertainty of measured heat release rates of $\pm 5\%$.

TGA was conducted under helium or airflow at a heating rate of 20 °C/min. Isothermal TGA was conducted in air only with a rapid initial heating rate of 50 °C/min and various holding temperatures for 30 min. Weight residual data at certain holding temperatures and holding times were collected from TGA curves to compare relative heat stability.

Depending on the nature of the filler distribution within the matrix, the morphology of nanocomposites can evolve from intercalated structure with a regular alternation of layered

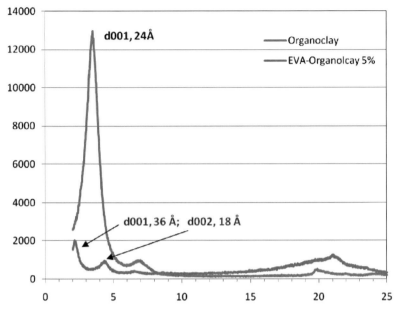

Figure 7.4 X-ray diffraction of organoclay and EVA–organoclay (5%) after melt-compounding.

silicates and polymer monolayers to exfoliated (delaminated) structure with layered silicates randomly and homogeneously distributed within the polymer matrix. Information on the nanocomposite morphology was obtained by transmission electron microscopy (TEM) and X-ray diffraction (XRD). Compounding was done on a twin-roll mill. Exfoliated silicate sheets were observed, together with small stacks of intercalated sheets. [39]. This structure may be described as a semi-intercalated semiexfoliated structure whose nature did not change with the vinyl acetate content of the EVA matrix, even larger numbers of stacks were observed for EVA with lower vinyl acetate content. There were no great differences in the morphology of these nanocomposites.

X-ray diffraction of a typical EVA–organoclay compound containing 5% organoclay is shown in Figure 7.4. The original diffraction peak (d001) at 24 Å disappeared after EVA compounding. The organoclay diffraction peaks shift to lower angle area after being incorporated into EVA. The d001 of the organoclay in the EVA compound is 36 Å. The basal spacing of the organoclay from 24 to 36 Å clearly demonstrates that the EVA molecules have entered the organoclay inner layer region. Because the layer thickness of the organoclay is nearly 10 Å, we can conclude that the inner layer distance is 26 Å. With consideration of the original inner layer spacing of the organoclay at 14 Å, a further 12-Å layer separation was realized by the EVA intercalation. In other words, a significant amount of EVA resides between the adjacent organoclay layers. The organoclay stacking order is regular, as evidenced by the appearance of the d002 diffraction at 18 Å. The insertion of EVA into the organoclay inner gallery will change the burning characteristics of the EVA material.

TGA is widely used to characterize the thermal stability of polymers. The mass loss of a polymer due to volatilization of products generated by thermal decomposition is monitored as a function of a temperature ramp. Nonoxidative decomposition occurs when the heating of the material is done under an inert gas flow such as helium or nitrogen, whereas the use of air or oxygen allows investigation of oxidative decomposition reactions. The experimental conditions highly influence the reaction mechanism of the degradation. EVA is known to decompose in two consecutive steps. The first is identical under oxidative and nonoxidative conditions. It occurs between 350 and 400 °C and is linked to the loss of acetic acid. The second step involves the thermal decomposition of the obtained unsaturated backbone either by further radical scissions (nonoxidative decomposition) or by thermal combustion (oxidative decomposition) (Figure 7.5).

In helium the EVA nanocomposite showed a negligible reduction in thermal stability compared with pure EVA or EVA filled with Na montmorillonite (microcomposite). In contrast, when decomposed in air, the same nanocomposite exhibited a rather large increase in thermal stability because the maximum of the second degradation peak was shifted 40 °C to higher temperature, whereas the maximum of the first decomposition peak remained unchanged [39] (Table 7.1). In this case the explanation for the improved thermal stability was char formation occurring under oxidative conditions. The char acted as a physical barrier between the polymer and the superficial zone where the combustion of the polymer occurred. Obviously, the optimum for thermal stabilization was already obtained at an organoclay level of 2.5–5.0 wt%, as indicated by the results in Table 7.1 on the temperature of the maximum decomposition rate for the main degradation peak for EVA nanocomposites.

Even with various flame-retardation testing methods, the cone calorimeter has been recognized as the main tool for understanding the burning of materials. The measuring principle is oxygen depletion, with a relationship between the mass of oxygen consumed from the air and the amount of heat released in the burning process. In a typical cone calorimeter experiment, a polymer sample plate in an aluminum dish is exposed to a defined heat flux (mostly 35 or 50 kW/m^2). Simultaneously, the properties of heat release rate, peak heat release rate (PHRR), time to ignition, total heat release, mass loss rate, mean CO yield, and mean specific extinction area are measured.

The flame retardant properties of EVA nanocomposites were determined by a cone calorimeter with a heat flux of 35 kW/m^2 (Figure 7.6). Under these conditions, simulating a developing fire scenario, the effect of 3 wt% organoclay was already evident. A decrease of the PHRR by 47%, as well as a shift toward longer time, was detected for a nanocomposite containing 5 wt% organoclay relative to the virgin EVA. Increasing the filler content to 10 wt% did not significantly improve the reduction of the PHRR any further. The decrease in PHRR indicated a reduction of burnable volatiles generated by the degradation of the polymer matrix. This reduction of PHRR clearly showed the flame retardant effect due to the presence of the organoclays and their "molecular" distribution throughout the matrix, as well as the polymer chains entrapped between the clay inner layer regions. Furthermore, the flame retardant properties were improved by the fact that the PHRR was spread over

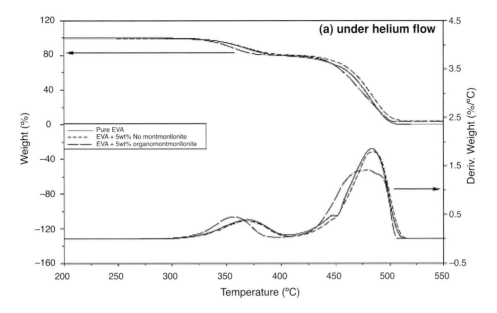

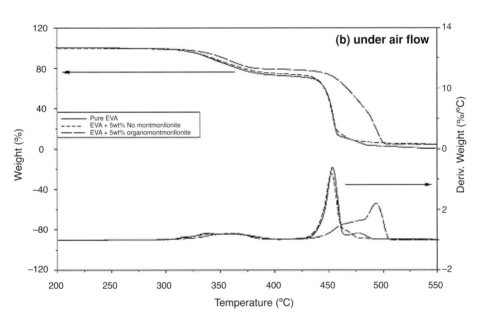

Figure 7.5 TGA of EVA, EVA microcomposite with 5 wt% Na montmorillonite, and EVA nanocomposite with 5 wt% organoclays under helium and air; heating rate 20 °C/min. EVA: Escorene UL 00328 with 28 wt% vinyl acetate content.

Table 7.1 *Temperature at the maximum degradation rate of the main decomposition peak (DTG) measured by TGA under air flow at 20 °C/min for EVA and EVA-based nanocomposites with different organoclay content*

Organoclay content (wt%)	Temperature at the degradation rate peak (main decomposition) (°C)
0	452.0
1	453.4
2.5	489.2
5	493.5

Note: EVA: Escorene UL 00328 with 28 wt% vinyl acetate content.

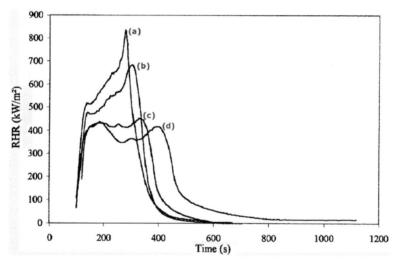

Figure 7.6 Heat release rated at heat flux 35 kW/m² for various EVA (Escorene UL 00328 with 28 wt% vinyl acetate content)–based materials: (a) pure EVA matrix and EVA matrix with 5 wt% Na montmorillonite; (b) EVA + 3 wt% organoclays; (c) EVA + 5 wt% organoclays; and (d) EVA + 10 wt% organoclays.

a much longer period of time. These properties were due to the formation of a char layer during nanocomposite combustion. This acted as an insulating and nonburning material and reduced the emission of volatile products (fuel) into the flame zone. The silicate layers of the organoclay played an active role in the formation of the char, but also strengthened it and made it more resistant to ablation.

Cone calorimeter experiments at a heat flux of 35 kW/m² showed that neat EVA was completely burned without any residue. In contrast, very early strong char formation was found for the EVA nanocomposite in an analogous cone calorimeter experiment; this char

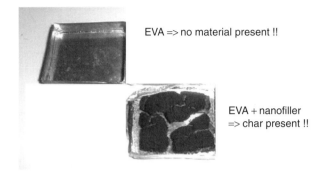

Figure 7.7 Samples of pure EVA and EVA nanocomposite with 5 wt% organoclays by cone calorimeter combustion after 200 s. EVA: Escorene UL 00328 with 28 wt% vinyl acetate content and heat
flux 35 kW/m^2; polymer plates of $100 \times 100 \times 3$ mm within aluminum dishes.

was stable and did not disappear by combustion (Figure 7.7). Finally, in the vertical burning
test UL94, the nanocomposite burned without producing burning droplets, a characteristic
feature that furthermore limits the propagation of a fire.

Solid state NMR spectroscopy on ^{13}C can reveal the chemical environmental change in
the burning process. Le Bras and co-workers [40] described the measurement method in
detail. EVA (Escorene UL 00112 with 12 wt% vinyl acetate content) and a nanocomposite
based on EVA (Escorene UL 00112) with 5 wt% organoclay were degraded by a cone
calorimeter with a heat flux of 50 kW/m^2. Samples were removed from the heat flux after
50, 100, 150, 200, and 300 s and the presence of EVA and the char formation were measured
by the shift (in ppm) of the signals relative to the standard tetramethysilane signal. The
following results were obtained [41].

Before cone irradiation of EVA and EVA nanocomposite:

- 33 ppm $\Rightarrow$ $-CH_2-$: polymer backbone of EVA
- 75 ppm $\Rightarrow$ $-CH_3$: acetate group of EVA
- 172 ppm $\Rightarrow$ $-C{=}O$: acetate group (small signal) of EVA.

After irradiation of EVA:

- 50 s: new signals at 130 ppm (char: aromatics and polyaromatics) and 180 ppm ($-C{=}O$
  with start of oxidation); EVA signals still present
- 150 s: no signals $\Rightarrow$ no organic material present.

After irradiation of EVA nanocomposite:

- 50 s: new signals at 130 ppm (char: aromatics and polyaromatics) and 180 ppm ($-C{=}0$
  with start of oxidation); EVA signals still present
- 100 s: char formation and EVA signals still present
- 200 s: char formation and EVA signals still present
- 300 s: no signals $\Rightarrow$ no organic material present.

The organoclay nearly doubled the survival of EVA under the regular cone irradiation testing condition. The formation of nanocomposites clearly promoted char formation and delayed the degradation of EVA.

Combinations of organoclay with ATH were also studied by cone calorimeter tests. ATH and organoclay were premixed before being compounded with EVA in a batch mixer at 145 °C for 15 min. Cone testing samples were prepared by compression molding and cut to fit the aluminum sample holder. With the same loading level of organoclay at 3%, two loading levels of ATH at 58% and 60% were used. A sample containing 65% ATH was made and tested for comparison purposes.

From cone testing (at heat flux 35 kW/m^2) data on HRR [Figure 7.8 (A)], the samples containing organoclays show significantly reduced values. More important, the secondary burning of the samples containing organoclay was reduced with higher ATH loading, such as the combination of 60% ATH with 3% organoclay. This was attributed to strong char formation by compounds containing organoclay. In contrast, the sample containing a higher amount of ATH does not have strong char formation and had great heat release and smoke generation associated with the secondary burning after the cracking of the original char. The rate of smoke production [Figure 7.8 (B)] for these test samples showed corresponding patterns such as the HRR graph. Smoke generation associated with secondary burning of the samples containing organoclay was nearly eliminated. These flame retardation benefits were also reported in fire tests of cables with nanocomposites as cable jackets.

Polyethylene is also a very important resin, used in flame retardant wire and cable applications because of its excellent electrical insulation and processing flexibility. However, polyethylene is a very poor char former in combustion processes because of its candlelike behavior. A char enhancer for polyethylene needs to be found. The use of organoclay was extended into polyethylene resins. Unlike EVA, polyethylene resins have no polarity, and forming nanocomposites creates a challenge. Maleic anhydride–grafted PE (PE-g-MA) was used as a compatibilizer in our attempt to increase polarity. An organoclay modified with dimethyldistearylammonium was used for PE nanocomposite.

A masterbatch containing 50 wt% organoclay predispersed in a mixture of PE and PE-g-MA was prepared using a 27-mm corotation twin-screw extruder at 500 rpm. The finished compound, containing 5 wt% organoclay, was prepared as a dilution of the masterbatch with LDPE in the same 27-mm twin-screw extruder at 500 rpm.

Cone calorimeter experiments at a heat flux of 35 kW/m^2 shows that the unfilled LDPE was completely burned without any residue. In contrast, a strong char was observed for the nano-LDPE [Figure 7.9 (A)]. In addition, there is a 15% reduction of PHRR [Figure 7.9 (B)] and 30 s delay in ignition for the nano-LDPE. It is worth noticing that PHHR reduction of the nano-LDPE is not as significant as with EVA containing organoclay. This is mainly due to the different polymer degradation mechanism in the burning process. Chain scission through radicals was considered the main mechanism. However, the good char foaming and reduced PHHR should provide a role for organoclays in LDPE-based flame materials.

Cable compounds must be flame retardant to achieve a low flame spread, defined by the widely used international cable fire test (IEC 60332–3-24) [42]. A compound containing

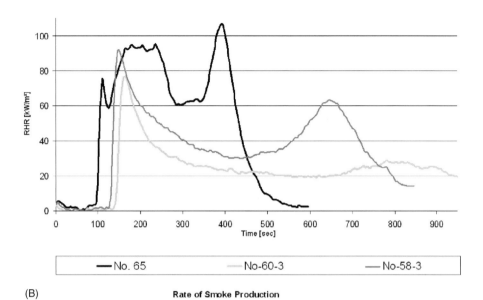

(A)

35 kW/m²

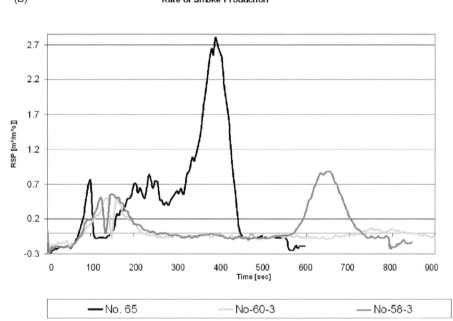

(B)

Rate of Smoke Production

Figure 7.8 (A) Heat release rate (HRR) of EVA-ATH compounds containing organoclay. 65: 65% ATH, 35%; 60–3: 60% ATH, 3% organoclay, 37% EVA; 58–3: 58% ATH, 3% organoclay, 39% EVA. EVA: Escorene UL 00328 with 28 wt% vinyl acetate content and heat flux 35 kW/m²; polymer plates of 100 × 100 × 3 mm within aluminum dishes. (B) Rate of smoke production (RSP) of EVA-ATH compounds containing organoclay. 65: 65% ATH, 35% organoclay; 60–3: 60% ATH, 3% organoclay, 37% EVA; 58–3: 58% ATH, 3% organoclay, 39% EVA. EVA: Escorene UL 00328 with 28 wt% vinyl acetate content and heat flux 35 kW/m²; polymer plates of 100 × 100 × 3 mm within aluminum dishes.

(A)

(B)   **35kW/m²**

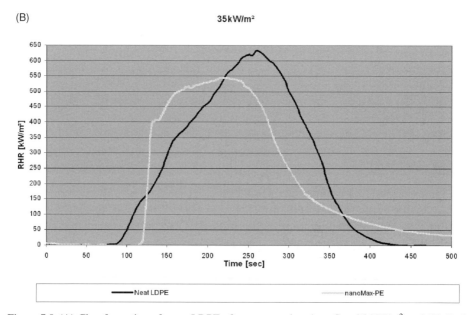

Figure 7.9 (A) Char formation of nano-LDPE after cone testing; heat flux 35 kW/m² and (B) Peak heat release rates of LDPE and nano-LDPE; heat flux 35 kW/m².

65 wt% of ATH and 35 wt% of a high filler level–accepting polymer matrix such as EVA must often be used for cable outer sheaths to ensure flame protection [43].

The performance of two compounds was compared: one made from 65 wt% ATH and 35 wt% EVA with 28% vinyl acetate content and a second from 60 wt% ATH, 5 wt% organoclays, and 35 wt% EVA with 28% vinyl acetate content. Both compounds were prepared on a BUSS Ko-kneader (46-mm screw diameter, L/D 11). Both were investigated

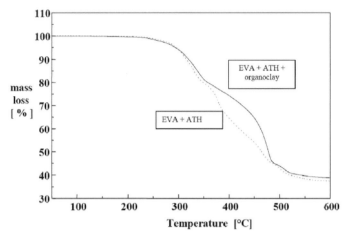

Figure 7.10 TGA in air of a compound with 35 wt% EVA and 65 wt% ATH in relation to a nanocomposite compound with 35 wt% EVA, 60 wt% ATH, and 5 wt% organoclays. EVA: Escorene UL 00328 with 28 wt% vinyl acetate content. ATH: Martinal OL 104 LE.

with TGA in air and a cone calorimeter at 50 kW/m^2 heat flux. TGA in air clearly showed a delay in degradation by the small amount of organoclays (Figure 7.10).

The char of the EVA/ATH/organoclay compound generated by a cone calorimeter was very rigid and showed very few small cracks; the char of the EVA/ATH compound was much less rigid (lower mechanical strength) and showed many big cracks. This could be why the PHRR for the nanocomposite was reduced to 100 kW/m^2, compared with 200 kW/m^2 for the EVA/ATH compound. To obtain the same decrease for the PHRR with the flame retardant filler ATH only, the ATH content must be increased to 78 wt%.

The great improvements in flame retardancy caused by the organoclays also opened the possibility of decreasing the level of ATH within the EVA polymer matrix. The content of ATH needed to maintain 200 kW/m^2 as a peak heat release rate could be decreased from 65 to 45 wt% by the presence of only 5 wt% organoclays within the EVA polymer matrix. Reduction in the total amount of these fillers resulted in improved mechanical and rheological properties of the EVA-based nanocomposite.

A coaxial cable containing an outer sheath made with EVA, ATH, and organoclay was studied under the large-scale fire test UL 1666 (riser test for cables) with a 145 kW burner in a two-story facility. There are many applications for indoor cables passing the UL 1666. This very severe fire test defines the following important points of measurement:

1. maximal temperature of fire gases at 12 ft: 850 °F
2. maximal height for flames: 12 ft

Compounds with halogenated flame retardants are often used to pass this test, but more and more flame retardant nonhalogen (FRNH) cables are demanded by the market for the

Table 7.2 *Fire performance of FRNH coaxial cables with EVA/ATH and EVA/ATH/organoclay outer sheaths*

UL 1666 requirements	EVA/ATH compound	EVA/ATH/organoclay compound
Maximal temperature at 12 feet: <850 °F	1930 °F	620 °F
Maximal flame height: <12 ft	12 ft	6 ft

Figure 7.11 Coaxial cable (1/2") with a nanocomposite-based outer sheath that passed the UL 1666 cable fire test.

riser test. Cables based on nanocomposite compounds demonstrate promising performance in this fire test.

An example of FRNH cables passing UL 1666 is shown in Figure 7.11. The outer sheath was based on a nanocomposite with an industrial EVA/ATH/organoclay composition. The analogous coaxial cable was tested with an outer sheath based on EVA/ATH. In both compounds the relation of polymer/filler was the same. Table 7.2 presents the results. The improved flame retardant properties were due to the formation of a char layer during the nanocomposite combustion. This insulating and nonburning char reduced the emission of volatile products from polymer degradation into the flame area and thus minimized the maximal temperature and height of the flames.

Heat stability has been a major concern for organoclays when they are used in plastics, because of processing temperature and residence time. Traditional organoclays with quaternary ammoniums such as dimethyldistearylammonium as exchange cations were considered, with heat stability only up to 200 °C [43]. Even with special selection of chemicals and refined processing, the heat stability can only be improved to 240 °C. This is mainly because of the Hofman elimination associated with the quaternary ammonium. Increasing the heat stability has been a hot topic with emphasis on the use of quaternary phosphonium chemistry.

We selected a primary amine (octadecylamine) to improve the heat stability of our organoclay because primary amines do not undergo Hofman elimination reactions upon heating. Octadecylamine was protonated by HCl and dissolved in water for the cation exchange

Table 7.3 *Weight loss by common organoclay and Nanomer I.31PS/I.30T*

Sample (DSC, 50 °C/min to ISO temperature)	Weight loss at 250 °C, 5 min	Weight loss at 300 °C, 5 min	Maximum compounding temperature (°C)
Nanoclay-1	8.0%	15%	220
Nanomer I.31PS/I.30T	0.8%	5%	280

Nanoclay-1

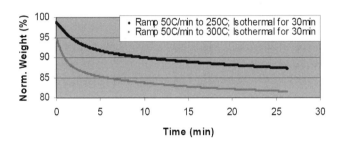

Nanomer I.31PS, I.30T

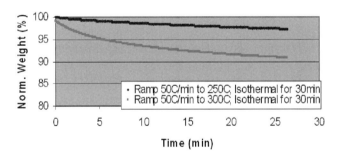

Figure 7.12 Isothermal TGA of common Nanoclay-1 and Nanomer I.31PS/I.30T.

reaction. Octadecylamine-modified organoclay has good hydrophobic properties, like other organoclays, allowing polymer to invade its interlayer resin to form nanocomposite materials.

Isothermal TGA was used to study the heat stability. We heated the sample at 50 °C/min to 250 or 300 °C and then held the temperature to record the weight loss of the organo-clay at these temperatures. This is similar to the heat experience of an organoclay in a polymer extrusion process. In Figure 7.12, we report the weight losses of nanoclay-1, which is traditional dimethyldistearylammonium-modified organoclay, and the weight loss of Nanomer I.31PS/I.30T, which is octadecylammonium-modified organoclay. The octadecylamine-modified organoclay has much more improved heat stability. Table 7.3

summarizes the heat stability results. For octadecylamine, the weight loss at the 250 °C isotherm is only 0.8% at 5 min compared with 8% weight loss of dimethyldisteary-lammonium-modified organoclay. At the 300 °C isotherm, octadecylamine-modified organoclay has 5% weight loss versus 15% weight loss of dimethyldistearylammonium-modified organoclay. Considering 5 min as the residence time of typical plastic processing, organoclay with octadecylamine can be used for processing at up to 280 °C, whereas organoclay with dimethyldistearylammonium can only be used at 220 °C. Organoclay with octade-cylamine may be suitable for use in engineering thermal plastics such as polyamides and polyesters to create new flame retardant compounds.

Other inorganic nanodimension particles such as layered doubled hydroxides, multiwall carbon nanotubes, POSS, and nanographite can also be potential nanoclay-sized flame retardant additives. Chemical compatibility of these materials is possible with surface modification reactions. Char formation is also observed in some plastic matrices. To have commercial value, general cost and easy processing have to be achieved.

7.6 Summary

Organoclays have been recognized as flame retardant additives in various plastic resins. The chemical modification of the montmorillonite inner-layer surface changes the nature of such inner-layer regions. Nanocomposites can be made with in situ polymerization and melt-compounding. It is possible to incorporate organoclays into plastics to form nanocomposites with regular processing equipment, and polymer chains can enter the inner-layer region in the nanocomposite formation process. The polymer matrix decomposition path was altered by phase segregation. In most cases, the entrapped polymers show slow burning characteristics compared to the bulk polymer matrix. The existence of the 1 nm–thick organoclay layer in a polymer matrix promotes char formation during the burning process. In addition, the nanocomposites demonstrate strong antidripping effects because of uniform dispersion in the polymer matrix.

Combination of traditional flame retardants with organoclay enables engineers to design new flame retardant compounds. These compounds have reduced burning heat, slower flame spread, and lower smoke generation. Commercially, organoclays have been used success-fully in low-smoke halogen-free flame retardant compounds by using combination with metal hydrates such as ATH and MDH. With the new development of organoclay chem-istry, flame retardant engineering plastics are possible and more environmental friendly flame retardant compounds are emerging.

References

1. M. Hirschler, Fire performance of organic polymers, thermal, decomposition, and chemical decomposition. *Polymeric Materials: Science and Engineering*, 83 (2000), 79–80.

2. V. Babrauskas, The generation of CO in bench scale fire tests and the prediction for real scale fires. *Fire and Materials*, 19 (1995), 205–13.
3. Amcol International Corporation, *Method of Manufacturing Polymer-Grade Clay for Use in Nanocomposites*, United States Patent 6,050,509 (2000).
4. A. Usuki, Synthesis of Nylon-6 clay hybrid. *Journal of Materials Research*, 8 (1993), 1179–84.
5. Amcol International Corporation, *Intercalates Formed with Polypropylene/Maleic Anhydride-Modified Polypropylene Intercalants*, United States Patent 6,462,122 (2002).
6. Y. Liang, G. Qian, J. Cho, V. Psihogios, and T. Lan, *Applications of Plastic Nanocomposites* (Clearwater Beach, FL: Additives, 2002).
7. NIST-IR 6312, *Interactions of Polymers with Fillers and Nanocomposites*, 1998.
8. G. Beyer, Nanocomposites offer new way forward for flame retardants. *Plastics Additives and Compounding*, 7 (2005), 32–5.
9. G. Camino, M. Zanetti, A. Riva, M. Braglia, and L. Falqui, Thermal degradation and rheological behaviour of EVA/montmorillonite nanocomposites. *Polymer Degradation and Stability*, 77 (2002), 299–304.
10. Y. Hu, Y. Tang, S. Wang, Z. Gui, and W. Fan, Preparation and flammability of ethylene-vinyl acetate copolymer/montmorillonite nanocomposites. *Polymer Degradation and Stability*, 78 (2002), 555–9.
11. G. Camino, R. Mulhaupt, M. Zanetti, and R. Thomann, Synthesis and thermal behaviour of layered silicate/EVA nanocomposites. *Polymer*, 42 (2001), 4501–7.
12. M. Zanetti, G. Camino, and R. Mulhaupt, Combustion behaviour of EVA/fluorohectorite nanocomposites. *Polymer Degradation and Stability*, 74 (2001), 413–17.
13. Y. Hu, H. Lu, Q. Kong, Y. Cai, Z. Chen, and W. Fan, Influence of gamma irradiation on high density polyethylene/ethylene-vinyl acetate/clay nanocomposites. *Polymers for Advanced Technologies*, 15 (2004), 601–5.
14. Y. Hu, H. Lu, Q. Kong, Z. Chen, and W. Fan, Gamma irradiation of high density poly (ethylene)/ethylene-vinyl acetate/clay nanocomposites: Possible mechanism of the influence of clay on irradiated nanocomposites. *Polymers for Advanced Technologies*, 16 (2005), 688–92.
15. U. Sundararaj and F. Zhang, Nanocomposites of ethylene-vinyl acetate copolymer (EVA) and organoclay prepared by twin-screw melt extrusion. *Polymer Composites*, 25 (2004), 535–42.
16. G. Camino, W. Gianelli, N. Dintcheva, S. Verso, and F. Mantia, EVA-montmorillonite nanocomposites: Effect of processing conditions. *Macromolecular Materials and Engineering*, 289 (2005), 238–44.
17. Y. Hu, Y. Tang, J. Wang, Z. Gui, Z. Chen, Y. Zhuang, W. Fan, and R. Zong, Influence of organophilic clay and preparation methods on EVA/montmorillonite nanocomposites. *Journal of Applied Polymer Science*, 91 (2005), 2416–21.
18. A. Morgan, L. Chu, and J. Harris, A flammability performance comparison between synthetic and natural clays in polystyrene nanocomposites: Thermal degradation of ethylene-vinyl acetate copolymer nanocomposites. *Fire and Materials*, 29 (2005), 213–29.
19. C. Wilkie, M. Costache and D. Jiang, Thermal degradation of ethylene/vinyl acetate copolymer nanocomposites. *Polymer* 46 (2005), 6947–58.

20. C. Wilkie, M. Costache, and B. Jang, The relationship between thermal degradation behavior of polymer and the fire retardancy of polymer/clay nanocomposites. *Polymer*, 46 (2005), 10678–87.
21. A. Frache, U. Constantino, A. Gallipoli, M. Nchetti, G. Camino and F. Bellucci, New nanocomposites constituted of polyethylene and organically modified ZnAl hydrotalcites. *Polymer Degradation and Stability*, 90 (2005), 586–90.
22. G. Nelson, R. Yngard, and F. Yang, Flammability of polymer–clay and polymer–silica nanocomposites. *Journal of Fire Sciences*, 23 (2005), 209–26.
23. M. Zammarano, M. Franceschi, S. Bellayer, J. Gilman and S. Meriani, Preparation and flame resistance properties of revolutionary self-extinguishing epoxy nanocomposites based on layered double hydroxides. *Polymer*, 46 (2005), 9314–28.
24. M. Zammarano, M. Franceschi, F. Mantovani, A. Minigher, M. Celotto, and S. Meriani, Flame resistance properties of layered-double-hydroxides/epoxy nanocomposites. In: *Proceedings of the 9th European Meeting on Fire Retardancy and Protection of Materials*, ed. M. Le Bras (Villeneuve D'Ascq, France: USTL Pub., 2003), pp. 17–19.
25. J. Jho, C. Hong, Y. Lee, J. Bae, B. Nam, G. Nam, and K. Lee, Tensile and flammability properties of polypropylene-based RTPO/clay nanocomposites for cable insulating material. *Journal of Applied Polymer Science*, 97 (2005), 2375–81.
26. C. Wilkie, Nanocomposite formation as a component of a fire retardant system. In: *Proceedings of the 10th European Meeting on Fire Retardancy and Protection of Materials*, ed. Bernhard Schartel (Berlin: Federal Institute for Materials Research and Testing [BAM], 2005), pp. 35–45.
27. S. Bourbigot, M. Le Bras, F. Dabrowski, J. Gilman, and T. Kashiwagi, PA-6 clay nanocomposite hybrid as char forming agent in intumescent formulations. *Fire and Materials*, 24 (2000), 201–8.
28. Y. Hu, Y. Tang, J. Xiao, J. Wang, L. Song, and W. Fan, PA-6 and EVA alloy/clay nanocomposites as char forming agents in poly(propylene) intumescent formulations. *Polymers for Advanced Technologies*, 16 (2005), 338–43.
29. Y. Hu, L. Song, Z. Lin, S. Xuan, S. Wang, Z. Chen, and W. Fan, Preparation and properties of halogen-free flame-retarded polyamide 6/organoclay nanocomposite. *Polymer Degradation and Stability*, 86 (2004), 535–40.
30. E. Leroy, L. Ferry, and J. Cuesta, Intumescence in ethylene-vinyl acetate copolymer filled with magnesium hydroxide and organoclays. In: *Fire Retardancy of Polymers*, ed. Michel Le Bras (London: The Royal Chemical Society, 2005), pp. 302–12.
31. R. Horrocks, B. Kondola, and S. Padbury, Interactions between nanoclays and flame retardant additives in polyamide 6, and polyamide 6,6 films. In: *Fire Retardancy of Polymers*, ed. Michel Le Bras (London: The Royal Chemical Society, 2005), pp. 223–37.
32. P. Whaley, J. Cogen, T. Lin, and K. Bolz, Nanocomposite flame retardant performance: Laboratory testing methodology. In: *Proceedings of the 53rd International Wire and Cable Symposium* (Philadelphia, PA: International Wire and Cable Symposium, 2004), pp. 605–11.
33. K. Shen and E. Olsen, eds., *Borates as FR in Halogen-Free Polymers: Proceedings of the 15th Annual BCC Conference on Flame Retardancy* (Stamford, CT: BCC Research, 2004).
34. K. Shen and E. Olsen, eds., *Recent Advances on the Use of Borates as Fire Retardants in Halogen-Free Systems: Proceedings of the 16th Annual BCC Conference on Flame Retardancy* (Stamford, CT: BCC Research, 2005).

35. N. Ristolainen, U. Hippi, J. Seppala, A. Nykanen, and J. Ruokolainen, Properties of polypropylene/aluminum trihydroxide composites containing nanosized organoclay. *Polymer Engineering and Science*, 45 (2005), 1568–75.

36. C. Wilkie and J. Zhang, Fire retardancy of polyethylene–alumina trihydrate containing clay as a synergist. *Polymers for Advanced Technologies*, 16 (2005), 549–53.

37. P. Cusack, M. Cross, and P. Hornsby, Effects of tin additives on the flammability and smoke emission characteristics of halogen-free ethylene-vinyl acetate copolymer. *Polymer Degradation and Stability*, 79 (2003), 309–18.

38. G. Beyer, M. Alexandre, C. Henrist, R. Cloots, A. Rulmont, R. Jerome, and P. Dubois. Preparation, morphology, mechanical and flame retardant properties of EVA/layered silicate nanocomposites. Presented at: World Polymer Congress, IUPAC Macro 2000, 38th Macromolecular IUPAC Symposium, Warsaw, 2000.

39. G. Beyer, M. Alexandre, C. Henrist, R. Cloots, A. Rulmont, R. Jerome, and P. Dubois, Preparation and properties of layered silicate nanocomposites based on ethylene vinyl acetate. *Macromolecular Rapid Communications*, 22 (2001), 643–6.

40. S. Bourbigot, M. Le Bras, R. Leeuwendal, K. Shen, and D. Schubert, Recent advances in the use of zinc borates in flame retardancy of EVA. *Polymer Degradation and Stability*, 64 (1999), 419–25.

41. G. Beyer, Flame retardancy of nanocomposites from research to technical products. *Journal of Fire Sciences*, 23 (2005), 75–87.

42. International Electrotechnical Commission (IEC), 60332–3-24, *Tests on Electrical Cables under Fire Conditions – Part 3–24: Test for Vertical Flame Spread of Vertically-Mounted Bunched Wires or Cables; Category C, 2000–10-0040.*

43. W. Xie, N. Whitely, W. Nathan, and W. Pan, Thermal stability of cationic surfactants modified layered silicates. *Proceedings of the NATAS Annual Conference on Thermal Analysis and Applications* 30 (2002), 137–42.

8

Flame retardant nanocomposites with polymer blends

J. ZHANG,[a] M. A. DELICHATSIOS,[a] A. FINA,[b] G. CAMINO,[b]
F. SAMYN,[c] S. BOURBIGOT,[c] AND S. NAZARÉ[d]

[a]*University of Ulster*
[b]*Politecnico di Torino, Sede di Alessandria*
[c]*ISP-UMET*
[d]*University of Bolton*

8.1 Introduction

Over the past decades, nanoclays have been widely used as additives to improve the strength as well as the fire performance of polymers, as evidenced by applications and a large number of studies reported in the literature (e.g., [1–9]).[1] The mechanism of action of nanoclays is now relatively well understood, despite some aspects remaining unclear, such as the phenomena controlling ignition time [10]. During the burning of polymer nanocomposites, a surface layer is formed on top of the virgin polymer, which acts as a mass and heat shield slowing down mass transfer of pyrolyzed gas to the surface, because less heat is transferred to unpyrolyzed material. Furthermore, in the presence of nanoparticles, the temperature at the surface of the surface layer increases far beyond the so-called ignition temperature of the polymer, which results in increased surface reradiation losses and, hence, decreased heat transfer to the solid. The formation of this surface layer has been observed in a number of studies using the cone calorimeter (e.g., [4–8]), where a significant reduction of the peak heat release rate (PHRR) compared with the corresponding pure polymer was observed for relatively thin samples. Zhang, Delichatsios, and Bourbigot [11] also studied the effect of the surface layer numerically, finding that the reduction in heat transfer at the interface of the surface layer and the virgin polymer is inversely proportional to the number of nanoparticles that remain on the surface after degradation of the polymer (if the concentration of nanoparticles is less than about 10%).

In recent years, nanoclays have also been used in polymer blends. Polymer blending is a useful and practical technique for achieving properties more advantageous than those of the single components. A clear advantage of polymer blends, as opposed to designing and synthesizing completely new monomers and polymers, is the reduction of research and development expense. Polymer blends often offer property profile combinations that are not easily obtained with new polymeric structures. These advantages have led to increased research interest, and comprehensive reviews of polymer blends can be found in [12–14].

[1] The authors acknowledge the EU for financially supporting the PREDFIRE-NANO project under Grant 013998 in the sixth Framework Program. The authors thank J. Hereid, M. Hagan, and M. McKee for conducting the cone calorimeter experiments and Dr. V. Mittal for inviting us to write this contribution.

Broadly speaking, polymer blends can be divided into three categories: miscible, partially miscible, and immiscible blends. A detailed classification was presented in [12] based primarily on the type of blends: elastomeric blends, engineering polymer blends, blends containing crystalline polymers, impact-modified blends, liquid crystalline polymer blends, and so forth. Polymer blends are now used in a wide range of applications, including the automobile industry, computer and other business machine housing, electrical components, appliances, consumer products, and construction and industrial applications [13]. Most of these applications require that the polymer blends have good fire properties, which is, unfortunately, not the case for most polymer blends, because of the nature of polymers (i.e., low flammability, high heat of combustion, and tendency to melt). Consequently, traditional fire retardants (FRs, mainly oxides) and, more recently, nanoclays are added to polymer blend systems in applications where good fire properties are required.

Although there has been extensive research (e.g., [1–9]) on the fire retardation effects of nanoclay on neat polymers, relatively few studies [15–31] have been conducted with polymer blend systems. In Table 8.1, the polymer blends as well as the tests conducted in [15–31] are presented. From this table, it is clear that the majority of these studies have been devoted to the morphology and thermal stability of polymer blend nanocomposites; very few studies actually focused on their fire performance. Nonetheless, these studies [15–31] have generally led to the conclusion that the addition of nanoclay to polymer blends can result in remarkable improvement in (a) mechanical properties, (b) compatibilization, (c) viscosity, (d) thermal stability, and (e) flammability.

In this chapter, we present a summary of fire retardant nanoclays used in polymer blends based on the authors' previous experience and the literature [15–31]. Because the main objective of this work is to study the fire retarding effects of nanoclays on polymer blends, we will focus on the properties affecting the fire performance of polymer blends: (a) dispersion of nanoclay, (b) rheology, (c) thermal stability, and (d) flammability (ignition, fire spread, and toxicity), whereas the effects of nanoclays on mechanical properties and compatibilization can be found easily in references listed in Table 8.1 (e.g., References [16, 17, 19–21] on compatibilization and [16, 18, 21, 25, 26, 29–31] on mechanical properties). A review of the mechanism by which nanoparticles organize in polymer blends is also available in [32].

This chapter is organized in the following way. First, we present some common techniques for characterizing the dispersion of nanoclays in polymer blends. The dispersion level has been shown to have a fundamental effect on the fire performance of polymer–clay nanocomposites (PCNs), as an exfoliated or intercalated polymer–clay system seems to enjoy reduced flammability. Second, the effects of nanoclays on the viscosity of polymer blends are discussed. With increased temperature in the condensed phase during combustion, most polymers (and hence polymer blends) have sufficiently low viscosity to flow under their own weight. This is highly undesirable, especially when the final products will be used in vertical orientation, because the melt can drip, having the potential to form a pool fire, which can increase fire spread. The results on thermal stability are presented next, followed by those for the cone calorimeter. The quantitative effects of nanoclays on the

Table 8.1 *Existing studies with polymer blend nanocomposites*

References	Blends	MP*	XRD	RH*	SEM	TEM	DSC	DTA	TGA	FTIR	LOI	UL94	CC*
Chuang et al. [15]	EVA/LLDPE												✓
Sinha Ray and Bousmina [16]	PC/PMMA	✓	✓		✓	✓	✓		✓	✓	✓		
Sinha Ray and Bousmina [17]	PC/PMMA		✓		✓	✓	✓		✓				
Lee et al. [18]	PMMA/SAN		✓	✓		✓	✓						
Xu et al. [19]	PEA/PMMA	✓	✓			✓	✓						
Chang et al. [20]	LDPE/EPDM				✓			✓	✓	✓	✓		✓
Haurie et al. [21]	LDPE/EVA							✓	✓		✓		✓
Lai, Li, and Liao [22]	PA6/ABS	✓	✓		✓	✓	✓		✓				
Yu et al. [23]	PLLA/PCL		✓		✓		✓		✓				
Elias et al. [24]	PP/PS			✓	✓	✓							
Scaffaro, Mistretta, and La Mantia [25]	PA6/HDPE	✓	✓	✓	✓	✓			✓	✓			
Acharya et al. [26]	EPDM/EVA	✓	✓		✓	✓	✓		✓				
Zhang et al. [27]	LDPE/EVA		✓	✓	✓				✓				✓
Gcwabaza et al. [28]	PP/PBS		✓	✓	✓	✓			✓				
Park et al. [29]	PC/PMMA & PC/SAN	✓	✓		✓	✓			✓				✓
Park et al. [30]	PS/PMMA	✓			✓	✓			✓				✓
Nayak and Mohanty [31]	PTT/m-LLDPE	✓	✓		✓	✓	✓		✓				

Notes: This table shows the base polymer blends and the tests conducted. MP: mechanical properties; RH: rheology; CC: cone calorimetry.

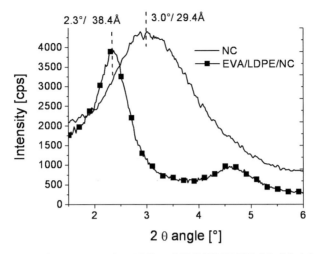

Figure 8.1 XRD pattern for neat nanoclay (NC) and EVA/LDPE/NC. Modified from Zhang *et al.* [27, p. 506]. With permission.

burning rates of an EVA/LDPE nanocomposite are examined using a methodology previously developed for neat polymer nanocomposites [11]. Finally, key findings from studies on the possible synergistic effects of nanoclays and other FRs are presented.

8.2 Morphological characterization

8.2.1 *Characterization of dispersion*

As shown in Table 8.1, the techniques commonly used for PCNs [such as X-ray diffraction (XRD), scanning electron microscopy (SEM), and transmission electron microscopy (TEM)] have also been used for polymer blend nanocomposites. Though useful, results from SEM and TEM are very qualitative because of the difficulties in defining the level of dispersion (such as exfoliation or intercalation). XRD, in comparison, provides more quantitative information on the dispersion of nanoclays. In this section, only the results from XRD tests are presented, but interested readers can find those with SEM and TEM in the relevant references in Table 8.1.

XRD involves illuminating samples with X-rays of a fixed wavelength, and based on the recorded intensity of the reflected radiation, calculating the *d*-spacing of the clay platelets. Studies [15–19, 22, 23, 25–29, 31] have shown a decrease of the scattering angle, 2θ (inversely proportional to the *d*-spacing of the clay platelets), in polymer blend nanocomposites in comparison to that in neat nanoclays.

Figure 8.1 shows an example of XRD analysis of an ethylene vinyl acetate/low-density polyethylene (EVA/LDPE) blend modified by an organically modified montmorillonite (OMMT, 5 wt%) [27]. Tests were conducted on a Philips X'Pert diffractometer using CuK$_\alpha$ radiation ($\lambda = 154{,}062$ Å) in continuous mode (step 0.02°, 1 s acquisition time). The neat

Table 8.2 *Compositions of the EVA/LDPE based materials*

Materials	EVA/LDPE (wt%)	NC (wt%)	FR (wt%)
EVA/LDPE	100	–	–
EVA/LDPE/nanoclay	95	5	–
EVA/LDPE/ATH	32	–	68
EVA/LDPE/ATH/nanoclay	32	5	63
EVA/LDPE/MH	32	–	68
EVA/LDPE/MH/nanoclay	32	5	63

nanoclay shows a diffraction signal centered at about 3.0° 2θ, reflecting an average interlayer distance of about 29.4 Å, which is a typical value for OMMT [33]. The compositions of the EVA/LDPE-based materials are presented in Table 8.2 for completeness, as they are used extensively throughout this chapter. The EVA/LDPE blend was modified with an organoclay (OMMT) and two FRs (aluminum trihydroxide, ATH, and magnesium hydroxide, MH). For EVA/LDPE/nanoclay, the corresponding signal is shifted to 2.3° 2θ, representing an average interlayer distance of 38.4 Å (an increase of 9 Å compared to the neat OMMT), suggesting the possible intercalation of polymer chains between the nanoclay platelets. This observation is also supported by the SEM micrographs of EVA/LDPE/NC as shown in Figure 8.2, where no large tactoids are detected in the polymer matrix, suggesting that the dispersion of the clay is submicrometric.

Acharya *et al.* [26] and Gcwabaza *et al.* [28] used XRD to study the effects of nanoclay (OMMT) loading on the morphology of polymer blend nanocomposites for an ethylene propylene diene terpolymer (EPDM)/EVA and a polypropylene/poly(butylene succinate) (PP/PBS) blend, respectively. Both studies showed an increase in the intensity of the diffraction peak due to nanoclay with an increase in the nanoclay loading, but the location (2θ) of the peak seems not to be affected by the increase of nanoclay loading.

8.2.2 *Viscosity*

To increase the viscosity of polymer blends, additives [such as traditional fire retardants (mainly oxides) and, more recently, nanoclays] are added to polymer blend systems. The present authors recently conducted dynamic rheological measurements for the EVA/LDPE nanocomposite, as reported in [27]. Figure 8.3 (a) and (b) compare the complex viscosity of the EVA/LDPE blend with and without nanoclay as a function of frequency and temperature, respectively. Measurements were carried out on 1 mm–thick samples using a Rheometrics RDA II Dynamic Analyzer rheometer. The frequency-sweep tests were conducted from 0.1 to 100 rad/s with constant temperature (140 °C) and strain amplitude (1%). For the temperature-sweep measurements, samples were heated from 300 to 530 °C (15 °C/min) under nitrogen with constant frequency (10 rad/s) and strain amplitude (10%). In both experiments, there is a significant increase of viscosity above that for the neat

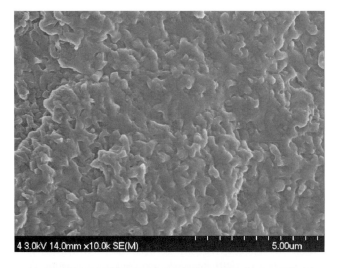

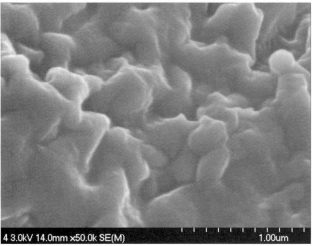

Figure 8.2 SEM micrographs of EVA/LDPE/nanoclay at two magnifications, 5 μm (top) and 1 μm (bottom).

blend with addition of 5 wt% nanoclay. For the frequency-sweep experiment in Figure 8.3 (a), EVA/LDPE/nanoclay shows an absolute shear thinning component of −0.54, compared with −0.35 for the neat blend, indicating significant polymer–clay and clay–clay interactions and therefore suggesting an intercalated nanocomposite system, which is also supported by the XRD analysis, where polybutylene/nanoclay shows a shift of 9 Å in *d*-spacing compared with the neat blend. The temperature-sweep results in Figure 8.3 (b) show that for the neat blend the viscosity increases to its peak at around 400 °C, which is probably because of deacetylation accompanied by slight foaming and partial cross linking of the EVA polymer. This peak is retained for EVA/LDPE/nanoclay, but its magnitude is

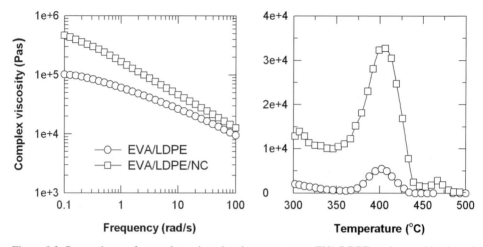

Figure 8.3 Comparison of complex viscosity between neat EVA/LDPE polymer blend and EVA/LDPE/NC as a function of (left) frequency and (right) temperature. For frequency-sweep experiments, the temperature is 140 °C and the stain at 1%; for temperature-sweep experiments, the frequency is 10 rad/s and the strain at 10%.

increased significantly, suggesting no melt dripping and hence reduced flammability in the presence of the nanoclay.

The results in Figure 8.3 are consistent with studies by other researchers [18, 25, 28], who also investigated the effects of nanoclay on the viscosity of polymer blends. Lee *et al.* [18] used three nanoclays, a natural (Cloisite Na[+]) and two organically modified MMTs (Cloisite 25A and Cloisite 15A), in a PMMA/poly(styrene-co-acrylonitrile) (SAN) blend and found that OMMTs increase the viscosity significantly, whereas the pristine MMT has a negligible effect. Similarly, in Reference [25], the viscosity of a polyamide-6/ high-density polyethylene (PA6/HDPE) blend was found to increase considerably with an OMMT (Cloisite 15A), especially in the low-frequency region. The effect of the nanoclay (C20A) loading on the viscosity of a PP/PBS blend was examined in [28], where a large increase of viscosity with nanoclay loading was observed. The results in [28] also indicated that there exists a critical loading (1% in this case) that corresponds to a transition from liquid to solid owing to the formation of a network or gel-like structure at higher C20A loadings.

8.3 Thermogravimetric analysis

Thermogravimetric analysis (TGA) is the most widely used thermal analysis technique, although other techniques such as differential scanning calorimetry (DSC) and differential thermal analysis (DTA) have also been used for polymer blends (see Table 8.1). In this section, we will limit our discussion to TGA, as its results (such as onset degradation temperature, degradation rate, and kinetic parameters) are most indicative of the fire performance of materials in fires.

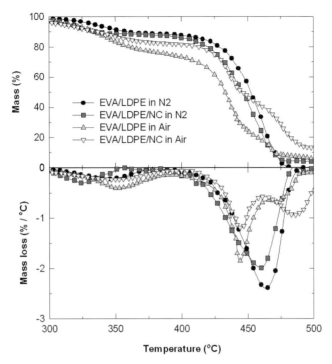

Figure 8.4 TGA/DTG profiles of the EVA/LPDE blend with or without nanoclay in air and nitrogen atmospheres (heating rate 10 °C/min).

Figure 8.4 shows the TGA weight loss (%) and degradation rate (DTG) (%/°C) of the neat EVA/LDPE blend and EVA/LDPE nanocomposite in air and under nitrogen. Under nitrogen, the profiles of EVA/LDPE and EVA/LDPE/nanoclay are similar, both showing two-stage degradation. The first step, occurring between 300 and 400 °C, is likely to correspond to the loss of acetic acid [34, 35], whereas the second step involves the thermal decomposition of the resulting backbone by further radical scissions [36]. The presence of the nanoclay accelerates the deacetylation of EVA, in agreement with results from Zanetti *et al.* [9]. The second weight loss step as well is affected by the presence of the nanoclay, the onset of the second weight loss step for EVA/LDPE/NC being early compared to that for EVA/LDPE, possibly related to catalytic cracking of polymer induced by the nanoclay, in agreement with results reported in the literature for PE and PP [37, 38].

The thermal-oxidative degradation is more complex because the presence of oxygen promotes additional degradation pathways. The degradation of both EVA/LDPE and EVA/LDPE/nanoclay involves three main decomposition steps, as shown in Figure 8.4. The two main weight loss stages observed in nitrogen remain in the thermo-oxidative atmosphere. However, there is an additional step toward the end of the test, which represents the oxidation of the carbonaceous residue formed in the reaction steps. The presence of the nanoclay seems to promote the thermo-oxidative stabilization of the neat blend through

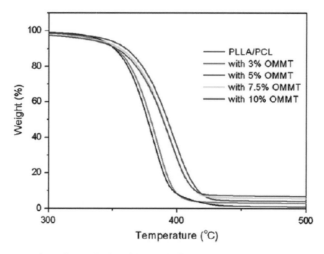

Figure 8.5 Thermogravimetric analysis of PLLA/PCL blend and nanocomposites with various amounts of OMMT. From Yu *et al.*, *Polymer* [23. p. 6443]. With permission.

the formation of a surface layer acting as a mass and thermal barrier to the unpyrolyzed material, resulting in an overall retarded weight loss. Although the maximum mass loss rate is reduced by the addition of nanoclay, the temperature at which it occurs is not affected.

The increase in the thermal stability of polymer blends with nanoclay in the thermo-oxidative atmosphere was also reported by other researchers (e.g., [19, 23, 28–31]). For example, Yu *et al.* [23] and Gcwabaza *et al.* [28] examined the effect of nanoclay (OMMT) loading on the TGA results for poly(L-lactide)/poly(ε-caprolactone) (PLLA/PCL) and PP/PBS blend. The results are shown in Figures 8.5 and 8.6, respectively. Both figures show that with the addition of a small amount of OMMT there is a noticeable increase in the thermal stability of the polymer blend. It can also be seen in Figures 8.5 and 8.6 that an optimized clay loading exists, above which there is no more improvement in thermal stability and, in some cases, even a decrease can be observed. This is because of the geometrical constraints that restrict exfoliation of large amounts of such high–aspect ratio silicate layers. Furthermore, for the PP/PBS blend in Figure 8.6 with higher nanoclay loadings (above 3%), the polymer blend nanocomposite shows two-step degradation, where stacked and intercalated layers are dispersed in the blend matrix. The first step may be related to chain scission, with some oxidation, and the second degradation step is due to the oxidative reactions [28].

8.4 Cone calorimetry

Flammability tests include (a) UL94, (b) limiting oxygen index (LOI), and (c) cone calorimeter tests. LOI is used to obtain the limit of oxygen concentration that sustains combustion, whereas UL94 studies the ignition from a small flame and subsequently the

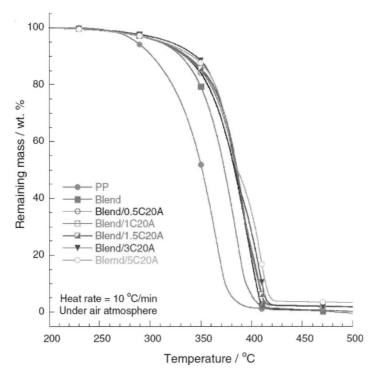

Figure 8.6 Thermogravimetric analysis of PP/PBS blend and nanocomposites with various amounts of OMMT in air atmosphere. From Gcwabaza *et al*. [28, p. 365]. With permission.

flame spread when the flame is removed. Although LOI and UL94 provide certain information regarding the ignition and burning behavior of a material in that particular configuration, they are less useful in determining the performance of materials in fires, in which other factors (such as heat transfer from flames and production of toxic gases) have to be considered. In comparison, the cone calorimeter test (ISO-5660), the most widely used fire test method, provides results that can usually be correlated to large-scale tests. In this section, the results of cone calorimetric studies are presented in terms of the effects of nanoclays on (a) time to ignition (TTI), (b) heat release rate (HRR), and (c) production of toxic gases (mainly smoke and carbon monoxide).

8.4.1 Time to ignition

Figure 8.7 compares the TTI between EVA/LDPE and EVA/LDPE/nanoclay. It can be observed that at higher heat fluxes the TTI is essentially not affected by the addition of the nanoclay. The differences at the lowest heat flux (20 kW/m^2) can be attributed to a carbonaceous layer formed on the surface prior to ignition, which poses uncertainties in measuring the time to ignition, as local ignition was observed in the experiments. A similar

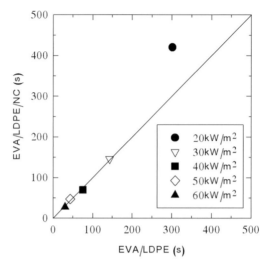

Figure 8.7 Comparison of TTI for EVA/LDPE and EVA/LDPE/nanoclay at different heat fluxes in the cone calorimeter. The data presented are average of duplicated tests at the same heat flux.

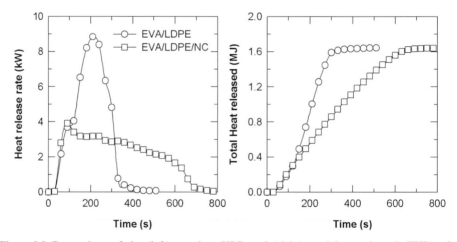

Figure 8.8 Comparison of the (left) transient HRR and (right) total heat released (THR) of a EVA/LDPE blend with or without nanoclay at 50 kW/m^2.

study by Chuang *et al.* [15] also revealed that there is no difference in the TTI for an EVA/linear LDPE blend when OMMT is added.

8.4.2 Heat release rate

Figure 8.8 (a) and (b) show, respectively, the transient HRR and total heat released (THR) for EVA/LPDE and EVA/LDPE/nanoclay at 50 kW/m^2. With the addition of 5 wt% nanoclay,

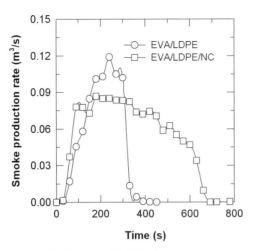

Figure 8.9 Comparisons of the SPR history at 50 kW/m^2.

the peak HRR observed for the neat blend is completely removed. This finding agrees with the one reported for a similar polymer blend system (EVA/LLDPE (linear LDPE)) [15]. The reduction of the steady and peak HRRs can be attributed to a surface layer formed during pyrolysis of nanocomposites, which reduces the heat transfer into unpyrolyzed material and increases the surface radiation heat loss. However, in the initial stage, there is no change in the HRR when the nanoclay is added, probably because there are not a sufficient number of nanoparticles to form the surface layer. With regard to the THR, Figure 8.8 (b) shows that EVA/LDPE/nanoclay has the same value as for the neat blend, implying that although it takes longer for EVA/LDPE/NC to burn, it burns completely.

8.4.3 Smoke, CO, and CO$_2$ production

Figure 8.9 presents a comparison of smoke production rates (SPR) between EVA/LDPE and EVA/LDPE/nanoclay at 50 kW/m^2. In the initial stage, the smoke production in the two materials is similar. However, after the formation of the surface layer, there is a steady decrease of the SPR for EVA/LDPE/nanoclay, which is likely due to the decreased HRR/MLR. As it takes longer for EVA/LDPE/nanoclay to burn completely, a more mean-ingful comparison would be the smoke yield, which can be calculated as the ratio of total smoke production (in g) to total mass lost (in g). The results are shown in Table 8.3, along with CO and CO$_2$ yields obtained in a similar manner. EVA/LDPE/nanoclay has higher smoke and CO yields and lower CO$_2$ yield than EVA/LDPE, indicating that the present nanoclay promotes production of smoke and CO. This finding is also consistent with the fact that slightly lower values of heat of combustion and CO$_2$ yield are found for EVA/LDPE/NC.

Table 8.3 *Comparisons of smoke, CO, and CO$_2$ yields between EVA/LDPE and EVA/LDPE/NC*

Material	Total heat released (MJ)	Total mass lost (g)	Heat of combustion (kJ/g)	Smoke yield (g/g)	CO yield (g/g)	CO$_2$ yield (g/g)
LDPE/EVA	1.62	50.4	32.2	0.0456	0.0181	2.44
LDPE/EVA/nanoclay	1.63	52.4	31.1	0.0771	0.0239	2.23

Note: The values shown here were calculated as averages of the tests at all the heat fluxes.

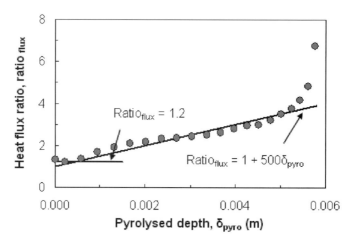

Figure 8.10 Derived heat flux ratio (ratio$_{flux}$) as a function of the pyrolyzed depth (δ_{pyro}) based on the mass loss data of the EVA/LDPE/nanoclay at 50 kW/m^2 in the cone calorimeter.

8.4.4 Mechanism of action of nanoparticles

To quantify the effects of nanoclay on the burning of polymers in the cone calorimeter, a methodology combining a numerical model and experimental data was developed and validated for a PA6 nanocomposite [11]. The main finding is that a heat flux ratio [the ratio of the heat flux at the interface of the nanolayer and virgin (unpyrolyzed) material to that at the surface when there is no surface layer] is proportional to the pyrolyzed depth (the thickness of the material that has been pyrolyzed), or equivalently to the number of nanoparticles accumulated on the surface. This proportionality is further supported by Figure 8.10, which plots the deduced heat flux ratio against the pyrolyzed depth for EVA/LDPE/nanoclay under an external heat flux of 50 kW/m^2. The correlation (shown as lines in Figure 8.10) obtained by linear fitting of the calculated heat flux ratio versus the pyrolyzed depth,

$$\text{ratio}_{flux} = 1.2 \text{ for } \delta_{pyro} < 4 \times 10^{-4} \text{ m} \tag{8.6.1a}$$

$$\text{ratio}_{flux} = 500\delta_{pyro} + 1 \text{ for } \delta_{pyro} >= 4 \times 10^{-4} \text{ m}, \tag{8.6.1b}$$

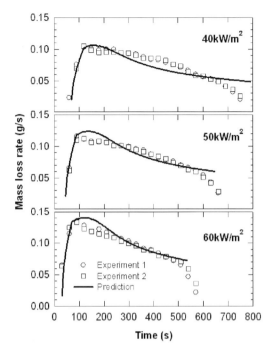

Figure 8.11 Comparison of the experimental and predicted MLRs at different heat fluxes using the correlation given by eq. (8.6.1) for EVA/LDPE/nanoclay. From Zhang *et al.* [44, p. 512]. With permission.

provides a means of predicting the mass burning rate under other heat fluxes. A detailed description of the methodology can be found in [11].

Figure 8.11 compares the predicted mass loss rates (MLRs) with the measured ones under several heat fluxes. Not only does the model predict correctly the trends of the experimental data, but the predicted MLRs are also in quantitative agreement with the measurements. Figure 8.11 implies, as noted in [11], that the correlation between ratio$_{flux}$ and δ_{pyro} (cf. eq. (8.6.1)) is independent of the external heat flux. It depends only on the thickness of the material that has pyrolyzed or, equivalently, on the amount of nanoclay accumulated on the surface after polymer pyrolysis. The result in Figure 8.11 demonstrates that the methodology is applicable not only to pure polymer nanocomposites but also to polymer blend nanocomposites.

8.5 Polymer blend nanocomposites combined with additional materials

In previous sections, we have shown that, although nanoclays can improve viscosity and reduce the peak HRR/MLR, their effects on ignition, the THR, and production of toxic gases are limited. Consequently, research has been conducted by including additional materials in polymer blend nanocomposites in an attempt to study the possible synergistic effect of

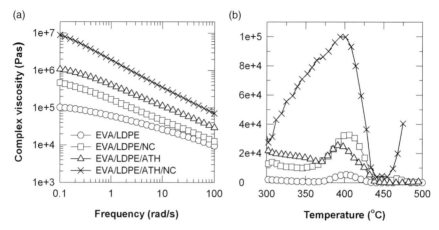

Figure 8.12 Complex viscosity of the EVA/LDPE polymer blend with nanoclay and /or FR (ATH) as a function of (a) frequency and (b) temperature. For frequency-sweep experiments, the temperature is 140 °C and the stain at 1%; for temperature-sweep experiments, the frequency is 10 rad/s and the strain at 10%.

nanoclays and other additives. The additives that have been studied include (a) oxides (ATH or MH) [15, 21, 27], (b) synthetic hydromagnesite (Hy) [21], and (c) nanosized hydroxyl aluminum oxalate (nano-HAO) [20]. This section summarizes the key findings of these studies.

8.5.1 Viscosity

Figure 8.12 (a) and (b) compare the viscosity as a function of frequency and temperature, respectively, for the EVA/LDPE blend modified with nanoclay and FR (ATH). In both experiments, there is a further increase in viscosity when ATH is added, in comparison to EVA/LDPE or EVA/LDPE/nanoclay. The absolute shear-thinning components of EVA/LDPE/ATH and EVA/LDPE/nanoclay are similar (around $\eta = -0.54$), indicating that the viscoelastic properties of the EVA/LDPE/ATH microcomposite with a filler loading of 68% are similar to those of the EVA/LDPE/nanoclay nanocomposite with a filler loading of only 5%. For EVA/LDPE/ATH/nanoclay, where 5 wt% of ATH was replaced with the same amount of nanoclay, there is a significant increase of viscosity.

8.5.2 TGA

A comparison of the TG curves is presented in Figure 8.13 for EVA/LDPE with nanoclay and/or FRs (ATH or MH). The inclusion of FRs lowers the weight loss onset temperature because of dehydration of ATH (to about 220 °C) or MH (to about 350 °C) [36]. A greater shift toward lower temperature is obtained with ATH, because of its lower thermal stability as compared with MH. The presence of nanoclay in EVA/LDPE/ATH/nanoclay causes a

Table 8.4 *Time to ignition (s) for EVA/LDPE-based materials at various heat fluxes*

	Heat flux (kW/m²)				
Materials	20	30	40	50	60
EVA/LDPE	301.5	143	75.5	43.5	30.5
EVA/LDPE/nanoclay	420.5	146	70	47.5	28
EVA/LDPE/ATH	357.5	172	110.5	74.5	52
EVA/LDPE/ATH/nanoclay	–	–	111	78	48
EVA/LDPE/MH	360	190.5	136.5	93	64.5
EVA/LDPE/MH/nanoclay	–	–	112.5	79.5	46.5

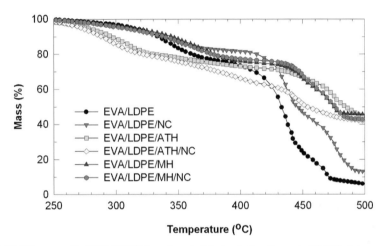

Figure 8.13 TG curves for EVA/LDPE-based materials in air (heating rate 10 °C/min).

slight destabilization relative to EVA/LDPE/ATH, whereas there are no significant differences between EVA/LDPE/MH/nanoclay and EVA/LDPE/MH. The reduction in thermal stability for EVA/LDPE/ATH/nanoclay compared with EVA/LDPE/ATH may be related to catalytic cracking of polymer induced by the nanoclay, as mentioned earlier. The efficiency of the cracking appears to be high when the nanoclay is combined with the oxides produced by dehydration of hydroxides.

8.5.3 Time to ignition

The effects of FRs on the TTI for the EVA/LDPE–based materials are shown in Table 8.4. There is a substantial increase in the TTI in the presence of FRs (ATH or MH), except at the lowest heat flux (20 kW/m²), where similar TTIs were observed for all materials.

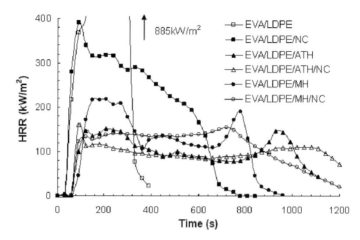

Figure 8.14 Heat release rate for EVA/LDPE-based materials. Modified from Zhang *et al.* [27, p. 509]. With permission.

The increases in ignition time caused by FRs occur because (a) the water released from the FRs hinders ignition by diluting the fuel in the gas phase below the flammability limit, and (b) FRs (ATH or MH) are completely transformed into oxides through an endothermic reaction prior to ignition and therefore the temperature of the substrate is lowered, which will increase the ignition time [39]. Partial replacement of FRs with nanoclay seems to increase the strength of the carbonaceous layer, as no ignition was observed for materials containing both FRs and nanoclay at 20 and 30 kW/m^2.

8.5.4 HRR

Figure 8.14 shows the HRR plots for the EVA/LDPE-based materials. As expected, there is a substantial reduction of the HRR with the addition of FRs. For FR-containing materials (FRCMs), the reduction in the HRR is much more substantial than that with nanoclay alone. Here, however, it is worthwhile to note that the reduction of the HRR by the FRCMs is also due partially to the reduced polymer content in the FRCMs. Partial replacement of FRs with nanoclay reduces the HRR/MLR further, probably because of an increase of the char strength in the presence of the nanoclay, which was also supported by the morphological and TGA results.

This results shown in Figure 8.14 are also similar to the ones reported by Haurie *et al.* [21], who studied the effect of synthetic hydromagnesite (Hy) combined with ATH and OMMT on the fire retardancy of an EVA/LDPE blend. Figure 8.15 shows a comparison of the HRR between Hy60 (30% LPDE + 10% EVA + 30% ATH + 30% Hy), Hy55 (33.75% LPDE + 11.25% EVA + 27.5% ATH + 27.5% Hy), and Hy/MMT50 (37.5% LPDE + 12.5% EVA + 30% ATH + 15% Hy + 5% OMMT). Comparing Hy/MMT50 with Hy55, where the filler content was reduced by 5 wt%, reductions in the PHHR and higher TTI

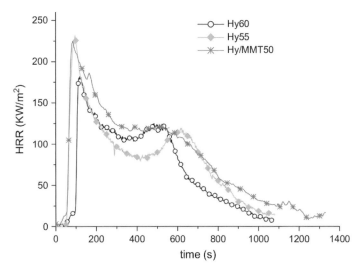

Figure 8.15 Heat release rate curves of hydromagnesite-containing composites: Hy60 (30% LPDE + 10% EVA + 30% ATH + 30% Hy), Hy55 (33.75% LPDE + 11.25% EVA + 27.5% ATH + 27.5% Hy) and Hy/MMT50 (37.5% LPDE + 12.5% EVA + 30% ATH + 15% Hy + 5% OMMT). From Haurie *et al*. [21, p. 1086]. With permission.

were found for the formulation containing OMMT. The improvement in thermal stability in the presence of OMMT in the flame retarding systems was attributed in [21] to the barrier effect exerted by the OMMT when dispersed in the polymer matrix. The increased stability provided by OMMT was also evidenced when the crusts of samples were examined after the test: the crusts for materials containing OMMT stayed intact, whereas those for materials without OMMT appeared broken [21].

In contrast to Figures 8.14 and 8.15, where the addition of nanoclay seems to have a limited effect on the reduction of the HRR compared with that of the polymer blends with FRs alone, a much more substantial reduction by nanoclay was reported in [15] for the EVA/LLDPE blend with ATH. Although no effect of nanoclay on the TTI was observed, the peak HRR was reduced from 320 to 190 kW/m^2 when nanoclay was added to EVA/LLDPE/ATH, with an even more significant decrease in the steady HRR (from around 250 to 60 kW/m^2). Considering that only 2.5 wt% nanoclay was used, the results in [17] indicate that the organoclay in combination with ATH seems to exert a synergistic effect on the fire retardancy of the polymer blend.

Chang *et al*. [20] used a nanosized hydroxyl aluminum oxalate (nano-HAO) with a dioctahedral 1:1 layered clay mineral (nanokaolin) to investigate their fire retardancy effects on a LDPE/ EPDM blend. In Table 8.5, the TTIs and HRRs of the composites containing 48 wt% nano-HAO and 12 wt% nanokaolin are compared with that of the composites containing 60 wt% nano-HAO alone. Whereas nano-HAO or nanokaolin has essentially no effect on the TTI, the peak and average HRRs are dramatically reduced with the addition

Table 8.5 *Effect of nanoclay (nanokaolin) and nanosized hydroxyl aluminum oxalate (nano-HAO) on the ignition and burning behavior of a LDPE/EPDM blend*

Materials	TTI (s)	PHRR (kW/m^2)	Ave. HRR (kW/m^2)
LDPE/EPDM	46	1487.6	818.6
LDPE/EPDM/Nano-HAO	43	307.7	124.7
LDPE/EPDM/Nano-HAO/Nanokaolin	44	270.4	106.0

Note: Modified from Chang *et al.* [20, p. 1207]. With permission.

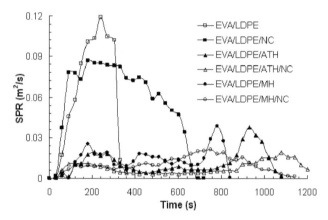

Figure 8.16 Smoke production rate curves for EVA/LDPE-based materials. Modified from Zhang *et al.* [27, p. 510]. With permission.

of FRs in comparison to those for the neat blend. A further reduction in the HRR was noted when nanokaolin was added, indicating an enhancement of the barrier property of the char layer in the presence of the nanoclay.

8.5.5 Toxic gases

Figure 8.16 shows a comparison of the transient SPR of all the EVA/LDPE-based materials. Significant reduction in SPR is achieved by FRs (ATH or MH) in comparison to that for the neat blend or that with nanoclay. Similar findings are reported in [40]. A comparison in Table 8.6 of the smoke yield [calculated as ratio of the total smoke production (in g) to the total mass loss (in g) where the water release due to degradation of the FRs is subtracted], however, indicates that FRCMs produce only marginally less smoke than the neat blend. This finding suggests the reduced smoke production rate by FRs (ATH or MH)

Table 8.6 *A summary of the cone calorimeter results for EVA/LDPE polymer blend with nanoclay (OMMT) and/or FRs (ATH and MH)*

Material	Total heat released (MJ)	Heat of combustion (kJ/g)	Smoke yield (g/g)	CO yield (g/g)	CO_2 yield (g/g)
LDPE/EVA	1.62	32.2	0.0456	0.0181	2.44
LDPE/EVA/nanoclay	1.63	31.1	0.0771	0.0239	2.23
LDPE/EVA/ATH	1.12	37.7	0.0376	0.0186	2.84
LDPE/EVA/ATH/nanoclay	1.17	37.5	0.0260	0.0186	2.86
LDPE/EVA/MH	1.13	39.5	0.0426	0.0169	2.93
LDPE/EVA/MH/nanoclay	1.27	38.2	0.0235	0.0179	2.56

is due primarily to the reduced polymer content and longer burning times. The CO and CO_2 yields by FRCMs are also similar to those for the neat blend, as shown in Table 8.6.

8.6 Conclusion

Polymer blending is useful in obtaining new materials with improved properties relative to their components, and polymer blends are widely used now in a large number of commercial applications. Polymer blends generally have poor fire properties, owing to the nature of polymers. To provide the required fire properties, traditional fire retardants, together with nanoclays, are used. In this chapter, we have summarized the main findings of previous studies on fire retardant polymer nanocomposites with polymer blends. The main conclusions are as follows:

(1) Studies using rheological measurements clearly show that nanoclay can increase the viscosity of polymer blends significantly. This is particularly relevant when the final products are used in a vertical orientation, as higher viscosity implies less tendency to drip upon heating. The dripping of melts has the potential to form a pool fire, which can increase fire spread.

(2) TGA studies show that the effect of nanoclays on the thermal stability of polymer blends depends on the environment. In pyrolytic environments, nanoclays reduce the stability of polymer blend systems, probably owing to catalytic cracking of polymer induced by nanoclay, which has also been observed for neat polymer nanocomposite systems. In comparison, in thermo-oxidative environments, an increase of the thermal stability caused by nanoclay has generally been observed. This has been attributed to the formation of a surface layer acting as a mass and thermal barrier to unpyrolyzed material. Studies investigating the effects of nanoclay loading on the degradation of polymer blends, however, indicate that there is an optimized loading level that can achieve the highest stability, and further increase in the nanoclay loading results in no

further improvement and, in some cases, a decrease in thermal stability owing to the geometrical constraints that restrict exfoliation of such high–aspect ratio silicate layers.

(3) Cone calorimeter studies indicate that nanoclay does not affect TTI, suggesting that there is no change in the properties of the polymer blend in the presence of a small amount (typically less than 5 wt%) of nanoclays. However, there is a substantial reduction of the HRR, especially the peak HRR. The mechanism of the nanoclay action in polymer blends appears to be the same as that in neat polymers, namely, a surface layer formed as a result of accumulation of nanoparticles after the polymer degrades prevents both heat transfer to the unpyrolyzed material and pyrolyzed gas escaping to the gas phase. The fact that the HRR in the initial stage is similar for materials with and without nanoclay suggests that the thickness of the surface layer is small (if any) just after ignition. The thickness of this layer (and its fire retardancy effect) increases with time as more nanoparticles are accumulated. This observation is further supported by applying a methodology developed for neat polymer nanocomposites [11] to an EVA/LDPE nanocomposite, where it is shown that the reduction of the heat flux at the interface of this surface layer and the unpyrolyzed material is inversely proportional to the thickness of the surface layer, in agreement with a previous study [11]. Studies also show that nanoclay has no effects on smoke suppression or reduction of CO production and, in some cases, actually promotes production of smoke and CO.

(4) The limited fire retardancy performance of nanoclay alone has led researchers to investigate the possible synergistic effects of nanoclay with other additives. The few studies available in the literature for multicomponent FR systems include combinations of nanoclay with (a) oxides (ATH or MH) [15, 21, 27], (b) synthetic hydromagnesite (Hy) [21], and (c) nanosized hydroxyl aluminum oxalate (nano-HAO) [20], where promising results are obtained, with noticeable improvement of the multicomponent systems in (a) viscosity, (b) thermal stability, and (c) fire performance (time to ignition, heat release rate, and production of toxic gases). These results highlight the necessity of further examining the synergistic effects between nanoclay and other FRs in the area of fire retardancy of polymer blend nanocomposites.

References

1. J. W. Gilman, T. Kashiwagi, A. B. Morgan, R. H. Harris, L. Brassell, M. VanLandingham, and C. L. Jackson, *Flammability of Polymer Clay Nanocomposites Consortium: Year One Annual Report*, Technical report NISTIR 6531. National Institute of Standards and Technology, 2000.
2. T. Kashiwagi, E. Grulke, J. Hilding, K. Groth, R. Harris, K. Butler, J. Shields, S. Kharchenko, and J. Douglas, Thermal and flammability properties of polypropylene/carbon nanotube nanocomposites. *Polymer*, 45 (2004), 4227–39.
3. T. Kashiwagi, R. H. Harris, Jr., X. Zhang, R. M. Briber, B. H. Cipriano, S. R. Raghavan, W. H. Awad, and J. R. Shields, Flame retardant mechanism of polyamide 6–clay nanocomposites. *Polymer*, 45 (2004), 881–91.

4. X. Liu and J. G. Quintiere, The thick and thin of burning nano-clay–nylon. In: *Proceedings of the 8th International Symposium on Fire Safety Science*, ed. D. T. Gottuck and B. Y. Latimer (Boston: Intl. Assoc. for Fire Safety Science, 2005), pp. 647–59.

5. C. A. Wilkie, Recent advances in fire retardancy of polymer–clay nanocomposites. In: *Proceedings of the 13th Annual BCC Conference on Flame Retardancy*, ed. M. Lewin (Norwalk: BCC Research, 2002).

6. N. J. Bok and A. W. Charles, The effect of clay on the thermal degradation of polyamide 6 in polyamide 6/clay nanocomposites. *Polymer*, 46 (2005), 3264–74.

7. H. Qin, Q. Su, S. Zhang, B. Zhao, and M. Yang, Thermal stability and flammability of polyamide 66/montmorillonite nanocomposites. *Polymer*, 44 (2003), 7533–8.

8. F. Dabrowski, S. Bourbigot, R. Delobel, and M. Le Bras, Kinetic modelling of the thermal degradation of polyamide-6 nanocomposite. *European Polymer Journal*, 36 (2000), 273–84.

9. M. Zanetti, G. Camino, D. Canavese, A. B. Morgan, F. J. Lamelas, and C. A. Wilkie, Fire retardant halogen–antimony–clay synergism in polypropylene layered silicate nanocomposites. *Chemistry of Materials*, 14 (2002), 189–93.

10. A. Fina, S. Bocchini, and G. Camino, Thermal behavior of nanocomposites and fire testing performance. In: *Fire and Polymers V, Materials and Concepts for Fire Retardancy*, ACS Symposium Series No. 10013, ed. C. A. Wilkie, A. B. Morgan, and G. L. Nelson (Washington, DC: Am. Chem. Soc., 2009), pp. 10–24.

11. J. Zhang, M. Delichatsios, and S. Bourbigot, Experimental and numerical study of the effects of nanoparticles on pyrolysis of a polyamide 6 (PA6) nanocomposite in the cone calorimeter. *Combustion and Flame*, 156 (2009), 2056–62.

12. L. M. Robeson, *Polymer Blends: A Comprehensive Review* (Germany: Hanser, 2007).

13. L. A. Utracki, ed., *Polymer Blends Handbook*, Vols. 1–2 (Dordrecht, the Netherlands: Kluwer Academic, 2003).

14. D. R. Paul and C. B. Bucknall, eds., *Polymer Blends Set: Formulation and Performance* (New York: Wiley, 2000).

15. T. H. Chuang, W. Guo, K. C. Cheng, S. W. Chen, H. T. Wang, and Y. Y. Yen, Thermal properties and flammability of ethylene–vinyl acetate copolymer/montmorillonite/polyethylene nanocomposites with flame retardants. *Journal of Materials Research*, 11 (2004), 169–74.

16. S. Sinha Ray and M. Bousmina, Compatibilization efficiency of organoclay in an immiscible polycarbonate/poly(methyl methacrylate) blend. *Macromolecular Rapid Communications*, 26 (2005), 450–55.

17. S. Sinha Ray and M. Bousmina, Effect of organic modification on the compatibilization efficiency of clay in an immiscible polymer blend. *Macromolecular Rapid Communications*, 26 (2005), 1639–46.

18. M. H. Lee, C. H. Dan, J. H. Kim, J. Cha, S. Kim, Y. Hwang, and C. H. Lee, Effect of clay on the morphology and properties of PMMA/poly(styrene-co-acrylonitrile)/clay nanocomposites prepared by melt mixing. *Polymer*, 47 (2006), 4359–69.

19. Y. Xu, W. J. Brittain, R. A. Vaia, and G. Price, Improving the physical properties of PEA/PMMA blends by the uniform dispersion of clay platelets. *Polymer*, 47 (2006), 4564–70.

20. Z. Chang, F. Guo, J. Chen, J. Yu, and G. Wang, Synergistic flame retardant effects of nanokaolin and nano-HAO on LDPE/EPDM composites. *Polymer Degradation and Stability*, 92 (2007), 1204–12.

21. L. Haurie, A. Fernández, J. Velasco, J. Chimenos, J. Lopez Cuesta, and F. Espiell, Thermal stability and flame retardancy of LDPE/EVA blends filled with synthetic hydromagnesite/aluminium hydroxide/montmorillonite and magnesium hydroxide/ aluminium hydroxide/montmorillonite mixtures. *Polymer Degradation and Stability*, 92 (2007), 1082–7.

22. S. M. Lai, H. C. Li, and Y. C. Liao, Properties and preparation of compatibilized nylon 6 nanocomposites/ABS blends. Part II – Physical and thermal properties. *European Polymer Journal*, 43 (2007), 1660–71.

23. Z. Yu, J. Yin, S. Yan, Y. Xie, J. Ma, and X. Chen, Biodegradable poly(L-lactide)/poly(ε-caprolactone)-modified montmorillonite nanocomposites: Preparation and characterization. *Polymer*, 48 (2007), 6439–47.

24. L. Elias, F. Fenouillot, J. C. Majesté, and P. Cassagnau, Morphology and rheology of immiscible polymer blends filled with silica nanoparticles. *Polymer*, 48 (2007), 6029–40.

25. R. Scaffaro, M. C. Mistretta, and F. P. La Mantia, Compatibilized polyamide 6/polyethylene blend–clay nanocomposites: Effect of the degradation and stabilization of the clay modifier. *Polymer Degradation and Stability*, 93 (2008), 1267–74.

26. H. Acharya, T. Kuila, S. K. Srivastava, and A. K. Bhowmick, Effect of layered silicate on EPDM/EVA blend nanocomposite: Dynamic mechanical, thermal, and swelling properties. *Polymer Composites*, 29 (2008), 443–50.

27. J. Zhang, J. Hereid, M. Hagen, D. Bakirtzis, and M. A. Delichatsios, Effects of nanoclay and fire retardants on fire retardancy of a polymer blend of EVA and LDPE. *Fire Safety Journal*, 44 (2009), 504–13.

28. T. Gcwabaza, S. Sinha Ray, W. W. Focke, and A. Maity, Morphology and properties of nanostructured materials based on polypropylene/poly(butylene succinate) blend and organoclay. *European Polymer Journal*, 45 (2009), 353–67.

29. S. Park, T. Kashiwagi, D. Stemp, J. Koo, M. Si, J. C. Sokolov, and M. H. Rafailovich, Segregation of carbon nanotubes/organoclays rendering polymer blends self-extinguishing. *Macromolecules*, 42 (2009), 6698–709.

30. S. Park, M. Si, J. Koo, J. C. Sokolov, T. Koga, T. Kashiwagi, and M. H. Rafailovich, Mode-of-action of self-extinguishing polymer blends containing organoclays. *Polymer Degradation and Stability*, 94 (2009), 306–26.

31. S. K. Nayak and S. Mohanty, Poly (trimethylene) terephthalate/m-LLDPE blend nanocomposites: Evaluation of mechanical, thermal and morphological behaviour. *Materials Science and Engineering A*, 527 (2010), 574–83.

32. F. Fenouillot, P. Cassagnau, and J. C. Majesté, Uneven distribution of nanoparticles in immiscible fluids: Morphology development in polymer blends. *Polymer*, 50 (2009), 1333–50.

33. M. A. Osman, M. Ploetze, and U. W. Suter, Surface treatment of clay minerals – Thermal stability, basal-plane spacing and surface coverage. *Journal of Materials Chemistry*, 13 (2003), 2359–66.

34. G. Camino, R. Sgobbi, S. Zaopo, S. Colombier, and C. Scelza, Investigation of flame retardancy in EVA. *Fire and Materials*, 24 (2000), 85–90.

35. N. S. Allen, M. Edge, M. Rodriguez, C. M. Liauw, and E. Fontan, Aspects of the thermal oxidation of ethylene vinyl acetate copolymer. *Polymer Degradation and Stability*, 6 (2000), 363–71.

36. G. Beyer, Flame retardant properties of EVA-nanocomposites and improvements by combination of nanofillers with aluminium trihydrate. *Fire and Materials*, 25 (2001), 193–7.

37. T. Manos, I. Y. Yusof, N. Papayannakos, and N. H. Gangas, Catalytic cracking of polyethylene over clay catalysts: Comparison with an ultrastable Y zeolite. *Industrial and Engineering Chemistry Research*, 40 (2001), 2220–25.

38. G. Tartaglione, D. Tabuani, G. Camino, and M. Moisio, PP and PBT composites filled with sepiolite: Morphology and thermal behaviour. *Composites Science and Technology*, 68 (2008), 451–60.

39. S. Bourbigot, M. Le Bras, R. Leeuwendal, K. K. Shen, and D. Schubert, Recent advances of zinc borates in flame retardancy of EVA. *Polymer Degradation and Stability*, 64 (1999), 419–25.

40. P. R. Hornsby and R. N. Rothon, Fire retardant fillers for polymers. In: *Fire Retardancy of Polymers*, ed. M. Le Bras, C. Wilkie, and S. Bourbigot (Cambridge: Royal Society of Chemistry, 2005), pp. 19–41.

9

Flame retardancy of polyamide/clay nanocomposites

YUAN HU, LEI SONG, AND QILONG TAI

University of Science and Technology of China

9.1 Introduction

Polymer/clay nanocomposites have received considerable attention during the past decade, both in industry and in academia, because of their attractive improvement of material properties relative to pure polymers and conventional polymer composites. The improvements include mechanical, thermal, flame retardant, and gas barrier performance [1–9]. It is believed that the improvements are mainly attributable to the nanometric size dispersion of the clay and the specific interfacial interaction between the polymer matrix and clay layers.

9.1.1 The structure and properties of clays

The clays commonly used in polymer nanocomposites belong to the family of 2:1 layered silicates or phyllosilicates. The crystal structure of the clay layers is made up of two tetrahedrally coordinated silicon atoms, which are fused to an edge-shared octahedral sheet of either aluminum or magnesium hydroxide. The layer thickness is about 1 nm and the lateral dimension of the layers may vary from 30 nm to several micrometers or even larger, depending on the particular silicate. There is a van der Waals gap between the layers, usually called a gallery or interlayer. Isomorphic substitution within the crystal structure of the layer (for example, Al^{3+} replaced by Mg^{2+} or by Fe^{2+}, or Mg^{2+} replaced by Li^+) generates negative charges that are counterbalanced by alkali and alkaline earth cations situated inside the interlayer. This type of clay is characterized by a moderate surface charge known as cation exchange capacity (CEC), and usually expressed as meq/100 g. Because the charge for each layer varies, the CEC value must be considered as an average.

Montmorillonite, hectorite, and saponite are the most widely investigated clays. Their structure and formula are shown in Figure 9.1 and Table 9.1, respectively. Montmorillonite is a very popular choice for nanocomposites because of its small particle size (<2 μm) and hence the easy diffusion of polymer into the clay particles. They also have high aspect ratios (10–2000) and high swelling capacity, which are essential for efficient intercalation of the polymer. The physical mixture of a polymer and a clay may not form a nanocomposite. This situation is analogous to that in a polymer blend, and separation into discrete phases takes place in most cases. Pristine clay usually contains hydrated Na^+ or K^+ ions. Obviously,

Table 9.1 *Chemical formula and characteristic parameters of commonly used 2:1 phyllosilicates [13]*

2:1 phyllosilicates	Chemical formula[a]	CEC (meq/100 g)	Particle length (nm)
Montmorillonite	$M_x(Al_{4-x}Mg_x)Si_8O_{20}(OH)_4$	110	100–150
Hectorite	$M_x(Mg_{6-x}Li_x)Si_8O_{20}(OH)_4$	120	200–300
Saponite	$M_x(Si_{8-x}Al_x)Si_8O_{20}(OH)_4$	86.6	50–60

[a] M = monovalent cation; x = degree of isomorphous substitution (between 0.3 and 1.3).

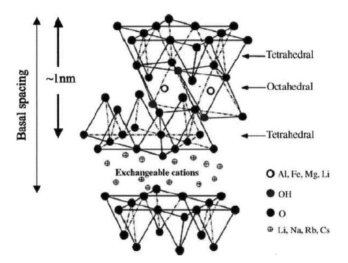

Figure 9.1 Structure of 2:1 phyllosilicates [13]. Reproduced from [13] by permission of Elsevier Science Ltd., UK.

the hydrophilic clay is immiscible with the hydrophobic polymer matrix. To make the clay miscible with the hydrophobic polymer matrix, the hydrophilic surface of the clay must be converted to a hydrophobic one, allowing the intercalation of many engineering polymers. Generally, this can be done by ion exchange reactions with primary, secondary, tertiary, and quaternary alkylammonium [10, 11] or alkylphosphonium [12] cations in the interlayer spacing, which can decrease the surface energy of the inorganic host and improve the wetting characteristics of the polymer matrix, resulting in a larger interlayer spacing. Thus, the organically modified clay (or organoclay) is more compatible with the organic polymers.

9.1.2 Polyamides

The polyamides are a class of polymers characterized by a carbon chain with amide (–CO–NH–) groups interspersed at regular intervals. Commercial polyamides were developed at Dupont with the aim of obtaining synthetic fibers. Poly(hexamethyleneadipamide), based

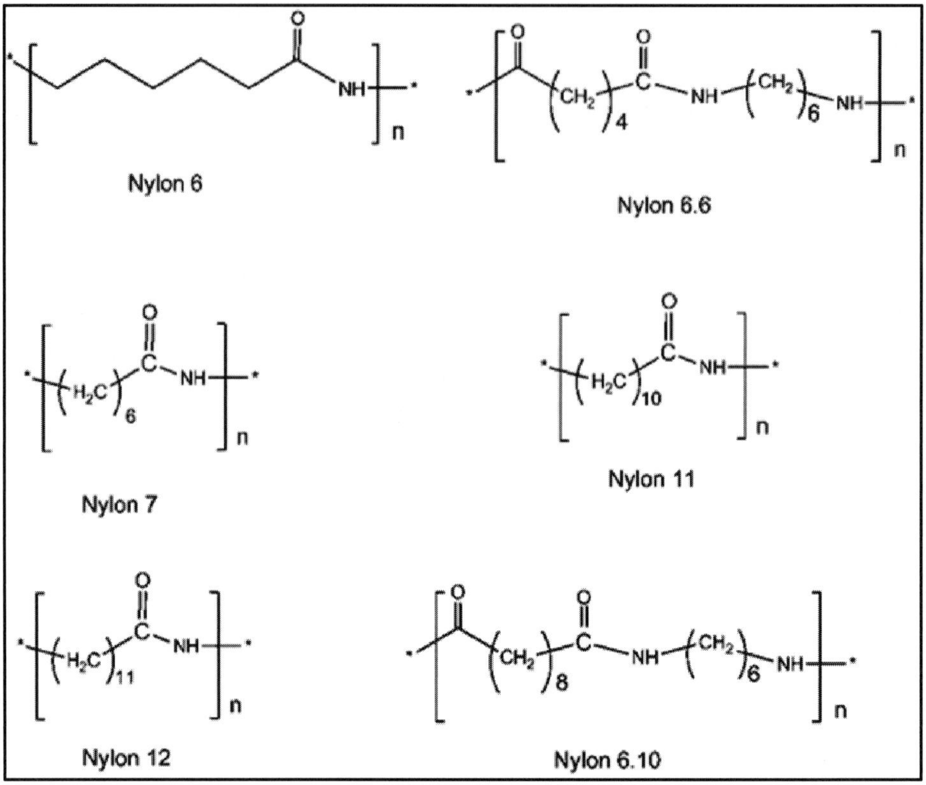

Figure 9.2 The structures of different nylons (or polyamides).

on hexamethylenediamine and adipic acid, was first scaled up in the United States in 1937–1938. The name "nylon" was first used by Dupont as the trade name for this product, but later was used as a common name for all the thermoplastic polyamides. They may be produced by the direct polymerization of amino acids or by reactions between diamine and dibasic acids. Generally, different nylons are identified by a numbering system that refers to the number of carbon atoms between successive nitrogen atoms in the main chain. Polyamides derived from amino acids are given by a single number. Polyamides derived from diamide and dibasic acids are referred to by two numbers, in which the first refers to the number of carbon atoms supplied by the diamide and the second to the number of carbon atoms contributed by the dibasic acid.

Nylon 6 and nylon 6.6 are the most commercially significant aliphatic nylons. About 80% of their usage is in synthetic fibers and the rest in engineering resins. In recent years, there has been fast growth of the production of engineering resins because of their excellent mechanical properties. Other commercial aliphatic nylons that have lower volumes than nylon 6 and nylon 6.6 are nylon 7, nylon 11, nylon 12, and nylon 6.10. The structures of these nylons (or polyamides) are presented in Figure 9.2. There are also various aromatic or aromatic–aliphatic nylons. As engineering resins, aliphatic nylons show high toughness over a range of temperature. Aliphatic nylons also show high fatigue resistance, abrasion resistance, lubricity, and good resistance to organic solvents. The structures of the different nylons are presented in Figure 9.2.

Nylons are often used in electrical products, such as electrical connectors, terminal blocks, small electrical housings, switch components, wire ties, and other industrial parts, where thermal stability and fire resistance are priorities. Because of this practical interest, the thermal stability and flame retardancy of nylons have been studied extensively. Halogen compounds are effective flame retardants for polyamides [14–17], but they have limited utility because their combustion products are toxic. Alternative flame retardants are used with polyamides, such as phosphorus- and nitrogen-containing substances. The most important organic nitrogen-containing flame retardants are melamine and its derivates [14, 18–20]. However, these traditional flame retardants are outside the scope of this chapter.

This chapter will highlight the major development of flame retardancy of polyamide/clay nanocomposites. Because nylon 6 and nylon 6.6 are the two most widely used polyamides, the content of this chapter will be based mainly on these two polyamides. Other less used polyamides such as nylon 12 and nylon 6.10 will be briefly discussed where appropriate. The different techniques used to prepare polyamids/clay nanocomposites, their thermal properties, and a possible flame retardant mechanism are also discussed.

9.2 Preparation approaches

With all of the methods discussed in this section, the primary goal is to break up the clay particle agglomerates and obtain nanocomposites with good dispersion. Because of the hydrophilic nature of the clay, it is hard to disperse the clay completely through the polyamide matrix. Thus, the clay should first be modified with an appropriate surfactant (for example, alkylammonium salt) via an ion exchange reaction to obtain organically modified clay, or *organoclay*. The role of the surfactant in the organoclay is to lower the surface energy of the clay layers and improve the compatibility between the clay and the polymer matrix. Generally, there are three main methods for polyamide/clay nanocomposite preparation: in situ polymerization, solution-blending, and melt-compounding.

9.2.1 In situ polymerization

In this method, the clay is swollen and intercalated within the liquid monomer or a monomer solution and then the polymer forms between the adjacent clay sheets. The in situ polymerization process usually yields polymer nanocomposites in which the clay is well dispersed. This can be attributed to the clay being exfoliated in the monomer (and/or solvent) prior to polymerization, or to the polymerization process causing expansion of the clay galleries.

One of the cornerstone studies, and probably the most important study in pioneering the polymer/clay field, was the work done by the Toyota group in which they polymerized nylon 6 in the presence of montmorillonite (MMT) clay [21]. Kojima *et al.* [22] prepared

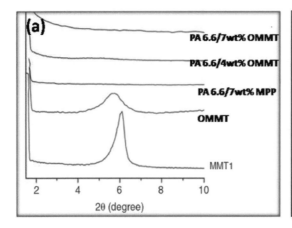 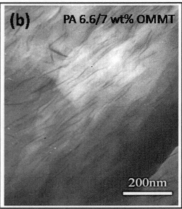

Figure 9.3 (a) The XRD patterns of MMT, OMMT, PA 6.6/OMMT nanocomposites; (b) TEM image of PA 6.6/7 wt% OMMT nanocomposites [27].

polyamide-6/clay composites by intercalating MMT with ε-caprolactam. The composites were obtained via the polymerization of ε-caprolactam, 6-aminocaproic acid, and the MMT at 260 °C for 6 h. Polyamide 6 (PA 6)/kaolin nanocomposites were also synthesized by first suspending kaolin filler in aqueous caprolactam, and then polymerizing caprolactam in situ at high temperature and pressure [23]. The kaolin mentioned is also a kind of clay, with chemical composition $Al_2Si_2O_5(OH)_4$ and layered structure. But it is different from montmorillonite, as kaolin is made up of one tetrahedral sheet linked through oxygen atoms to one octahedral sheet of alumina octahedra [24]. Since then, the in situ polymerization strategy has been used with a variety of polymers. Recently, a novel method using water as a predispersant of pristine sodium montmorillonite (Na-MMT) and ε-caprolactam monomers was developed by Yang and co-workers [25] for the synthesis of the exfoliated monomer casting polyamide 6 (MCPA6)/MMT nanocomposites via in situ anionic ring-opening polymerization. The results showed that the MMT sheets were completely exfoliated at low MMT loading (below 1.5 wt%). Furthermore, the exfoliated monomer casting PA 6/MMT and the intercalated PA 6/OMMT nanocomposites using Na-MMT with water dispersion and OMMT with acetone dispersion were also synthesized via in situ anionic ring-opening polymerization [26]. The effects of Na-MMT and OMMT on PA 6 were compared in detail, and the results indicated that incorporation of the exfoliated Na-MMT improved tensile modulus, strength, and thermal stability in comparison with those of intercalated OMMT.

Hu's group recently reported the preparation of PA 6.6/OMMT (organic montmorillonite) nanocomposites based on adipic acid, hexamethylendiamine, and OMT modified by protonated aminocaproic acid via in situ polymerization [27]. The XRD patterns are shown in Figure 9.3 (a). According to Figure 9.3 (a), the average basal spacing of MMT and OMMT was 1.45 and 1.54 nm, respectively. The PA 6.6/OMMT nanocomposites containing 4 and 7 wt% OMMT showed no diffraction peak, indicating that exfoliated structures were formed. The TEM image [Figure 9.3 (b)] for PA 6.6/7 wt% OMMT showed that an exfoliated morphology had been formed in the nylon 6.6 matrix and the clay layers were well dispersed in the polymer matrix.

Tarameshlou *et al.* [28] reported a novel method for preparation of PA 6.6/OMMT nanocomposites, using in situ interfacial polycondensation of an aqueous hexamethylene-diamine and a nonaqueous adipoyl chloride in dichloromethane solutions containing varied amounts of OMMT. Exfoliated or highly intercalated PA 6.6/OMMT nanocomposites were obtained in their studies. Similar interfacial polycondensation was employed to prepare PA 6.6/Na-MMT nanocomposites [29].

For other polyamide/clay nanocomposites, PA 11/OMMT nanocomposites were prepared by Zhang, Yu, and Fu [30] via in situ intercalative polymerization. The polymerization involved two steps: first the cation exchange of MMT with 11-aminolauric acid, and then the polymerization of 11-aminolauric acid in the presence of OMMT. The crystallization rate of the PA 11 matrix was obviously improved by the nanoscale-dispersed MMT clay. A similar in situ polymerization of exfoliated PA11/MMT nanocomposites containing different amounts of MMT was performed by Yang and co-workers [31]. First, 11-aminoundecanoic acid was mixed with organoclay in a vessel, and then the mixtures were heated to 100–150 °C to drive off the water and heated again to 230–260 °C for 4 h to achieve polymerization.

In situ polymerization usually yields nanocomposites that are well dispersed at both micro- and nanoscale. But for industrial applications, the commercial PA 6 nanocomposites from Unitika and Toyota seem to be the only example of an in situ process that has been scaled up. If one were going to take advantage of existing capital equipment, the expenses of developing a new process and even engineering changes for handling the nanocomposite polymerization might present a large commercial hurdle to commercial application.

9.2.2 Solution-blending

Solution-blending is based on a solvent system in which the polymer is soluble and the clay is swellable, such as water *or N,N*-dimethylformamide, acetone, benzene, or toluene for pristine clay or organoclay. The clay is first swollen in a solvent, and then the polymer is added to the clay solution, and the polymer chains diffuse, intercalate, and displace the solvent molecules previously accommodated within the galleries. Upon solvent removal, the nanocomposites are obtained. The structure of the nanocomposites obtained depends on molecular factors such as surface energy and interactions between the intermolecular and kinetic factors, including the solvent evaporation rate and shear mixing. This method has been used widely with water-soluble polymers to produce intercalated nanocomposites, such as poly(ethylene oxide) (PEO) [32], poly(vinyl alcohol) (PVA) [33], and poly(2-vinyl pyridine) (PVP) [34] intercalated into the clay galleries. Examples with nonaqueous solvents are nanocomposites of poly(1-caprolactone) (PCL)/clay [35] and poly(lactide) (PLA)/clay [36] with chloroform as a cosolvent and high-density polyethylene with xylene and benzonitrile [37]. However, there are a few reports of polyamide/clay nanocomposites prepared using a solution intercalation method.

Ma and co-workers [38] have investigated this method in attempts to prepare nanocomposites with PA 6–based polymer. These nanocomposites were filled with OMMT modified by three different swelling agents, namely, *n*-dodecylamine, 12-aminolauric acid, and 1,12-diaminodecane. Upon treatment, the interlayer spacing of the OMMT was increased. In order to obtain the PA 6–based nanocomposites, the PA 6 was dissolved in formic acid in the presence of different amounts of OMMT. After deposition in deionized water, the nanocomposites were recovered in a vacuum at 40 °C for 24 h.

PA 6/Fe-OMT nanocomposite fibers have been produced using a similar method. The composites were prepared by dissolving Fe-OMT in *N,N*-dimethylformamide (DMF), mixing it with a formic acid solution of PA 6, and then electrospinning the solutions at a positive voltage of 14 kV with a working distance of 10 cm [39]. A variety of aromatic polyamides and clay hybrids were prepared by Zulfiqar *et al.* [40–44] using solution-blending. In their studies, different concentrations of MMT modified with dodecylamine were blended with a polyamide solution of *N,N'*-dimethylacetamide (DMAc), resulting in complete dispersion of clay throughout the matrix.

For the solution-blending method, the driving force for the polymer chains' intercalation into the clay from solution is the entropy gained by desorption of solvent molecules, which compensates for the decreased entropy of the confined, intercalated chains [45]. Thus, intercalation using this method may only occur for certain polymer/solvent pairs. For example, polystyrene can be used with various solvents, so it is easy to choose a solvent compatible simultaneously with the clay and the polymer [46, 47], whereas the polyamides may require acidic solvents, which could react with some organic treatments.

The commercial feasibility of the method must take into account that large amounts of solvents may be required, though they can usually be recovered during devolatilization. However, the solvent requirements may limit the production rates, produce some contamination, and probably result in environmental pollution. Clearly, the solvent method for preparing polyamide/clay nanocomposites is not a very good option on the industrial scale.

9.2.3 Melt-compounding

Melt-compounding or melt-blending involves mixing clay by annealing it, statically or under shear, with polymer above the softening point of the polymer. This has great advantages in comparison to other approaches to the preparation of polyamide/clay nanocomposites, and is the most widely used and published method. On one hand, melt-compounding is environmentally friendly because of the absence of organic solvents,in contrast to the in situ polymerization solution-blending. On the other hand, it is compatible with current industrial processes, such as extrusion, roll mixing, batch mixing, and injection-molding. Therefore, this method is easily industrialized and has a lower cost of capital equipment.

Recently, many groups have reported different polyamide/clay nanocomposites prepared using melt-compounding, with or without flame retardant additives. For example,

Bourbigot, Devaux, and Flambard [48] reported a new route for preparing PA 6 textiles with permanent flame retardancy. First, an exfoliated PA 6/clay hybrid was prepared via melt-compounding, and then the textiles were obtained by melt-spinning. The resultant hybrid textiles showed a 40% reduction in peak heat release rate (PHRR) as compared with pure PA 6. In another report, Samyn *et al.* [49] prepared PA 6/OMMT clay nanocomposites exhibiting different morphologies by melt-compounding. The results showed that the PA 6 nanocomposites exhibited significant reduction in PHRR irrespective of the nanomorphology (exfoliation, intercalation, and presence of tactoids). PA 6 nanocomposites containing bentonite with different organic modifiers were also prepared by melt-blending, and TEM observation showed that the low-cost bentonite dispersed well in the polymer matrix [50]. Some researchers prepared PA 6 or PP/OMMT nanocomposites by melt-blending and injection-molding [51], respectively. The results suggested that the morphology of the nanocomposites depended on the nature of the polymer matrix but was not affected by the preparation method. Hu's group reported several kinds of flame retardant PA 6/OMMT nanocomposites with either halogenated antimony or halogen-free flame retardant additives prepared using melt-blending methods [6, 52].

Liu, Wu, and Berglund [53] prepared PA 6.6/clay nanocomposites via melt compounding using a new kind of organophilic clay, which was obtained through co-intercalation of epoxy resin and quaternary ammonium into Na-MMT. Because of the strong interaction between epoxy and the PA 6.6 matrix, the clay layers dispersed homogeneously and nearly exfoliated in the matrix. Yang, Guo, and Yu [54] prepared exfoliated PA 6.6/MMT nanocomposites with methyl methacrylate (MMA) as co-intercalation agent. A series of nanocomposites containing MMT and MMA were obtained via melt-compounding. The MMA could diffuse into OMMT galleries and attract PA 6.6 molecules through hydrogen bonding, and thus facilitate exfoliation of the clay layers.

McNally and co-workers [55] prepared PA 12/4 wt% fluoromica nanocomposites via melt-compounding. Two different fluoromicas were used in their experiments: virgin fluoromica with formula $Na[Mg_{2.5}Si_4O_{10}F_2]$, and organo-fluoromica modified with quaternary tallow ammonium chloride. The XRD and TEM results showed that both intercalated and exfoliated structures were obtained. The tensile modulus and strength of the PA 12/organofluoromica system were, respectively, 27% and 50% greater than those of neat PA 12. Hocine, Mederic, and Aubry [56] investigated the mechanical properties of PA 12/OMMT clay nanocomposites prepared by melt-compounding as a function of clay volume fraction. The results clearly showed that Young's modulus, the yield stress, the strain at break, and the stress at break all increased strongly with volume fraction up to a solid volume fraction threshold at ~1%, meaning that there was significant improvement in mechanical properties.

This solvent-free method is much preferred for industrialization of polyamide nanocomposites because of its high efficiency and low cost and the possibility of avoiding environmental hazards.

9.3 Flame retardancy of polyamide/clay nanocomposites

9.3.1 Thermal stability

Thermogravimetric analysis (TGA) is widely used to evaluate the thermal stability of the polymer/clay nanocomposites. The weight loss of a nanocomposite through the volatilization generated by thermal decomposition is monitored as a function of temperature. Thermal decomposition occurs when the materials are heated under an inert gas such as nitrogen or helium, whereas thermo-oxidative decomposition is always performed in air. Generally, the incorporation of clay into a polymer matrix was found to enhance thermal stability by providing a heat insulator and a mass transport barrier to the volatile products generated during decomposition.

The thermal decomposition of PA 6 is a complex process and can lead to many different products, according to Levchik, Weil, and Lewin [57]. Recent research found that the main volatile product of the thermal decomposition of PA 6 was caprolactam. Additional products were mainly oligomers containing nitrile and vinyl chain ends, formed as a result of depolymerization [58]. Pramoda *et al.* [59] have studied the thermal degradation of PA 6 and PA 6/clay nanocomposites using TG-FTIR. They found that the presence of clay did not appear to have any effect on the degradation, as seen from an analysis of gases evolved under nitrogen. The major evolved gases were cyclic monomers, hydrocarbons, CO_2, CO, NH_3, and H_2O for both PA 6 and the PA 6/clay nanocomposites.

The typical TGA curves of PA 6 and PA 6/OMMT clay nanocomposites both under nitrogen and in air are shown in Figure 9.4 [60] and Figure 9.5 [61]. Under pyrolytic conditions, PA 6 and PA 6/OMMT degraded in one step, corresponding to one large peak in the mass loss rate for each of the three samples. The thermal stability of the nanocomposites did not vary significantly from that of virgin PA 6, except in having a small earlier mass loss starting at about 350 °C. This small mass loss could be attributed to thermal degradation of the organic modified component of the clay. Under thermo-oxidative conditions (Figure 9.5), the PA 6 degraded in a two-step process, consisting primarily of the decomposition of the bulk polymer followed by the degradation of the char layer through chain scission volatilization. The PA 6/clay nanocomposites also degraded in two steps, and the temperature of initial degradation onset was relatively unaffected by clay addition. According to some other investigations [60, 62, 63], no significant changes were observed in the onset temperature of degradation. However, it is interesting that Pramoda *et al.* [59] observed that the degradation onset temperature was 12 °C higher for PA 6 with 2.5% clay loading than for virgin PA 6, and the onset temperature for the higher clay loading remained unchanged. It seems that the degradation onset temperature of the PA 6/clay nanocomposites cannot be obviously increased by the introduction of clay. However, the char yields at 500 and 800 °C both consistently increased with increasing clay content. From the inserted diagram in Figure 9.5, the char decomposed more slowly and at higher temperature with increasing clay concentration. The results showed that increasing clay concentration generally increased the thermal stability of the char. But the concentration of clay appeared to have little effect on the onset temperature and rate of degradation.

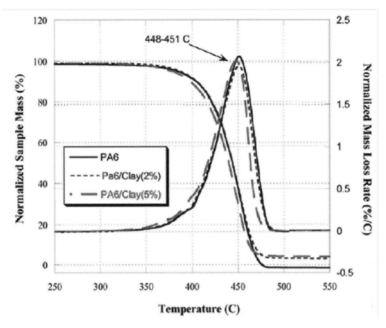

Figure 9.4 TG and DTG curves of PA 6 and PA 6/clay nanocomposites (2 wt%, 5 wt%) in N_2 at 10 °C/min [60]. Reproduced from [60] by permission of Elsevier Science Ltd., UK.

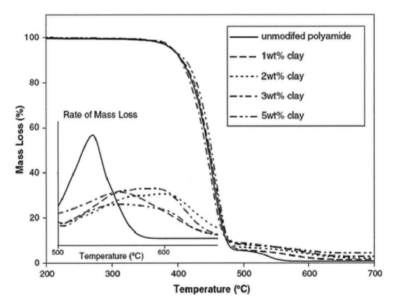

Figure 9.5 TGA curves of PA 6/clay nanocomposites with different clay loading in air at 10 °C/min. Insert shows the rate change of mass loss between 500 and 600 °C [61]. Reproduced from [61] by permission of Elsevier Science Ltd., UK.

Table 9.2 *The TGA data of PA 6.6 and PA 6.6/OMMT under nitrogen atmosphere [71]*

Sample	$T_{5\%}$ (°C)	Residue at 780 °C (wt%)
PA 6.6	350	4.18
PA 6.6/4 wt% OMMT	383	10.68
PA 6.6/7 wt% OMMT	411	15.29

The thermal decomposition of PA 6.6 is quite complex, and the products decomposed are affected by the temperature [57]. It was first reported that PA 6.6 eliminates cyclopentanone as the main decomposition product, but also some hydrocarbons, nitriles, and vinyl fragments [56]. Wiloth [64] later suggested that the main mechanisms of thermal decomposition of PA 6.6 were based on the tendency of the adipic acid fragment to undergo cyclization. With the development of modern test methods, the thermal decomposition of PA 6.6 has been extensively investigated by mass spectrometry, IR spectrometry, evolved-gas analysis, and so forth. PA 6.6 fragments, hexamethylenediamine, ammonium, CO_2, cyclopentanone, and many other products were detected [65–70].

For PA 6.6/clay nanocomposites, it seems that thermal stability depends on the method of preparation. Table 9.2, reported by Hu's group [71], showed the TGA data under nitrogen for PA 6.6 and PA 6.6/OMMT nanocomposites prepared by condensation polymerization, including the onset degradation temperature at 5 wt% loss of the sample ($T_{5\%}$) and the fraction of material that is nonvolatile at 780 °C (char residue). The thermal stability of PA 6.6 was obviously enhanced when the nanocomposites were formed. The $T_{5\%}$ of the nanocomposites increased from 350 to 411 °C, and the content of OMMT increased from 0 to 7 wt%. The weight percentage of char residue of samples at 780 °C also increased with increasing clay loading. In another report, the researchers prepared PA 6.6/OMMT nanocomposites and PA 6.6/MMT microcomposites via melt-blending, both 5 wt% clay loading [72]. It was found that the nanocomposite had a higher decomposition temperature (434.2 °C under nitrogen and 447.8 °C in air) than the microcomposite (425.1 °C under nitrogen and 442.3 °C in air). Moreover, the decomposition temperature for the nanocomposites was 10 °C lower than that for the pure PA 6.6 under nitrogen atmosphere but 7 °C higher in air. The results indicated that the addition of clay could accelerate the decomposition of the PA 6.6 matrix under nitrogen. This may be caused by the catalysis of water in MMT. In air, the degradation of samples was mainly oxygenolysis and the barrier effect of the silicate layers was dominant because of the formation of carbonaceous silicate char on the surfaces of the nanocomposites. Therefore, the nanocomposites had high thermal stability in air. To further understand the barrier effect, an isothermal oxidation experiment was performed in air at 360 °C (Figure 9.6). The volatilization rate of either the nanocomposite or the microcomposite was faster than that of pure PA 6.6 at the beginning; thereafter, the weight loss slowed down in both composites relative to the pure PA 6.6, but the nanocomposite had

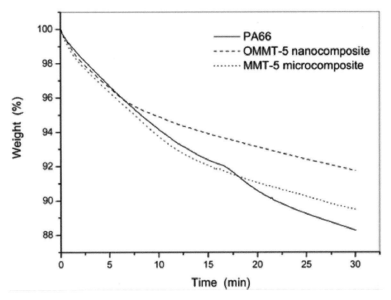

Figure 9.6 Isothermal TGA curves of pure PA 6.6, OMMT-5 nanocomposite, and MMT-5 micro-composite (air, 360°C) [72]. Reproduced from [72] by permission of Elsevier Science Ltd., UK.

higher thermal stability than the microcomposite. The results showed that the exfoliation of the silicate layers in a polymer matrix could enhance the barrier effect.

9.3.2 Cone calorimeter test

Without question, the cone calorimeter test is one of the most important and effective bench-scale methods for evaluating the flammability properties of polymeric materials. In this method, sample plates 100×100 mm in size are investigated under forced flaming conditions. Generally, several important parameters can be obtained from the cone test, including time to ignition (t_{ign}), heat release rate (HRR), peak heat release rate (PHRR), total heat release (THR), mass loss rate (MLR), and specific extinction area (SEA); PHRR is thought to be the most important parameter for evaluating the fire safety of natural or synthetic polymeric materials [73]. Cone calorimetry can provide other data such as smoke release and CO production for fire hazard evaluation. It has been reported that polymer/clay nanocomposites exhibit improved flame retardancy as compared with virgin polymer through flammability characterization by cone calorimetry. Sharp reductions in PHRR and decreases in MLR were observed in most cases [1, 74–76]. In contrast, microcomposites showed no change in PHRR [77].

Gilman evaluated the flammability of both nylon 6 and nylon 12-clay nanocomposites by cone calorimetry [78]. Detailed data are shown in Table 9.3 and the HRR plots for nylon 6 and exfoliated nylon 6 nanocomposites are presented in Figure 9.7 (a). A 63% reduction in the PHRR in comparison with pure nylon 6 could be clearly observed for the

Table 9.3 *Cone calorimeter data [78]*

Sample	PHRR (Δ%) (kW/m^2)	Mean HRR (Δ%)(kW/m^2)	Mean H_c(MJ/kg)	Mean SEA(m^2/kg)	Mean CO yield (kg/kg)
Nylon 6	1,010	603	27	197	0.01
Nylon 6/clay nanocomposites (2% exfoliated)	686 (32%)	390 (35%)	27	271	0.01
Nylon 6/clay nanocomposites (5% exfoliated)	378 (63%)	304 (50%)	27	296	0.02
Nylon 12	1,710	846	40	387	0.02
Nylon 12/clay nanocomposites (2% exfoliated)	1,060 (38%)	719 (15%)	40	435	0.02

Note: Heat flux: 35 kW/m^2.

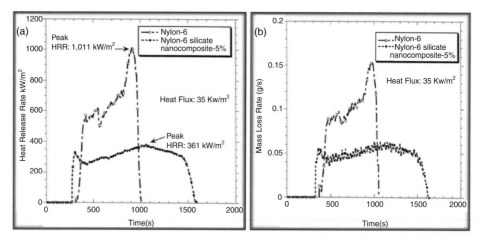

Figure 9.7 (a) Comparison of the HRR plot for nylon 6 and nylon-6 silicate nanocomposites (mass fraction 5%) at 35 KW/m^2 heat flux; (b) the mass loss rate data for nylon 6 and nylon-6 clay nanocomposites [78]. Reproduced from [78] by permission of Elsevier Science Ltd., UK.

nanocomposites containing 5% clay. The cone calorimetry data showed that both the PHRR and average HRR were reduced significantly for polyamide nanocomposites with a low clay mass fraction (2% to 5%). However, the heat of combustion (H_c), SEA, and carbon monoxide yields were unchanged. These data tend to demonstrate that the improvement in flame retardancy did not arise from the flame reaction in the gas phase but rather the thermal decomposition and combustion process in the condensed phase. It was believed

Table 9.4 *Cone calorimeter results [72] of PA 6.6/OMMT nanocomposites at a heat flux of 35kW/m²*

Sample	PA 6.6	OMMT-2	OMMT-5	OMMT-10
PHRR	802.4	496.1	335.5	209.4
Average EHC (MJ/kg)	0.18	0.18	0.19	0.19
THR (MJ/m²)	249.2	245.0	247.5	247.8
Mean SEA (m²/kg)	2.02	3.20	3.32	3.16
Mean CO yield (kg/kg)	0.0001	0.0002	0.0002	0.0002
Mean CO_2 yield (kg/kg)	0.01	0.01	0.01	0.01
Ignition time (s)	169	163	139	152

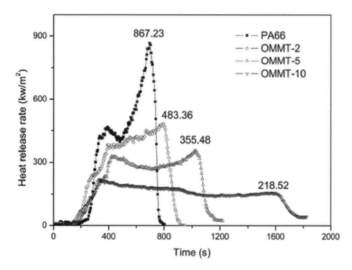

Figure 9.8 HRR plots for pure PA 6.6 and PA 6.6/OMMT nanocomposites at a heat flux of 35 KW/m² [72] (reproduced from [72] by permission of Elsevier Science Ltd., UK).

that the MLR was responsible for the lower HRR of the nanocomposites. Figure 9.7 (b) shows that the MLR for nylon-6 nanocomposite was significantly reduced compared with that for pure nylon-6. The two sets of data essentially mirror the HRR data.

In a recent study, Zhang and co-workers [72] reported the fire properties of PA 6.6/OMMT nanocomposites with different OMMT contents via melt-blending in a two-screw extruder. The HRR curves for pure PA 6.6 and the three nanocomposites with different clay content are shown graphically in Figure 9.8. The cone calorimetry data are presented in detail in Table 9.4. The PHRR of PA 6.6 was obviously reduced by the addition of clay and decreased with increasing content of clay.

Nevertheless, irrespective of the large reduction in PHRR or MLR of polyamide/clay nanocomposites, the THR was almost similar to that for the neat polyamides. For instance,

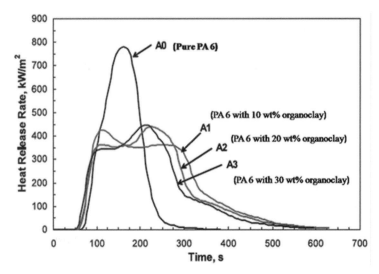

Figure 9.9 Effect of clay loading rate on HRR curves for PA6 nanocomposites at 50 KW/m^2 [79] (reproduced from [79] by permission of IOP Publishing Ltd., UK).

the mean heat of combustion is 27 MJ/kg for PA 6 and 27 MJ/kg for PA 6/clay nanocomposites. In PA 6.6, the total heat of combustion is 249.2 MJ/m^2 for pure PA 6.6, 245 MJ/m^2 for PA 6.6/clay nanocomposites with 2% clay, and 247.5 MJ/m^2 for PA 6.6/clay nanocomposites with 5% clay. This is because the nanocomposites do not self-extinguish until most of the fuel has been burnt; that is, the fuel burns slowly but completely. From Table 9.4, the values of EHC, THR, average CO yield, and average CO$_2$ yield for the nanocomposites were similar to those for pure PA 6.6. The specific extinction area of the nanocomposites had a 60% increase relative to that of pure PA 6.6. These results were similar with the PA 6 nanocomposites discussed previously.

Highly filled PA 6/clay nanocomposites were prepared using melt-blending by Dasari *et al.* [79] and the flammability properties were characterized. As shown in Figure 9.9, the PHRR was significantly reduced from that for pure PA6 when 10 wt% organoclay was added. But, with increasing clay loading, a slight reduction in PHRR was observed. The THR was somewhat higher than for the pure polymer because of the organic surfactant in the clay. According to the investigations, the THR for pure PA 6 is ∼82 MJ/m^2, whereas for A1-A3, tit is as high as 94, 100, or 105 MJ/m^2.

In conclusion, the addition of a small amount of clay into polyamides can reduce the PHRR significantly from that for pure polymer, irrespective of intercalated or exfoliated nanocomposites. However, the THR of the nanocomposites is similar to that for the pure polymer. In some samples, the THR of the nanocomposites is even higher than that of pure polymer. Furthermore, the ignition of the nanocomposites usually occurs earlier than for pure polyamide. This can be rationally interpreted as follows. On one hand, large amounts of organic surfactants are present in modified clay, which thermally decompose earlier than polyamides, resulting in shorter ignition time. On the other hand, the polyamide matrix of the nanocomposites and the organic surfactants in modified clay will be burnt out during combustion. Thus, the nanocomposites have a somewhat higher THR than pure polymer. A decrease in PHRR indicated a reduction of combustible volatile products generated by the degradation of the polyamide matrix. The increase in flame retardancy due to the presence of clays and their nanometer distribution throughout the polymer matrix was clearly shown. Moreover, the flame retardancy was improved by the fact that the heat release was spread over a much longer period of time.

All data in this chapter show that when small concentrations of clay are well dispersed in polyamides, the PHRR as measured in a cone calorimeter is strongly decreased in comparison to that for virgin polymer. This decrease is often explained by the formation of a protective barrier on the surface of combusting polymer matrix. Recently, it became evident that the barrier could have been formed by a migration of the clay to the surface of the polymer during combustion or pyrolysis. Moreover, the migration of clay was also observed at lower temperature; that is, it could occur far below the pyrolysis temperature [80, 81]. The migration in nanocomposites has been studied in a number of publications by Lewin, Tang, and co-workers [82–87]. They proposed many causes for the migration, including temperatures and viscosity gradients of the polymer matrix melt, propulsion of the clay to the surface by gas bubbles from the decomposed polymer or surfactants, and the difference in surface free energy between the polymer matrix and polymer/clay composites [88].

9.3.3 Limiting oxygen index and UL-94 tests

Limiting oxygen index (LOI) indicates relative flammability by measuring the minimum concentration of oxygen in a precisely controlled nitrogen-oxygen mixture that will just support the flaming combustion of a specimen. The LOI test evaluates the ease with which combustion is extinguished for downward-burning materials. Thus, this test describes the effectiveness of flame retardant additives at early stages of combustion.

The UL-94 test developed by Underwriters Laboratory is another widely used flammability testing methodology, which is a standard technique for selecting materials used in electronic equipment in many countries. The UL-94 vertical test gives insight into the ease with which a polymer can burn upward or self-extinguish. These two tests give very different information from cone calorimetry, but are all important to fully and properly evaluate the flame retardant behavior of a given polymeric material.

With few exceptions, nanocomposites do not self-extinguish until most of the fuel has been burnt. That is, they burn slowly but completely. Therefore, polymer/clay nanocomposites are unable to meet the fire safety standard of UL-94 and LOI tests when used alone. Improvements on LOI or UL-94 tests can be observed in some systems that combine clay and conventional flame retardants. In this section, the issue will be discussed briefly.

Researchers have explored many combinations of clay with various halogenated non-halogenated flame retardants (FRs). This idea is to combine the barrier effect of the clay with the chemical action of conventional flame retardants.

Table 9.5 *The LOI and UL-94 results of flame-retarded PA 6/organoclay nanocomposites [6]*

Sample	Composition	UL-94 test	LOI
PA 6	Pure nylon 6	Burning	21
PA 6–1	PA6 + OMMT 2 wt%	Burning	21.5
PA 6–2	PA6 + MH 8 wt% + RP 5 wt%	V-0	29
PA 6–3	PA6 + OMMT 2 wt% + MH 6 wt% + RP 5 wt%	V-0	31

Table 9.6 *UL-94 results of organoclay-filled PA 11 with intumescent FR [90]*

Sample	PA 11 (wt%)	FR^a (wt%)	OMMT (wt%)	UL-94	t_1/t_2
PA 11–1	100	0	0	Failed	21.22/ N/A
PA 11–2	70	30	0	V-0	0/1.75
PA 11–3	75	20	5	V-1	4.59/19.95
PA 11–4	72.5	20	7.5	V-0	1.53/5.47

^{a} Exolit OP 1230 (FR, Clariant), an intumescent FR additive based on metal phosphinate.

Hu's group prepared flame-retarded PA 6/OMMT nanocomposites using magnesium hydroxide (MH) and red phosphorus (RP) as flame retardants and OMMT as a synergist via melt-compounding [6]. As shown in Table 9.5, the LOI increased by 0.5 with the addition of 2 wt% OMMT clay to PA 6. However, the LOI increased by 2.0 with the combination of 2 wt% OMMT with MH and RP. This suggested a synergistic effect on the LOI values between the OMMT clay and MH and RP flame retardants.

Isitman, Gunduz, and Kaynak [89] investigated the effect of a nanoclay in a flame-retardant glass fiber–reinforced PA 6 system based on phosphorus compounds and zinc borate. They found that substitution of nanoclays for a certain fraction of the flame retardant significantly reduced the PHRR and delayed ignition in cone calorimetry. Furthermore, improvements of the LOI values and maintained UL-94 ratings of the nanocomposites were observed in comparisons with samples without nanoclay. For example, the sample containing 10 wt% FRs had a LOI value of 24.9, whereas the LOI value of the sample with a combination of 5 wt% FRs and 5 wt% nanoclay was as high as 29.1. The improved flame retardancy was attributed to the formation of a strong and consolidated barrier owing to the reinforcement of the char by nanoscale char layers.

Flame retardant PA 11 and PA 12 nanocomposites with low concentrations of OMMT, carbon nanofibers, and nanosilica were prepared by Lao *et al.* via a twin-screw extruder [90]. Table 9.6 shows the UL-94 results of PA 11/OMMT nanocomposites filled with commercial intumescent flame retardant. The sample consisting of 70% PA 11 and 30% FR composites obtained a V-0 rating in the UL-94 test. Although 10% of FR was replaced by 5% OMMT, a V-1 rating was obtained in the UL-94 test. When 7.5% OMMT was added into the system, the sample PA-4 passed the UL-94 V-0 test. Under this condition, the total flame retardation is 27.5%. The results indicated that the combination of OMMT with the metal phosphinate led to an improvement on the UL-94 test.

Improvements in LOI value and UL-94 ratings were also found in some other polymer/clay nanocomposites. For example, Ribeiro *et al.* [91] found a synergistic effect between MMT clays and intumescent FRs (ammonium polyphosphate and pentaerythritol) in an ethylene–butyl acrylate matrix. The value of LOI was found to increase from 21 for the intumescent system to 30 for the system containing 3 wt% natural clay. Wang *et al.* [92] prepared a series of novel epoxy/clay nanocomposites with different amounts of phosphorous-containing organoclay. LOI was 11 units higher than that of the neat epoxy, indicating that a significant enhancement of flame retardancy was obtained from the nanocomposites containing 5 wt% organoclay. Zhang *et al.* [93] incorporated Fe-montmorillonite (Fe-MMT) into ethylene vinyl acetate/magnesium hydroxide (EVA/MH) formulations. The authors claimed a synergistic effect on flame retardancy between the Fe-MMT and MH, as the incorporation can evidently enhance the LOI value and thermal stability of the EVA/MH composites. Meanwhile, many other researchers have also found synergistic effects of clay in combination with other conventional flame retardants, such as aluminum hydroxide, magnesium hydroxide/red phosphorus, aluminum phosphinate [94–96].

In conclusion, it is unfortunate that despite impressive flame retardancy results with clays, almost all the nanocomposites fail to pass the LOI or UL-94 test when the clays are used alone. When the clays are combined with conventional flame retardants, the results are usually more promising.

9.4 Flame retardant mechanism

It appears that the flame retardant mechanism for the polyamide/clay nanocomposites discussed in this chapter relates to the formation of a continuous protective carbonaceous char layer that acts as a heat shield. The mechanism is similar to that for other kinds of polymer/clay nanocomposites.

During combustion, the polyamide matrix is heated to thermal degradation temperature and volatile thermal degradation products are generated in it. Because the boiling points of most volatiles are much lower than the decomposition temperature of the polyamide matrix, the volatiles are superheated as they are generated. Thus bubbles nucleate and grow below the heated polymer surface and are released into the gas phase. These bubbles agitate the melt-polymer surface and can interfere with the formation of a carbonaceous char layer and heat transfer barrier [97]. The reduced flammability of polyamide/clay nanocomposites can best be explained by enhanced char formation in the condensed phase. Specifically, the presence of nanoscale clay layers in the polymer matrix retards the vigorous bubbling and facilitates the formation of a continuous protective char layer. The clay and carbonaceous

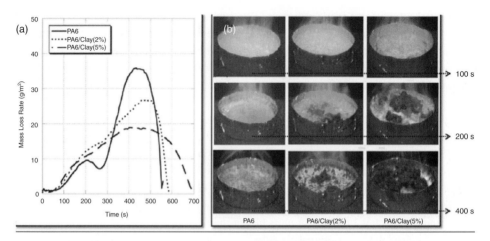

Figure 9.10 (a) Effects of clay content on mass loss rates of PA 6 and PA 6 clay nanocomposites in N₂ at 50 kW/m^2; (b) selected video images at 100, 200, and 400 s in nitrogen at 50 kW/m^2 [60]. Reproduced from [60] by permission of Elsevier Science Ltd., UK.

char layer on the burning surface can insulate the underlying polymeric substrate and thus slow the heat and mass transfer between the gaseous and condensed phases.

A detailed study of the char formation process in PA 6/clay nanocomposites was performed by Kashiwagi *et al.*, using a NIST radiative gasification apparatus [60].

Figure 9.10 (a) shows the mass loss rate data for PA 6 and PA 6 nanocomposites containing 2 and 5 wt% clay. Video images of the three samples taken at different times are shown in Figure 9.10 (b). The times at which images were taken corresponded to particular events in the pyrolysis of the PA 6 nanocomposites. The 100-s image corresponds to events shortly after ignition. The 200-s reflects the start of the plateau of steady burning behavior. The following images at 400 s relates to the peak MLR, which could be related to the peak heat release rate.

According to Kashiwagi *et al.* [60], pure PA 6 revealed small bubbles of degradation products followed by the appearance of many large bubbles during the nonflaming gasification tests. At the end of the test, a very thin black coating was left over the bottom of the container. The amount of char residue was less than 1% of the initial sample mass. The gasification behavior of the PA 6/clay (2%) sample was initially similar to that of the PA 6 sample, except that it appeared more viscous. Around 150 s, several small, dark floccules appeared on the surface and grew with time, as shown in Figure 9.10 (b). The mass of the residue was about 2% of the initial sample mass. The PA 6/clay (5%) sample appeared more viscous than the PA 6 sample during the gasification test, but it still formed numerous bubbles. Around 150 s, many carbonaceous floccules appeared near the perimeter of the sample, moved to the center, and formed larger, rough floccules. The mass of the residue at the end of the test was about 5% of the initial sample mass. At the end of the test, the PA 6/clay (5%) left more carbonaceous floccule residues than the PA 6/clay (2%) sample.

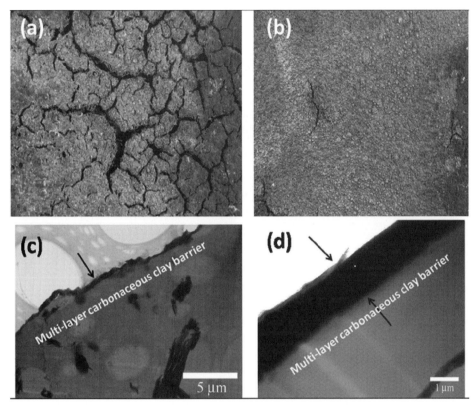

Figure 9.11 Digital photographs of the residues collected after a combustion test at a heat flux of 50 kW/m² for (a) PA 6 organoclay (90/10) and (b) PA 6 organoclay (70/30); and TEM micrographs of cross sections beneath the top surfaces for (c) PA 6 organoclay (90/10) and (d) PA 6 organoclay (70/30) [79] (reproduced from [79] by permission of IOP Publishing Ltd., UK).

In summary, significant reduction in the peak heat release rate for the PA 6/clay nanocomposites was achieved by the formation of protective floccules on the polymer surface, which shielded the PA 6 from external thermal radiation and feedback from the flame. That is, the carbonaceous floccules acted as thermal insulation.

As has been discussed, the reduction in the flammability of the nanocomposites was mainly attributed to the barrier effect of carbonaceous char during the combustion. To obtain in-depth knowledge of the protective barrier stability and uniformity under fire condition, Dasari *et al.* [79] prepared densely filled PA 6/organoclay nanocomposites and identified whether a critical composition was needed to form a stable char with no apertures or cracks. The HRR curves of the PA 6 nanocomposites with 10, 20, and 30 wt% organoclay were shown previously in Figure 9.9. The nanocomposites showed significant reduced HRR compared to pure PA 6; but among the nanocomposites, the difference was not significant once a critical composition was reached. Digital images of char residue collected after the cone calorimeter test are shown in Figure 9.11 (a, b). At a low loading of

10 wt% [Figure 9.11 (a)], discrete islandlike structures with many cracks were observed. In contrast, at a higher loading of 30 wt% [Figure 9.11 (b)], the residues appeared to be solid and continuous. TEM examinations [Figure 9.11(c, d)] at the cross section beneath the surface of the char were also performed to investigate the uniformity and thickness of the heat insulator. They revealed that at 10 wt% clay loading, the carbonaceous char is very thin (about 0.2–0.4 μm), nonuniform, and broken [Figure 9.11 (c)], whereas at higher loadings (30 wt%), the char is relatively heavily packed (about 1.8–2.0 μm) and continuous [Figure 9.11 (d)]. Therefore, it seems that the formation of a multilayered carbonaceous silicate barrier is important and necessary to lower the heat release rate and mass loss rate.

9.5 Summary and outlook

Polymer/clay nanocomposites may be considered environmentally friendly alternatives to some traditional flame retardant polymers. Apart from the thermal stability and/or flame retardant behavior, polymer/clay nanocomposites have many other excellent properties. Moreover, polymer/clay nanocomposites can be prepared and processed with normal techniques used for polymers, such as extrusion, injection-molding, and casting. As discussed in this chapter, most of the clays were treated with organic surfactants or modifiers to enhance the compatibility of clay layers with polyamides and achieve homogeneous dispersion. However, most of the surfactants are alkylammonium compounds and they are thermally unstable and decompose, usually, from ∼180 °C. But all the polyamides require much higher processing temperatures. The thermal decomposition of the organic-modified components is unavoidable and impairs the thermal stability of the nanocomposites. Generally, the presence of small molecular surfactants has a negative effect on flame retardant behavior. The decomposition of organic surfactants may cause quick release of fuel and affect the flammability properties by causing earlier ignition than the pure polyamides. So it is very important to develop new routes to producing polyamide/clay nanocomposites with more thermally stable clay modifiers.

Polyamide/clay nanocomposites exhibited much lower peak heat release rates (PHRRs) than pure polyamides. However, the total heat release (THR) of the nanocomposites was always unchanged and the nanocomposites did little to improve the limiting oxygen index (LOI) and UL-94 V test. Thus, we think it unlikely that clay will be used as a single flame retardant system for polyamides. However, clay is useful as a component of such systems. In fact, combinations of clay with various conventional flame retardants, such as inorganic, halogen, and phosphorus systems, have been investigated and reported in many publications. The combination is effective and inspiring, as it can not only decrease the HRR but also enhance the LOI or UL-94 rating of the polymers. More and more researchers have understood the advantages of using clay as a synergist with other flame retardants, and we believe that these nanocomposites will greatly advance the development of flame retardants.

References

1. J. G. Zhang and C. A. Wilkie, Polyethylene and polypropylene nanocomposites based on polymerically-modified clay containing alkylstyrene units. *Polymer*, 47 (2006), 5736–43.
2. J. W. Gilman, W. H. Awad, R. D. Davis, J. Shields, R. H. Harris, C. Davis, A. B. Morgan, T. E. Sutto, J. Callahan, P. C. Trulove, and H. C. DeLong, Polymer/layered silicate nanocomposites from thermally stable trialkylimidazolium-treated montmorillonite. *Chemistry of Materials*, 14 (2002), 3776–85.
3. G. Beyer, Flame retardancy of nanocomposites – from research to reality – Review. *Polymers and Polymer Composites*, 13 (2005), 529–37.
4. J. W. Gilman, R. H. Harris, J. R. Shields, T. Kashiwagi, and A. B. Morgan, A study of the flammability reduction mechanism of polystyrene-layered silicate nanocomposite: Layered silicate reinforced carbonaceous char. *Polymers for Advanced Technologies*, 17 (2006), 263–71.
5. B. N. Jang, M. Costache, and C. A. Wilkie, The relationship between thermal degradation behavior of polymer and the fire retardancy of polymer/clay nanocomposites. *Polymer*, 46 (2005), 10,678–87.
6. S. Lei, H. Yuan, Z. H. Lin, S. Y. Xuan, S. F. Wang, Z. Y. Chen, and W. C. Fan, Preparation and properties of halogen-free flame-retarded polyamide 6/organoclay nanocomposite. *Polymer Degradation and Stability*, 86 (2004), 535–40.
7. M. Fermeglia, M. Ferrone, and S. Pricl, Computer simulation of nylon-6/organoclay nanocomposites: Prediction of the binding energy. *Fluid Phase Equilibria*, 212 (2003), 315–29.
8. J. Brus, M. Urbanova, I. Kelnar, and J. Kotek, A solid-state NMR study of structure and segmental dynamics of semicrystalline elastomer-toughened nanocomposites. *Macromolecules*, 39 (2006), 5400–5409.
9. D. Sikdar, D. R. Katti, and K. S. Katti, A molecular model for epsilon-caprolactam-based intercalated polymer clay nanocomposite: Integrating modeling and experiments. *Langmuir*, 22 (2006), 7738–47.
10. D. Garcia-Lopez, I. Gobernado-Mitre, J. F. Fernandez, J. C. Merino, and J. M. Pastor, Influence of clay modification process in PA6-layered silicate nanocomposite properties. *Polymer*, 46 (2005), 2758–65.
11. E. Erdmann, M. L. Dias, V. Pita, F. Monasterio, D. Acosta, and H. A. Destefanis, *Effect of the Organoclay Preparation on the Extent of the Intercalation/Exfoliation and the Barrier Properties in Polyamide-6/Montmorillonite Nanocomposites*, ed. D. S. Dos Santos (Rio de Janeiro: Trans Tech Publications Ltd., 2006), pp. 78–84.
12. A. Akelah, A. Rehab, T. Agag, and M. Betiha, Polystyrene nanocomposite materials by in situ polymerization into montmorillonite-vinyl monomer interlayers. *Journal of Applied Polymer Science*, 103 (2007), 3739–50.
13. S. S. Ray and M. Okamoto, Polymer/layered silicate nanocomposites: A review from preparation to processing. *Progress in Polymer Science*, 28 (2003), 1539–1641.
14. S. V. Levchik and E. D. Weil, Combustion and fire retardancy of aliphatic nylons. *Polymer International*, 49 (2000), 1033–73.
15. BP Amoco Corporation, *Flame Retardant Anti-drip Polyamide Compositions*, United States Patent Application 20010053819 A1 (2001).
16. DSM IP ASSETS BV, *Flame Retardant Polyamide Composition*, European Patent application 1479728 A1 (2004).

17. DSM N.V. (Heerlen, NL), *Flame Retardant Polyamide Composition*, United States Patent Application 6037401 (2000).
18. U. Braun, B. Schartel, M. A. Fichera, and C. Jager, Flame retardancy mechanisms of aluminium phosphinate in combination with melamine polyphosphate and zinc borate in glass-fibre reinforced polyamide 6,6. *Polymer Degradation and Stability*, 92 (2007), 1528–45.
19. DSM NV, *Halogen-Free Flame-Retardant Thermoplastic Polyester or Polyamide Composition*, European Patent Application 0996678 A1 (2000).
20. J. H. Koo, S.-C. Lao, W. Yong, C. Wu, C. Tower, G. E. Wissler, L. A. Pilato, and Z. Luo, *Material Characterization of Intumescent Flame Retardant Polyamide 11 Nanocomposites*. American Institute of Aeronautics and Astronautics Inc., 2008.
21. A. Usuki, Y. Kojima, M. Kawasumi, A. Okada, Y. Fukushima, T. Kurauchi, and O. Kamigaito, Synthesis of nylon 6-clay hybrid. *Journal of Materials Research*, 8 (1993), 1179–84.
22. Y. Kojima, A. Usuki, M. Kawasumi, A. Okada, T. Kurauchi, and O. Kamigaito, Synthesis of nylon-6-clay hybrid by montmorillonite intercalated with epsilon-caprolactam. *Journal of Polymer Science, Part A: Polymer Chemistry*, 31 (1993), 983–6.
23. T. Kyu, Z. L. Zhou, G. C. Zhu, Y. Tajuddin, and S. Qutubuddin, Novel filled polymer composites prepared from in situ polymerization via a colloidal approach. 1. Kaolin/nylon-6 in situ composites. *Journal of Polymer Science, Part B: Polymer Physics*, 34 (1996), 1761–8.
24. W. A. Deer, R. A. Howie, and J. Zussman, *An Introduction to the Rock-Forming Minerals*, 2nd ed. (Harlow, Longman, 1992).
25. A. D. Liu, T. X. Xie, and G. S. Yang, Synthesis of exfoliated monomer casting polyamide 6/Na+-montmorillonite nanocomposites by anionic ring opening polymerization. *Macromolecular Chemistry and Physics*, 207 (2006), 701–7.
26. A. D. Liu, T. X. Xie, and G. S. Yang, Comparison of polyamide-6 nanocomposites based on pristine and organic montmorillonite obtained via anionic ring-opening polymerization. *Macromolecular Chemistry and Physics*, 207 (2006), 1174–81.
27. L. Song, Y. Hu, Q. L. He, and F. You, Study on crystallization, thermal and flame retardant properties of nylon 66/organoclay nanocomposites by in situ polymerization. *Journal of Fire Sciences*, 26 (2008), 475–92.
28. M. Tarameshlou, S. H. Jafari, H. A. Khonakdar, M. Farmahini-Farahani, and S. Ahmadian, Synthesis of exfoliated polyamide 6,6/organically modified montmorillonite nanocomposites by in situ interfacial polymerization. *Polymer Composites*, 28 (2007), 733–8.
29. Z. S. Kalkan and L. A. Goettler, In situ polymerization of polyamide 66 nanocomposites utilizing interfacial polycondensation. II. Sodium montmorillonite nanocomposites. *Polymer Engineering and Science*, 49 (2009), 1825–31.
30. Q. Zhang, M. Yu, and Q. Fu, Crystal morphology and crystallization kinetics of polyamide-11/clay nanocomposites. *Polymer International*, 53 (2004), 1941–9.
31. X. K. Zhang, G. S. Yang, and J. P. Lin, Synthesis, rheology, and morphology of nylon-11/layered silicate nanocomposite. *Journal of Polymer Science, Part B: Polymer Physics*, 44 (2006), 2161–72.
32. P. Aranda and E. Ruizhitzky, Poly(ethylene oxide)–silicate intercalation materials. *Chemistry of Materials*, 4 (1992), 1395–1403.
33. Y. H. Yu, C. Y. Lin, J. M. Yeh, and W. H. Lin, Preparation and properties of poly(vinyl alcohol)–clay nanocomposite materials. *Polymer*, 44 (2003), 3553–60.

34. J. Ma, D. Gao, B. Lu, Y. Chu, and J. Dai, Study on PVP/C-MMT nanocomposite material via polymer solution-intercalation method. *Materials and Manufacturing Processes* 34 (2007), 715–20.

35. G. Jimenez, N. Ogata, H. Kawai, and T. Ogihara, Structure and thermal/mechanical properties of poly(epsilon-caprolactone)–clay blend. *Journal of Applied Polymer Science*, 64 (1997), 2211–20.

36. N. Ogata, G. Jimenez, H. Kawai, and T. Ogihara, Structure and thermal/mechanical properties of poly(l-lactide)–clay blend. *Journal of Polymer Science, Part B: Polymer Physics*, 35 (1997), 389–96.

37. H. G. Jeon, H. T. Jung, S. W. Lee, and S. D. Hudson, Morphology of polymer/silicate nanocomposites – High density polyethylene and a nitrile copolymer. *Polymer Bulletin*, 41 (1998), 107–13.

38. C. C. M. Ma, C. T. Kuo, H. C. Kuan, and C. L. Chiang, Effects of swelling agents on the crystallization behavior and mechanical properties of polyamide 6/clay nanocomposites. *Journal of Applied Polymer Science*, 88 (2003), 1686–93.

39. Y. Cai, F. Huang, Q. Wei, L. Song, Y. Hu, Y. Ye, Y. Xu, and W. Gao, Structure, morphology, thermal stability and carbonization mechanism studies of electrospun PA6/Fe-OMT nanocomposite fibers. *Polymer Degradation and Stability*, 93 (2008), 2180–85.

40. S. Zulfiqar, Z. Ahmad, and M. I. Sarvar, Preparation and properties of aramid/layered silicate nanocomposites by solution intercalation technique. *Polymers for Advanced Technologies*, 19 (2008), 1720–28.

41. S. Zulfiqar, A. Kausar, M. Rizwan, and M. I. Sarwar, Probing the role of surface treated montmorillonite on the properties of semi-aromatic polyamide/clay nanocomposites. *Applied Surface Science*, 255 (2008), 2080–86.

42. S. Zulfiqar, M. I. Sarwar, I. Lieberwirth, and Z. Ahmad, Morphology, mechanical, and thermal properties of aramid/layered silicate nanocomposite materials. *Journal of Materials Research*, 23 (2008), 2296–304.

43. S. Zulfiqar, Z. Ahmad, M. Ishaq, and M. I. Sarwar, Aromatic-aliphatic polyamide/montmorillonite clay nanocomposite materials: Synthesis, nanostructure and properties. *Materials Science and Engineering A*, 525 (2009), 30–36.

44. S. Zulfiqar, M. Rafique, M. S. Shaukat, M. Ishaq, and M. I. Sarwar, Influence of clay modification on the properties of aramid-layered silicate nanocomposites. *Colloid and Polymer Science*, 287 (2009), 715–23.

45. R. A. Vaia and E. P. Giannelis, Lattice model of polymer melt intercalation in organically-modified layered silicates. *Macromolecules*, 30 (1997), 7990–99.

46. Y. Q. Li and H. Ishida, Solution intercalation of polystyrene and the comparison with poly(ethyl methacrylate). *Polymer*, 44 (2003), 6571–7.

47. L. Torre, G. Lelli, and J. M. Kenny, Synthesis and characterization of sPS/montmorillonite nanocomposites. *Journal of Applied Polymer Science*, 100 (2006), 4957–63.

48. S. Bourbigot, E. Devaux, and X. Flambard, Flammability of polyamide-6/clay hybrid nanocomposite textiles. *Polymer Degradation and Stability*, 75 (2002), 397–402.

49. F. Samyn, S. Bourbigot, C. Jama, and S. Bellayer, Fire retardancy of polymer clay nanocomposites: Is there an influence of the nanomorphology? *Polymer Degradation and Stability*, 93 (2008), 2019–24.

50. D. Garcia-Lopez, I. Gobernado-Mitre, J. F. Fernandez, J. C. Merino, and J. M. Pastor, Properties of polyamide 6/clay nanocomposites processed by low cost bentonite and different organic modifiers. *Polymer Bulletin*, 62 (2009), 791–800.

51. A. Frache, O. Monticelli, S. Ceccia, A. Brucellaria, and A. Casale, Preparation of nanocomposites based on PP and Pa6 by direct injection molding. *Polymer Engineering and Science*, 48 (2008), 2373–81.
52. Y. Hu, S. F. Wang, Z. H. Ling, Y. L. Zhuang, Z. Y. Chen, and W. C. Fan, Preparation and combustion properties of flame retardant nylon 6/montmorillonite nanocomposite. *Macromolecular Materials and Engineering*, 288 (2003), 272–6.
53. X. Liu, Q. Wu, and L. A. Berglund, Polymorphism in polyamide 66/clay nanocomposites. *Polymer*, 43 (2002), 4967–72.
54. Q. Q. Yang, Z. X. Guo, and J. Yu, Preparation and characterization of polyamide 66/montmorillonite nanocomposites with methyl methacrylate as cointercalation agent. *Journal of Applied Polymer Science*, 108 (2008), 1–6.
55. T. McNally, W. R. Murphy, C. Y. Lew, R. J. Turner, and G. P. Brennan, Polyamide-12 layered silicate nanocomposites by melt blending. *Polymer*, 44 (2003), 2761–72.
56. N. A. Hocine, P. Mederic, and T. Aubry, Mechanical properties of polyamide-12 layered silicate nanocomposites and their relations with structure. *Polymer Testing*, 27 (2008), 330–39.
57. S. V. Levchik, E. D. Weil, and M. Lewin, Thermal decomposition of aliphatic nylons. *Polymer International*, 48 (1999), 532–57.
58. H. Bockhorn, A. Hornung, U. Hornung, and J. Weichmann, Kinetic study on the non-catalysed and catalysed degradation of polyamide 6 with isothermal and dynamic methods. *Thermochimica Acta*, 337 (1999), 97–110.
59. K. P. Pramoda, T. Liu, Z. Liu, C. He, and H.-J. Sue, Thermal degradation behavior of polyamide 6/clay nanocomposites. *Polymer Degradation and Stability*, 81 (2003), 47–56.
60. T. Kashiwagi, R. H. Harris, X. Zhang, R. M. Briber, B. H. Cipriano, S. R. Raghavan, W. H. Awad, and J. R. Shields, Flame retardant mechanism of polyamide 6-clay nanocomposites. *Polymer*, 45 (2004), 881–91.
61. R. J. Varley, A. M. Groth, and K. H. Leong, The role of nanodispersion on the fire performance of organoclay-polyamide nanocomposites. In *5th Asian/ναισαλαρτσυA)5− 6MXXA(σλαιρεταM ετισοπμοX νο εχνερεφοX* (Hong Kong: Elsevier Science Ltd., 2006) pp. 2882–91.
62. F. Dabrowski, S. Bourbigot, R. Delobel, and M. Le Bras, Kinetic modelling of the thermal degradation of polyamide-6 nanocomposite. *European Polymer Journal*, 36 (2000), 273–84.
63. B. N. Jang and C. A. Wilkie, The effect of clay on the thermal degradation of polyamide 6 in polyamide 6/clay nanocomposites. *Polymer*, 46 (2005), 3264–74.
64. F. Wiloth, Zur thermischen Zersetzung von Nylon 6.6. III. Messungen zur Thermolyse von Nylon 6.6 und 6.10. *Makromolekulare Chemie*, 144 (1971), 283–307.
65. P. R. Hornsby, J. Wang, R. Rothon, G. Jackson, G. Wilkinson, and K. Cossick, Thermal decomposition behaviour of polyamide fire-retardant compositions containing magnesium hydroxide filler. *Polymer Degradation and Stability*, 51 (1996), 235–49.
66. Y. Nagasawa, M. Hotta, and K. Ozawa, Fast thermolysis/FT-IR studies of fire-retardant melamine–cyanurate and melamine–cyanurate containing polymer. *Journal of Analytical and Applied Pyrolysis*, 33 (1995), 253–67.
67. U. Bahr, I. Luderwald, R. Muller, and H. R. Schulten, Pyrolysis field desorption mass spectrometry of polymers. III. Aliphatic polyamides. *Angewandte Makromolekulare Chemie*, 120 (1984), 163–75.
68. I. Luderwald and F. Mertz, Über den thermischen Abbau von Polyamiden der Nylon-Reihe. *Angewandte Makromolekulare Chemie*, 74 (1978), 165–85.

69. D. H. MacKerron and R. P. Gordon, Minor products from the pyrolysis of thin films of poly(hexamethylene adipamide). *Polymer Degradation and Stability*, 12 (1985), 277–85.
70. J. M. Andrews, F. R. Jones, and J. A. Semlyen, Equilibrium ring concentrations and the statistical conformations of polymer chains: 12. Cyclics in molten and solid nylon-6. *Polymer*, 15 (1974), 420–24.
71. L. Song, Y. Hu, Q. L. He, and F. You, Study of nylon 66-clay nanocomposites via condensation polymerization. *Colloid and Polymer Science*, 286 (2008), 721–7.
72. H. L. Qin, Q. S. Su, S. M. Zhang, B. Zhao, and M. S. Yang, Thermal stability and flammability of polyamide 66/montmorillonite nanocomposites. *Polymer*, 44 (2003), 7533–8.
73. V. Babrauskas and R. D. Peacock, Heat release rate: The single most important variable in fire hazard. *Fire Safety Journal*, 18 (1992), 255–61.
74. S. P. Su and C. A. Wilkie, Exfoliated poly(methyl methacrylate) and polystyrene nanocomposites occur when the clay cation contains a vinyl mononer. *Journal of Polymer Science, Part A: Polymer Chemistry*, 41 (2003), 1124–35.
75. H. Y. Ma, Z. B. Xu, L. F. Tong, A. G. Gu, and Z. P. Fang, Studies of ABS-graft-maleic anhydride/clay nanocomposites: Morphologies, thermal stability and flammability properties. *Polymer Degradation and Stability*, 91 (2006), 2951–9.
76. Y. B. Cai, Y. Hu, L. Song, L. Liu, Z. Z. Wang, Z. Chen, and W. H. Fan, Synthesis and characterization of thermoplastic polyurethane/montmorillonite nanocomposites produced by reactive extrusion. *Journal of Materials Science*, 42 (2007), 5785–90.
77. J. W. Gilman, T. Kashiwagi, M. Nyden, J. E. T. Brown, C. L. Jackson, S. Lomakin, E. P. Giannelis, and E. Manias, Flammability studies of polymer layered silicate nanocomposites: Polyolefin, epoxy and vinyl ester resins. In: *Chemistry and Technology of Polymer Additives*, ed. S. Al-Malaika, A. Golovoy, and C. A. Wilkie (Malden: Blackwell Science, 1999), pp. 249–65.
78. J. W. Gilman, Flammability and thermal stability studies of polymer layered-silicate (clay) nanocomposites. *Applied Clay Science*, 15 (1999), 31–49.
79. A. Dasari, Z. Z. Yu, Y. W. Mai, and S. Liu, Flame retardancy of highly filled polyamide 6/clay nanocomposites. *Nanotechnology*, 18 (2007), 445602–10.
80 M. Lewin, Surface barrier formation in the pyrolysis and combustion of nanocomposites. *Recent Advances in Flame Retardancy of Polymeric Materials*, 12 (2002), 84–96.
81. M. Lewin, Some comments on the modes of action of nanocomposites in the flame retardancy of polymers. *Fire and Materials*, 27 (2003), 1–7.
82. M. Lewin, Reflections on migration of clay and structural changes in nanocomposites. *Polymers for Advanced Technologies*, 17 (2006), 758–63.
83. M. Lewin, A. Mey-Marom, and R. Frank, Surface free energies of polymeric materials, additives and minerals. *Polymers for Advanced Technologies*, 16 (2005), 429–41.
84. Y. Tang, M. Lewin, and E. M. Pearce, Effects of annealing on the migration behavior of PA6/clay nanocomposites. *Macromolecular Rapid Communications*, 27 (2006), 1545–9.
85. M. Zammarano, J. W. Gilman, M. Nyden, E. M. Pearce, and M. Lewin, The role of oxidation in the migration mechanism of layered silicate in poly(propylene) nanocomposites. *Macromolecular Rapid Communications*, 27 (2006), 693–6.
86. J. Hao, M. Lewin, C. A. Wilkie, and J. Wang, Additional evidence for the migration of clay upon heating of clay-polypropylene nanocomposites from X-ray photoelectron spectroscopy (XPS). *Polymer Degradation and Stability*, 91 (2006), 2482–5.

87. Y. Tang and M. Lewin, Maleated polypropylene OMMT nanocomposite: Annealing, structural changes, exfoliated and migration. *Polymer Degradation and Stability*, 92 (2007), 53–60.
88. M. Lewin, Reflections on migration of clay and structural changes in nanocomposites. *Polymers for Advanced Technologies*, 17 (2006), 758–63.
89. N. A. Isitman, H. O. Gunduz, and C. Kaynak, Nanoclay synergy in flame retarded/glass fibre reinforced polyamide 6. *Polymer Degradation and Stability*, 94 (2009), 2241–50.
90. S. C. Lao, C. Wu, T. J. Moon, J. H. Koo, A. Morgan, L. Pilato, and G. Wissler, Flame-retardant polyamide 11 and 12 nanocomposites: Thermal and flammability properties. *Journal of Composite Materials*, 43 (2009), 1803–18.
91. S. P. S. Ribeiro, L. R. M. Estevão, C. Pereira, J. Rodrigues, and R. S. V. Nascimento, Influence of clays on the flame retardancy and high temperature viscoelastic properties of polymeric intumescent formulations. *Polymer Degradation and Stability*, 94 (2009), 421–31.
92. W. S. Wang, H. S. Chen, Y. W. Wu, T. Y. Tsai, and Y. W. Chen-Yang, Properties of novel epoxy/clay nanocomposites prepared with a reactive phosphorus-containing organoclay. *Polymer*, 49 (2008), 4826–36.
93. Y. Zhang, Y. Hu, L. Song, J. Wu, and S. L. Fang, Influence of Fe-MMT on the fire retarding behavior and mechanical property of (ethylene–vinyl acetate copolymer/magnesium hydroxide) composite. *Polymers for Advanced Technologies*, 19 (2008), 960–66.
94. B. Schartel, U. Knoll, A. Hartwig, and D. Putz, Phosphonium-modified layered silicate epoxy resins nanocomposites and their combinations with ATH and organo-phosphorus fire retardants. *Polymers for Advanced Technologies*, 17 (2006), 281–93.
95. L. Yang, Y. Hu, H. D. Lu, and L. Song, Morphology, thermal, and mechanical properties of flame-retardant silicone rubber/montmorillonite nanocomposites. *Journal of Applied Polymer Science*, 99 (2006), 3275–80.
96. M. Modesti, A. Lorenzetti, S. Besco, D. Hreja, S. Semenzato, R. Bertani, and R. A. Michelin, Synergism between flame retardant and modified layered silicate on thermal stability and fire behaviour of polyurethane–nanocomposite foams. *Polymer Degradation and Stability*, 93 (2008), 2166–71.
97. T. Kashiwagi, Polymer combustion and flammability – Role of the condensed phase. Presented at: Twenty-Fifth Symposium (International) on Combustion, Pittsburgh, 1994, pp. 1423–37. Available at: http://www.fire.nist.gov/bfrlpubs/fire95/art104.html.

10

Self-extinguishing polymer–clay nanocomposites

SEONGCHAN PACK[a] AND MIRIAM H. RAFAILOVICH[b]

[a]*Samsung Cheil Industries, Inc.*
[b]*State University of New York at Stony Brook*

10.1 Introduction

Thermodynamically, the introduction of a solid particle into a polymer matrix either decreases or increases the interfacial energy, depending on the degree of interaction between polymer chains and solid surfaces.* If strong absorption of the polymer chains on the surfaces takes place, the system can be approached through minimization of the interfacial energy, reducing the energy factors. Furthermore, the minimization of interfacial energy can be optimized by increasing the interfacial area of solid particles. Therefore, in order to maximize reduction of the interfacial energy, the solid particles need a large aspect ratio, making both layered silicates and carbon nanotubes (CNTs) good candidates [1]. In particular, layered silicates cation-exchanged with organophilic surfactants can be delaminated into a single silicate sheet in a polymer matrix and remain as nanosheets with aspect ratio 100–1000. Because of this unique delamination of organophilic silicates, polymer–organoclay nanocomposites are of great interest in industry and academia. Numerous research groups have characterized and predicted the microstructures of polymer/organoclay nanocomposites using advanced techniques [2–4].

Furthermore, the addition of organoclays can compatibilize immiscible polymer blends. Si *et al.* [5] demonstrated that the addition of Cloisite clays could prevent polymer blends from undergoing further phase segregation because the clays segregate into the interfaces between the blends. They argued that the segregation of the clays could minimize the interfacial tension while forming *in situ grafts* on the segregated clay surfaces with the polymer chains. Hence, the addition of functionalized clays in polymer blends can enhance compatibility and thereby improve mechanical properties [5–6]. In addition to intercalation or exfoliation within a polymer matrix, the thermal stability of polymer–clay nanocomposites also depends on the degree of heat dissipation. Pack *et al.* [7] showed that clay nanoparticles could order into long ribbonlike structures, whereas Kashiwagi *et al.* [8] blended nanotubes directly into the matrix and then demonstrated that the entanglements of the nanotubes were well formed during combustion. In both cases the improvement in thermal properties, such as heat release rate (HRR) and mass loss rate (MLR) [7–10], was attributed to

* This work was supported by NSF-MRSEC. The authors thank Professor Chad Korach for the use of the instrumented-indentation at Stony Brook University.

enhanced thermal conduction within the matrix by the extended tubelike structures. Despite this improvement, most nanocomposites still cannot be rendered self-extinguishing unless conventional flame retardant (FR) agents are added. The FR agents can be classified further into two basic types, halogenated and nonhalogenated. The improved dispersion obtained with functionalized clays also improves the dispersion of the FR [7, 9], but optimal results were shown to be obtained from blends containing nanoparticle mixtures, where synergy could be obtained between the particles that were interfacially active and those that improved the conductivity [10–11]. In this chapter, we try to explain, using the data from complementary experiments, some of the factors that lead to flame retardance and self-extinguishing behavior in nanocomposites.

10.2 Flame retardant formulations and nanoparticle synergy

10.2.1 Halogenated flame retardant formulations

The addition of halogenated FR particles is a conventional method for engineering flame retardant polymeric materials and composites. Because of their efficacy in quenching heat in either gas-phase or condensed-phase reactions, halogenated FR agents have been developed for use either as additives or copolymerized onto the polymer chains [12–14]. Most of the halogenated FR agents are based on bromine or chlorine compounds, which can produce hydrogen halides that react with either hydrogen or hydroxyl radicals during combustion. Therefore, the rates of chain reactions can be reduced because the hydrogen halides inhibit exothermic oxidation reactions. [15]. In spite of the fact that good flame retardancy is obtained in the gas phase when halogenated FRs are added, in many cases more than 30–40 wt% of FR particles must be added to the end products for the materials to pass standard flame tests. Because the FR is often in crystalline form, when too many halogenated FR particles are added, the polymer matrix can be embrittled. For the past decade, the addition of nanoparticles, such as functionalized clays and carbon nanotubes, has been of great interest in the field of flame retardancy because they have been shown to produce rigid chars once the polymer–nanoparticles matrix decomposes at high temperature [7–11]. In this manner the formation of chars can act as a heat barrier, which can retard propagation of the heat fronts, and lead to a large reduction in MLR and HRR.

Although the effects on the MLR and HRR can be explained by good dispersion of nanoparticles in a melt polymer matrix [9], to our knowledge there have been no reports where the addition of clay nanoparticles alone has produced a self-extinguishing nanocomposite from an otherwise flammable polymer matrix. Recently, Si *et al.* reported that the addition of FR in combination with organoclays could render even highly flammable polymers, such as acrylic resins, self-extinguishing according to the UL-94 V0 standard [9]. They demonstrated that neither the FR nor the clays could achieve this result, and hence, they postulated that a synergistic effect occurred. Using transmission electron microscopy on thin sections, they demonstrated that the FR, in this case decabrome, adsorbed onto the clay surfaces, and because the clays were exfoliated in this matrix, the dispersion of the

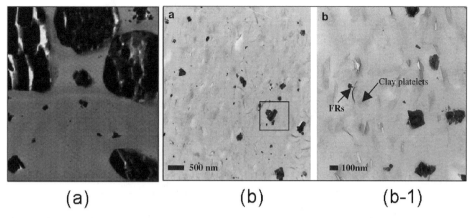

Figure 10.1 TEM images of PMMA and the FR blend 20% decabrome and 5% antimony trioxide (a) without the addition of clays and (b) and with the addition of 5% Cloisite 20A. (b-1) Higher magnification of section delineated by the box in (b) where the clay platelets are marked. Reproduced from Ref. [7] with permission from Elsevier.

FR was greatly improved (Figure 10.1). As a result of the improved dispersion, only 5% Cloisite clay, together with 20% decabrome and 5% antimony trioxides, was needed to pass the UL-94 V0 protocol with a sample 1.6 mm thick. Wang, Echols, and Wilkie reported that a concentration of more than 40% dibromostyrene in polypropylene was needed to achieve the same result in the absence of clay [16].

10.2.2 Nonhalogenated flame retardant formulations

Because of environmental concerns regarding the use of halogenated FR derivatives [17], the demand for an alternative FR to replace the halogenated FR formulations has increased. One set of candidates is a group of aryl phosphates, whose chemical structures and applications for flame retardancy have been reviewed extensively in the literature [18, 19]. Some commonly used aryl phosphates and their properties are listed in Table 10.1.

These phosphate esters can serve either as flame retardants or as plasticizers. For example, because poly(vinyl chloride) (PVC) can decompose before its viscosity is sufficiently low for melt-blending, aryl phosphates have been used to lower the processing temperature and prevent decomposition. Aromatic resins such as polycarbonate (PC) are good char formers and flame retardants, according to the UL-94 V2 criteria. Addition of small amounts of the aryl phosphates (less than 5 wt%) allows one to achieve a grade of UL-94 V0 [18]. On the other hand, styrenic resins and polyolefins (e.g., polystyrene or polyethylene), which are not good char formers and not flame retardants, are immiscible with most phosphate esters, such as resorcinol bis(diphenyl phosphate) (RDP) and bisphenol A bis(diphenyl phosphate) (BDP). This results in poor processing conditions because the phosphates phase-separate from the styrenic matrix during melt-blending at high processing temperatures ($\sim$200 °C).

Table 10.1 *The common aryl phosphates: Chemical structures and their properties*

Name	Chemical structure	Phosphorus content %	Viscosity at 25 °C (centipoise)	Decomposition temperature (at 10 wt%)
Triphenyl phosphate		9.50%	Solid	198 °C
Resorcinol bis(diphenyl phosphate)		10.70%	12,450 cps	325 °C
Bisphenol A bis(diphenyl phosphate)		8.90%	600 cps	375 °C

Recently, Pack *et al.* have demonstrated that this problem can be overcome when RDP is directly absorbed onto unfunctionalized sodium clay surfaces [10]. They showed that the interactions between RDP and clays were sufficiently strong so that the RDP was able to penetrate the galleries between clay platelets, allowing the clays to exfoliate or intercalate in a large class of matrices. This method of functionalization was shown to be simpler and more cost-effective than those that functionalized the clays with di-tallow molecules, and in certain cases the same synergistic effects were observed with halogenated FR.

In contrast to the Cloisite clays, addition of phosphate-coated clays was able to improve the flame retardant response in certain classes of polymers. For example, addition of 5% RDP-coated clays to PC was able to achieve a UL-94 V1 rating, whereas the addition of Cloisite 20A failed. Similarly, addition of 10% RDP-coated clays to HIPS achieved a rating of V1, whereas no effect was observed with Cloisite clays.

Aryl phosphates were also adsorbed onto halloysite clay tubes, with similar results. In Figures 10.2 (a, b) it can seen that the neat halloysites are not well dispersed in the PC matrix, but segregate into large clusters. However, the halloysites treated with 0.5 wt% RDP are seen to be uniformly dispersed within the matrix. Furthermore, the nanocomposites with coated tubes achieved a UL-94 V1 rating, whereas those with uncoated tubes did not. The halloysite clay nanotubes have the potential advantage of being able to disperse more of the RDP, which can coat the outside of the tube as well as filling the inside space. The large aspect ratio of the clay nanotubes allows them to function in a manner similar to the clay platelets. Hence, the synergistic effects reported for clay platelets were also observed with the coated halloysite nanotubes. Figure 10.2 (c, d) shows TEM images of halloysite

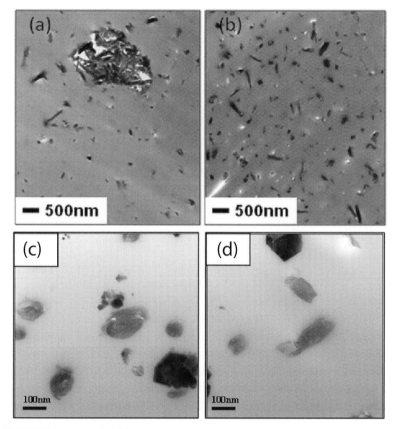

Figure 10.2 TEM images: (a) PC/neat halloysites (95/5 wt%), (b) PC/RDP halloysites (95/5 wt%), (c) PMMA/DB/AO/untreated halloysites (70/20/5/5 wt%), and (d) PMMA/DB/AO/RDP halloysites (70/20/5/5 wt%). Reproduced from Ref. [7] with permission from Elsevier.

nanotubes dispersed in a PMMA matrix where 20% decabrom and 5% AO were added. From the figure the dispersion of the FR formulation is much better in the matrix where the halloysite tubes were coated than in the matrix with uncoated nanotubes. Furthermore, the matrix with the coated tubes achieved a UL-94 V0 rating, similar to that for the PMMA nanocomposite, where Cloisite 20A and the same FR were used.

10.3 Flame retardant polymer blends with nanoparticles

Most engineered plastics mix with more than two polymers. It is hard to achieve flame retardant formulations because the microstructures of polymer blends are continuously changing during heating, which can result in phase separation. Even though the functionalized clays can be exfoliated and/or intercalated in homopolymers, it is necessary to consider the interfaces in polymer blends. Recently, Si *et al.* [5] showed that when Cloisite clays

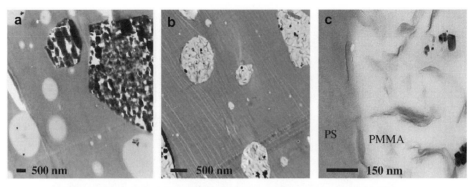

Figure 10.3 TEM images: (a) PS/PMMA/DB/AO(70/30/15/4 wt%), (b) PS/PMMA/DB/AO/Cloisite 20A(70/30/15/4/5 wt%), high magnification, (c) PS/PMMA/DB/AO/Cloisite 20A(70/30/15/4/5 wt%). Reproduced from Ref. [7] with permission from Elsevier.

were added to polymer blends, the polymers could be adsorbed onto the clay interfaces, forming in situ grafts. Because of the large aspect ratio of the platelets, the grafts were not stable in any of the phases, but segregated to the domain interfaces, resulting in partial compatibilization. Further annealing of the blends showed that the effect was not simply a phase-pinning effect, where surface interactions reduced polymer mobility and hence lowered the rate of phase segregation. Rather, upon annealing, the domains decreased, and compatibilization increased, as would be expected when the positioning the clays at the interfaces resulted in a reduction of the overall energy of the system. In the absence of clays, the advancing heat front of a flame induces phase segregation in a polymer blend, further destabilizing the material against degradation. The presence of the clays at the interface prevents segregation and induces certain synergies, as will be described, which facilitate the formation of flame retardant polymer blends.

10.3.1 Microstructures

PS/PMMA/FR blends with organoclays

PS and PMMA are immiscible polymers that phase-segregate when blended. Addition of the halogenated FR formulation forms yet another phase, which segregates into large domains distinct from the PS and the PMMA domains, which are clearly seen in the TEM images shown in Figure 10.3 (a). TEM images of the same blend composition, but with the addition of 5% Cloisite 20A clay, are shown in Figure 10.3 (b), where we can see that the dispersion of the FR is greatly improved. In Figure 10.3 (c), a higher-magnification image, we can see individual clay platelets segregated to the PS and PMMA domain interfaces, stabilizing the blend. Furthermore, we see that the improved dispersion of the FR results from the adsorption of the FR onto the clay platelet surfaces, which are exfoliated or intercalated within the PMMA domains.

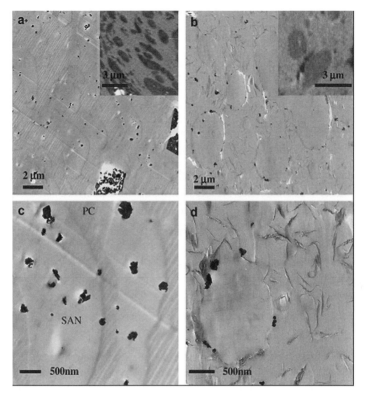

Figure 10.4 Scanning transmission X-ray microscopy (STXM) images taken at 286.8eV, the absorption of the energy of SAN24. SAN24 is dark, which is shown in the inserted images. TEM images: (a) PC/SAN24/DB/AO (50/50/15/4 wt%) and (b) PC/SAN24/DB/AO/Cloisite 20A (50/50/15/4/ 5 wt%). High magnification: (c) PC/SAN24/DB/AO (50/50/15/4 wt%) and (d) PC/SAN24/DB/AO/ Cloisite 20A (50/50/15/4/5 wt%). Reproduced from Ref. [7] with permission from Elsevier.

PC/SAN24/FR blends with organoclays

Si *et al.* also studied PC/SAN blends and observed a similar effect, where the addition of as little as 3% Cloisite 20A resulted in compatibilization of the blend. On the other hand, even though PC is an intrinsically better flame retardant polymer than either PS or PMMA, addition of the FR formulation with or without the clays produced blends with inferior flame retardant response. In Figure 10.4, we show the TEM images of a PC/SAN24/FR blend with and without the addition of Cloisite 20A clay. Because the electron contrast between the two polymers is not sufficient to distinguish the separate phases, scanning transmission X-ray microscopy (STXM) images [20] are also shown in the insets of the lower-magnification images in Figure 10.4 (a, b). Higher-magnification images are shown in Figure 10.4 (c, d). In Figure 10.4 (a), we see that in the absence of clay, most of the FR particles are segregated inside the SAN phase. When 5% Cloisite 20A clay is added, the FR is again observed to adsorb onto the clay platelet surfaces and segregate to the blend interfaces, together with the clay platelets. The results of DMA analysis on the blends are shown in Figure 10.5, where the addition of Cloisite clays to the blend is shown to increase the storage modulus by nearly 65%, and results in a single peak in tan δ, indicating excellent compatibilization of the blend with a single glass transition temperature T_g. When the FR was added to the clay, the degree of compatibilization was decreased, as evidenced by the multiple peak structures observed in the tan δ trace and the significant reduction of the storage modulus from 3.7 to 2.7 GPa.

Nonhalogenated FR formulation

An alternate method for achieving similar results was recently proposed by Pack *et al.* [10], who showed that RDP-coated clays can also compatibilize polymer blends, and are even more effective than Cloisite 20A clays when the blends contain styrenic components. In this case, the FR did not compete with the polymers for the clay surfaces, because both polymers also adsorbed onto the RDP-coated surfaces. As a result, addition of these clays improved the mechanical properties of the blends and resulted in a material that was able to pass the UL-94 V1 flame test.

PS/PMMA/FR/Cloisite 20A with multiwall carbon nanotubes (MWCNTs)

It was shown that the addition of organoclays, such as Cloisite 20A, can enhance the compatibilization of the blends. However, when the blend was also required to be self-extinguishing, significantly higher concentrations of clay were required. Furthermore, because the blends also contained 15–18% of the FR formulation and at least 10% by weight of clays, the total amount of fillers was sufficiently large to affect the mechanical properties of the material. Hence Pack *et al.* [11] proposed to use combinations of nanoparticles in an attempt to explore new synergies, which could reduce the overall filler composition, while improving the degree of compatibilization and the self-extinguishing properties. They experimented with adding long and short multiwalled carbon nanotubes [*l*-MWCNTs (2–40 μm in length) and *s*-MWCNTs (< 1 μm in length)] together with Cloisite 20A clays and the halogenated FR formulation. The results are shown in Figure 10.6, where the *l*-MWCNTs and *s*-MWCNTs are shown in Figure 10.6 (a, b). The longer tubes are seen to form an interconnected network, whereas the shorter tubes remain distinct. Figure 10.6 (e, f) shows TEM images of thin cross sections obtained from the blends PS/PMMA/DB/AO/Cloisite 20A/*l*-MWCNT (70/30/15/4/3/2 wt%) and PS/PMMA/DB/AO/Cloisite 20A/*s*-MWCNT (70/30/15/4/3/2 wt%), respectively. Comparing the images with the cross section of the unfilled blend in Figure 10.6 (c), it is clear that both filled blends have superior dispersion of the FR (dark, electron-dense regions) material in the PMMA phase. The histogram plot in Figure 10.6 (d) shows that, even though the images appear superficially similar, the size of the dispersed particles is significantly smaller in the blend containing the *s*-MWCNT. Further detail is provided in the higher-resolution images shown as inserts into Figure 10.6 (e, f). Here one can see that (1) the clay platelets are segregated at the phase domain boundaries and (2) the FR particles are clustered around the

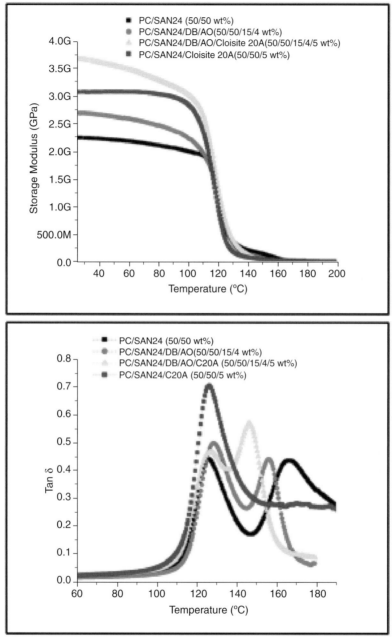

Figure 10.5 (top) Storage modulus and (bottom) tan δ in PC/SAN24 (50/50 wt%), PC/SAN24/DB/AO (50/50/15/4 wt%), PC/SAN24/DB/AO/Cloisite 20A (50/50/15/4/5 wt%), and PC/SAN24/Cloisite 20A (50/50/5 wt%). Reproduced from Ref. [7] with permission from Elsevier.

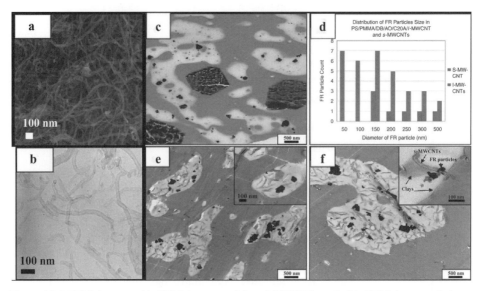

Figure 10.6 TEM images: (a) SEM images of long-MWCNTs and (b) TEM image of short-MWCNTs. (c) PS/PMMA/DB/AO (70/30/15/4 wt%). (d) Distribution of FR particle size in PS/PMMA/DB/AO/Cloisite 20A/*l*-MWCNT and *s*-MWCNT, (e) PS/PMMA/DB/AO/Cloisite 20A/*l*-MWCNT (70/30/15/4/3/2 wt%), and (f) PS/PMMA/DB/AO/Cloisite 20A/*s*-MWCNT (70/30/15/4/3/2 wt%). Scale bar = 500 nm. Reproduced from Ref. [11] with permission from the American Chemical Society.

MWCNT, rather than the clay platelets. These results indicate that the FR is preferentially adsorbed onto the MWCNT surfaces, and hence no longer competes with the polymer in the formation of the in situ grafts on the clay platelet surfaces. As a result, the efficiency of the platelets in compatibilizing the blend is restored, and both blends are flame retardant with half the concentration of clay. It is interesting to note that the *s*-MWCNT blend satisfies the UL-94 V0 requirement, whereas the blend containing the *l*-MWCNT only satisfies UL-94 V2, which may be because of improved FR particle dispersion. In addition, close examination of the TEM images shows that the *s*-MWCNT are dispersed in both the PS and PMMA phases, whereas the *l*-MWCNT are observed mostly in the PS phase.

These differences may be explained by the fact that the entanglements of the *l*-MWCNTs prevent the tubes from diffusing through the matrix as the phase separation progresses, whereas the shorter tubes can adjust to the changes in the domain distribution during the melt-blending process.

The difference of dispersion can affect the rheological properties of blends. In Figure 10.7 (A, B) we show the DMA data obtained from the different polymer blends.

From the figures we can see that the addition of FR particles lowers the modulus of the PS/PMMA blend by nearly 26%, because the FR particles are also immiscible into the blend. However, the addition of Cloisite 20A (3 wt%) somewhat restores the modulus of the

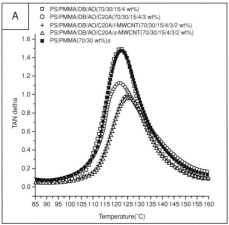
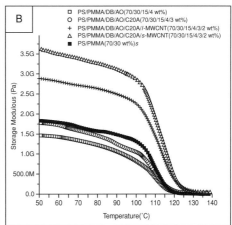

Figure 10.7 (A) Storage Modulus and (B) tan δ versus temperature curves: PS/PMMA/DB/AO (70/30/15/4 wt%) – unfilled square, PS/PMMA/DB/AO/Cloisite 20A (70/30/15/4/3 wt%) – unfilled circle, PS/PMMA/DB/AO/Cloisite 20A/*l*-MWCNT (70/30/15/4/3/2 wt%) – unfilled cross, PS/PMMA/DB/AO/Cloisite 20A/*s*-MWCNT(70/30/15/4/3/2 wt%) – unfilled black triangle, and PS/PMMA (70/30 wt%) – filled square. Reproduced from Ref. [11] with permission from the American Chemical Society.

unfilled blend, which could result from slightly improving the compatibilization, thereby reducing the interfacial tension at the interfaces. Significant improvement is observed when the MWCNTs are added. In particular, the added *s*-MWCNTs can give the blend the highest modulus compared to that in the added *l*–MWCNTs blend because of the different degree of dispersion in the PS/PMMA/FR/C20A blend. It is also confirmed by the frequency dependence of G'. From Figure 10.8 we can see that the PS/PMMA/FR/C20A blend with the s-MWCNTs behaves as a liquid-like state, even at 30 Hz. However, starting at about 1 Hz the dynamic response of the *l*–MWCNTs case is independent of frequency until 60 Hz, indicating the formation of an entangled gel-like state because of the entangled substructure of the MWCNT network.

10.3.2 Small angle X-ray scattering

According to these TEM images of polymer–clay nanocomposites, the presence of FR particles can affect the compatibilization of blends. The degree to which the clays compatibilize the blends depends on the ability to the polymer to adsorb onto the clay surfaces, and hence is correlated with the extent of exfoliation or intercalation. To observe whether the clays are exfoliated or intercalated, we conducted SAXS on PMMA/DB/AO/Cloisite20A, PS/PMMA/Cloisite 20A, PS/PMMA/ DB/AO/Cloisite 20A, PC/SAN24/Cloisite 20A, and PC/SAN24/DB/AO/Cloisite20A nanocomposites and compared the results with SAXS data

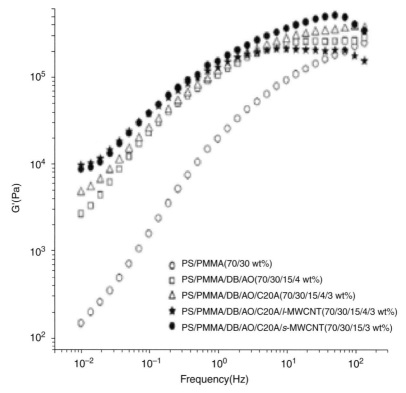

Figure 10.8 *G'* versus frequency curves: PS/PMMA (70/30 wt%) – unfilled circle, PS/PMMA/DB/AO (70/30/15/4 wt%) – unfilled square, PS/PMMA/DB/AO/Cloisite 20A (70/30/15/4/3 wt%) – unfilled triangle, PS/PMMA/DB/AO/Cloisite 20A/*l*-MWCNT (70/30/15/4/3/2 wt%) – filled star, and PS/PMMA/DB/AO/Cloisite 20A/*s*-MWCNT (70/30/15/4/3/2 wt%) – filled circle. Reproduced from Ref. [11] with permission from the American Chemical Society.

from pure Cloisite 20A clays. In Figure 10.9, we can see that the Cloisite 20A clays have two peaks at $q = 0.2319$ Å^{-1} and $q = 0.4638$ Å^{-1}, where the d spacing ($d = 2\pi/q$) is 2.70 nm and 1.35 nm in two directions, [001] and [002]. When the clays are largely exfoliated, as in the case of PMMA/Cloisite20A nanocomposites, no diffraction peaks can be observed. A peak is observed, though, at $q = 4.10$ nm in the PMMA/FR/C20A blend, indicating that the FR is hindering the exfoliation. This value is still larger than that observed in the pure clays, indicating that intercalation, rather than exfoliation, is occurring. This is consistent with the TEM images described previously, where adsorption of the FR onto the clay platelets competed with their ability to interact with the polymer matrix.

The TEM images in Figure 10.1 also showed the result that the FR particles were absorbed onto the clay surfaces, which could reduce the interfacial area available to adsorb

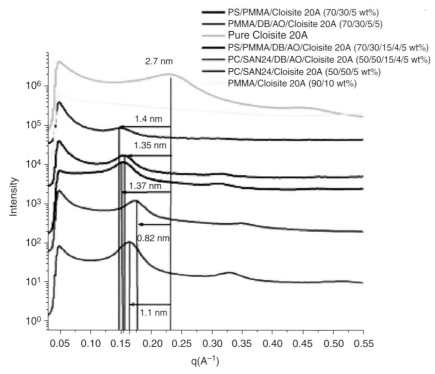

Figure 10.9 SAXS spectra of pure Cloisite 20A – (line), PMMA/C20A (90/10 wt%) – (line), PMMA/DB/AO/Cloisite 20A (70/20/5/5 wt%) – (line), PS/PMMA/DB/AO/Cloisite 20A (70/30/15/4/5 wt%) – (line), PS/PMMA/C20A (70/30/5/ wt%) – (line), PC/SAN24/DB/AO/Cloisite 20A (50/50/15/4/5 wt%) – (line), and PC/SAN24/C20A (50/50/15/4/5 wt%) – (line). Reproduced from Ref. [7] with permission from Elsevier.

the polymer chains. Hence, this competitive absorption can hinder the process of exfoliation in the PMMA matrix. In the case of the PS/PMMA/DB/AO/Cloisite20A blends, a high degree of intercalation is present, for which the interlayer spacings obtained from the spectra in Figure 10.9 are listed in Table 10.2. Addition of the FR is observed to decrease the interlayer spacing, indicating that less polymer has been adsorbed. A similar result is also obtained from the PC/SAN24 blend, for which the change in interlayer spacing is also tabulated in Table 10.2. The observed decrease in spacing is consistent with decreased adsorption, and as was shown previously with DMA measurements in Figure 10.5, with decreased degree of compatibilization. Hence, the strong adsorption of the FR onto the clay platelets, which greatly improves their dispersion, comes at the cost of a decrease in intercalation and compatibilization.

Table 10.2 d_{001} *of nanocomposites with FR and Cloisite 20A*

Sample	d_{001} (nm)
Cloisite 20A (C20A)	2.75
PMMA/Cloisite 20A	–
PMMA/DB/AO/C20A (70/20/5/5 wt%)	4.1
PS/PMMA/C20A (70/30/5 wt%)	4.07
PS/PMMA/DB/AO/C20A (70/30/15/4/5 wt%)	4.05
PC/SAN24/C20A (50/50/5 wt%)	3.8
PC/SAN24/DB/AO/C20A(50/50/15/4/5 wt%)	3.52

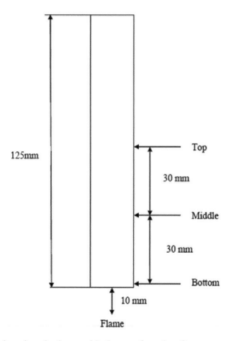

Figure 10.10 Scheme for burning the bar and indexes of cutting for cross sections. Reproduced from Ref. [7] with permission from Elsevier.

10.3.3 The effect of heat absorption on morphology

PS/PMMA/FR/C20A

In order to determine whether FR adsorption interferes with the ability of the clay platelets to stabilize the phase-segregated structures upon heating, thin cross sections were obtained from PS/PMMA blend test slabs exposed to the UL-94 flame test apparatus, as shown in Figure 10.10. The localization of the clays at interfaces does not seem to be interrupted by the addition of FR particles, and the internal microstructures remain stabilized against further

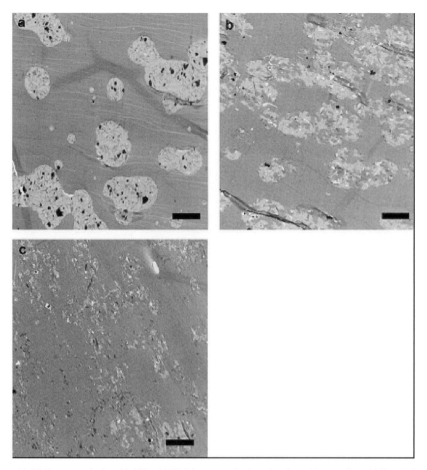

Figure 10.11 Low-resolution (4,800×) TEM images of a burning bar: (a) top, (b) middle, (c) bottom of PS/PMMA/DB/AO/Cloisite 20A(70/30/15/4/5 wt%). Scale bar = 2 μm. Reproduced from Ref. [7] with permission from Elsevier.

phase separations when heat is applied. The TEM images are shown in Figure 10.11. In Figure 10.11 (A), we can see that the morphology, taken furthest from the UL-94 V0 flame, is unchanged, where both the FR particles and the clays are seen in the PMMA domains. A closer examination shows that the microsized PMMA domains are well dispersed in the matrix, and the clays are also well dispersed either inside the PMMA domains or at the interfaces between the PS/PMMA phases. In the TEM image cross-sectioned closer to the flame, the PMMA domains become less visible, without coalescing. Similarly, the FR particles become smaller and fewer in number. The images are consistent with the PMMA polymer degrading, without further progression of the phase segregation, while the products enter the vapor phase, together with the FR particles. Because the FR reaction occurs in the vapor phase, simultaneous volatilization of the two products will synergize and increase

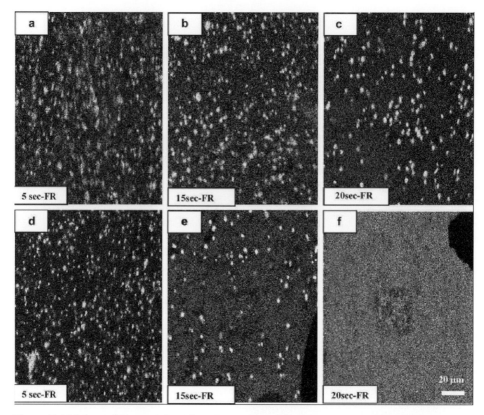

Figure 10.12 Maps of Br particles (dots) in a series of burning times at 650 °C in the oven: (a)–(c) from the PS/PMMA/DB/AO (70/30/15/4) samples, (d)–(f) from PS/PMMA/DB/AO/Cloisite 20A (70/30/15/4/3) samples. Reproduced from Ref. [7] with permission from Elsevier.

the efficiency of the FR formulation. This is further illustrated in Figure 10.12, where we mapped the distribution of Br, using EDAX, in the samples with the same formulation. In the figure, the samples were subjected to heat treatment for 5 min at 650 °C rather than to exposure to flame. From the figure we can see that the Br is consumed significantly faster in the samples containing clay, where a large reduction is already evident after 15 s and no bromine is observed after 20 s. On the other hand, only a small reduction in the Br particles is observed in the sample without Cloisite 20A clay even after 20 s.

From Figure 10.12 (a–c) we can see that most of the Br particles still remain in the blend without the clays after 20 s. However, in the case of the blend with the clays, from Figure 10.12 (d–f) we can see that the Br particles are well dispersed, and most of the Br particles disappear after 15 s. Therefore, the addition of clays can accelerate the volatiliza-tion of the halogenated particles, which were strongly absorbed onto the clay surfaces with the polymer chains. These results also support the conclusion that the FR particles

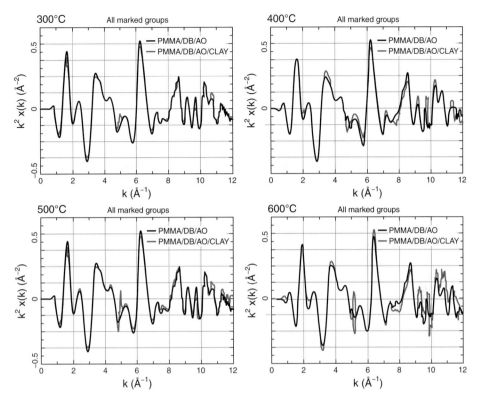

Figure 10.13 EXFAS data on Br particles in PMMA/DB/AO (70/20/5 wt%) burnt samples without or with Cloisite 20A(5 wt%) as a function of temperature. Reproduced from Ref. [7] with permission from Elsevier.

mainly react with the volatile products in the gas phase, which may be confirmed by the fact that there are no changes in the nearest-neighbor configuration of the decabromine molecules in the EXAFS data shown in Figure 10.13, obtained from a PMMA/FR sample.

PS/PMMA/FR/Cloisite20A with either the l-MWCNTs or the s-MWCNTs

Each sample was exposed to the UL-94 V0 flame for 20 s and then cross-sectioned for TEM images, which were taken from the top (20 mm from the middle) and middle (30 mm from the flame) of the each sample. The TEM images of the *l*-MWCNTs and the *s*-MWCNTs blends are shown in Figure 10.14 (a, b) and Figure 10.14 (c, d), respectively. From the figures we can see that the degree of dispersion of MWCNTs differs between the *l*-MWCNTs and the *s*-MWCNTs. Most of the *l*-MWCNTs are only seen in the PS matrix, whereas the s-MWCNTs are seen in both phases. In particular, in Figure 10.14 (b)

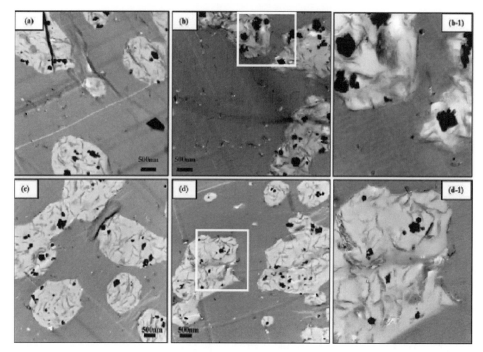

Figure 10.14 TEM images of two series of *l*- and *s*-MWCNTS annealing bars: (a) top, (b) middle of PS/PMMA/DB/AO/Cloisite 20A/*l*-MWCNTs (70/30/15/4/3/2 wt%) bar and (c) top, (d) middle of PS/PMMA/DB/AO/Cloisite 20A/*s*-MWCNT(70/30/15/4/3/2 wt%). Scale bar = 500 nm. (b-1) and (d-1) Higher magnification TEM images of (b) and (d). (e) The mean size of FR particles in each section in both the *l*-MWCNTs and the *s*-MWCNTs. Reproduced from Ref. [11] with permission from the American Chemical Society.

we can see that the clays are concentrated around the FR particles, and a few *l*-MWCNTs segregate at the phase interfaces. However, in the case of the *s*-MWCNTs blend, many carbon nanotubes are seen in the PMMA domains containing the clays. Furthermore, the *s*-MWCNTs also segregate into the interfaces in the same way that the clays do.

Therefore, this different distribution of MWCNTs could cause the two types of MWCNTs to be entangled differently in the melt matrix. Considering with the TEM images of both MWCNTs, we postulate an evolution of the microstructures in Scheme 10.1. We show that the *l*-MWCNTs cannot be entangled with one another in PMMA domains because they are too long to reside there. However, the *s*-MWCNTs can be dispersed in both the domains and the matrix because of their length. As temperature increases, a strong entanglement of *l*-MWCNTs is formed only in the PS matrix. On the other hand, because the *s*-MWCNTs are well dispersed in both phases, the entanglement networks are well established in the whole matrix. Therefore, the degree of entanglement of the *s*-MWCNTs can

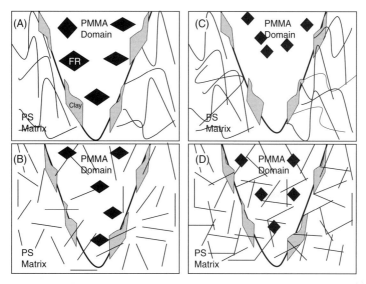

Scheme 10.1 A model of the dispersion of either *l*-MWCNTs or *s*-MWCNTs in a PS/PMMA/FR/C20A blend.

be much higher than that of the *l*-MWCNTs. As a result, the modulus of the *s*-MWCNT blend is higher than for the *l*-MWCNT blend, which is shown in the previous section.

Furthermore, the clays can be also entangled with the MWCNTs once the blends decompose at high temperatures. In the case of the *l*-MWCNT blend, the clays at the interfaces have relatively many contacts with the *l*-MWCNTs. However, in the case of the *s*-MWCNT blend, because the physical contact with the clays at the interfaces is less, the entanglements of *s*-MWCNTs may provide higher heat conductivity, which can be responsible for higher heat dissipation. The different heat conductivity can lead to different rates of volatilization of the FR particles, which can be confirmed in Figure 10.15. In the figure, Br is a primary indicator of the distribution of FR particles because of its higher electron density: the dots and the bright dots correspond to the Br-rich areas in the EDAX map and in the SEM image, respectively. These areas are also representative of the PMMA domains, because the TEM images show that most of the FR particles are in the domains. From the figure we can see that the Br particles are seen in both the *l*-MWCNT and *s*-MWCNT samples after 5 s combustion. However, most of the Br particles in the *s*-MWCNT blend are volatile after 15 s. In contrast, in the case of the *l*-MWCNT blend, the Br particles are still visible in the 15 s image. These different results can be attributed to the smaller Br particles in the *s*-MWCNT samples, which could be more easily volatilized into the flame zone. This different rate of volatilization of the FR particles may support the model, in that the better dispersion of the *s*-MWCNTs can create entanglement networks in which heat transfer is faster, so that the FR particles are released earlier to the flame zone than the FR particles in the *l*-MWCNTs blend.

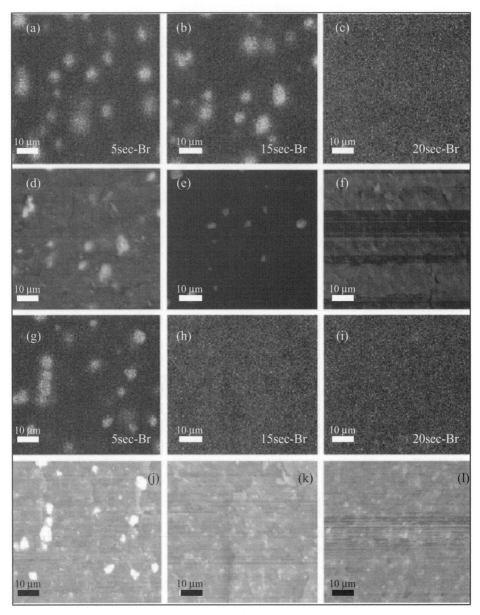

Figure 10.15 Br Mapping and SEM images: (a)–(f) PS/PMMA/DB/AO/Cloisite 20A/*l*-MWCNT (70/30/15/4/3/2) samples, (g)–(l) from the PS/PMMA/DB/AO/Cloisite 20A/*s*-MWCNT (70/30/15/ 4/3/2) samples. Scale bar = 10 μm. Reproduced from Ref. [11] with permission from the American Chemical Society.

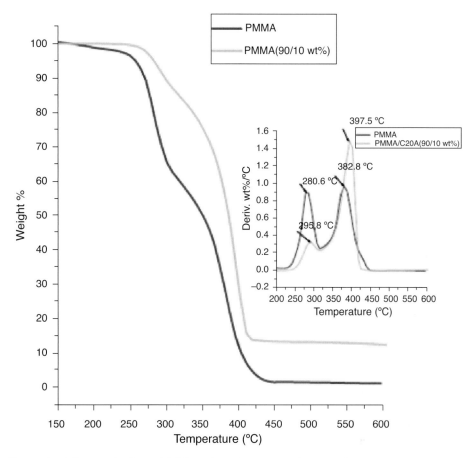

Figure 10.16 Curves of TGA and DTG (an insert image) as a function of temperature from PMMA and PMMA/clay nanocomposites. Reproduced from Ref. [7] with permission from Elsevier.

10.3.4 Thermogravimetric analysis

To determine an order of decomposition in polymer–clay nanocomposites containing the FR particles, performed thermogravimetric analysis (TGA) on the PMMA homopolymers and PS/PMMA blends with and without inorganic additives.

PMMA and PMMA/Cloisite 20A

In Figure 10.16, we show the TGA data for PMMA and PMMA with 10% Cloisite 20A nanocomposite. From the figure we can see that pure PMMA has two main decomposition temperatures, the first around 250–325 °C for degradation of the unsaturated end groups, and the second at 350–450 °C for the depolymerization of random scission [21, 22]. In the insert, the derivatives of the spectra are plotted. The first peak in the derivative is significantly reduced when the clays are added, which may indicate that the end groups are

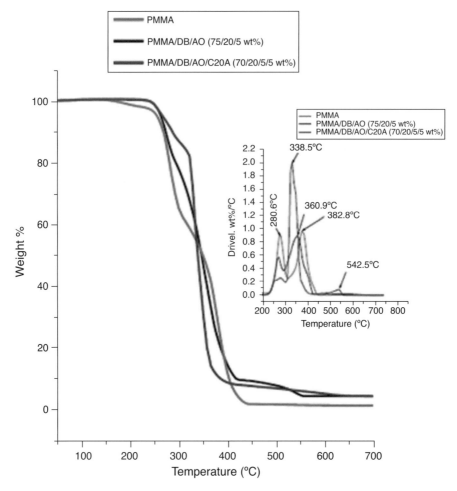

Figure 10.17 Curves of TGA and DTG (an insert image) as a function of temperature from PMMA and PMMA/DB/AO without and with clay nanocomposites. Reproduced from Ref. [7] with permission from Elsevier.

more stable thermally than those of pure PMMA. The second peak ($T = 382.8\,°\text{C}$) is greatly increased and then observed at the higher temperature ($T = 397\,°\text{C}$). This improvement of thermal stability could result from the fact that the PMMA was strongly absorbed on the clay surfaces. Hence, for the depolymerization of random scission to occur, more heat is needed to decouple the polymer chains from the clay surfaces.

PMMA/FR/Cloisite 20A

In Figure 10.17 we plot the TGA data of the PMMA homopolymer when both the FR particles and the clays are present and the nanocomposite was graded as V0 under the UL-94 criteria. In the insert, the derivatives of the spectra are plotted, which accentuate the areas of fastest degradation and allow the determination of the temperatures with greater precision. From the figure it can be seen that in the absence of clays the sample has several decomposition peaks, meaning that the different components decompose at different temperatures and times. As a result, the gas-phase reactions with the FR particles may not be effective in the flame zone. However, the TGA trace of PMMA/FR/Cloisite 20A nanocomposite shows that most of the mass of the nanocomposite decomposes at the same time and temperature, which is more clearly seen in the derivative. Based on the TGA data, we can then make a tentative model for the synergy that exists between the FR and the clays. The TGA data show that in the absence of clays the decomposition temperatures are different for the PMMA and the FR components. Because the FR reaction occurs mainly in the gas phase, both components must be vaporized at the same time in order for the hydrogen halide reactions to occur and extinguish the flame. The addition of clays to the PMMA/FR system decreases the amount of the first peak and increases the second peak, indicating a synergistic effect in which all components vaporize in the same temperature ranges. This factor may explain the higher yield of gas phase combustion products from the PMMA/FR/Cloisite 20A, which was reported in the mass spectrometry–GC analysis [8].

PS/PMMA and PS/PMMA/Cloisite 20A

The effect of Cloisite 20A on the decomposition mechanism of PS/PMMA polymer blends is seen in the TGA analysis shown in Figure 10.18. For a comparison with the experimental TGA trace, we also plot a calculated TGA spectrum of PS/PMMA/C20A if PS and PMMA/Cloisite 20A were to go into the vapor phase separately. From the figure we can see that the addition of clays increases the onset temperature of pure PMMA to $T = 397.2\ °C$, which is a result of the average decomposition temperature of the two components (PMMA, $T = 382.5\ °C$, and PS, $T = 414.5\ °C$). This increase can affect the overall TGA trace of the PS/PMMA blend when the clays are added. From the figure we can see that an expected TGA trace from the combination of [PS] with [PMMA/C20A] (black solid line) is best fit to the experimental TGA trace among the other calculated traces: solid line for the individually added [PS]/[PMMA]/[C20A] blend and solid line for the combination of [PS/PMMA] with Cloisite 20A. This result is consistent with the TEM images, in which the PMMA polymer chains are strongly absorbed onto the clay surfaces, and most of the clays segregate in the PMMA domains. Therefore, the distinct phases, PS and PMMA/C20A, could decompose individually. Furthermore, it is interesting that the decomposition temperature of the PS/PMMA/C20A appears in the temperature ranges of FR decomposition.

PS/PMMA/FR and PS/PMMA/FR/Cloisite 20A

Because we have seen that, in the case of PMMA/FR/Cloisite 20A, all components go into the flame zone at the same temperature and time, the addition of both FR particles and clays to the PS/PMMA blend may lead to similar decomposing processes. To investigate the decomposition mechanism, in Figure 10.19, we plot the TGA experimental and calculated traces. From the figure we can see that the simulation of a combination PS/PMMA

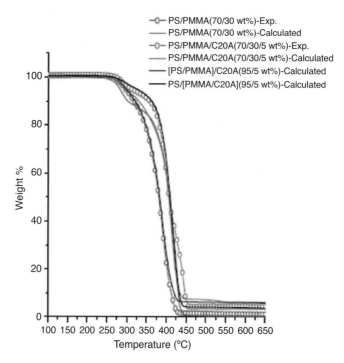

Figure 10.18 Experimental and calculated curves of TGA from PS/PMMA and PS/PMMA/clay nanocomposites. Reproduced from Ref. [7] with permission from Elsevier.

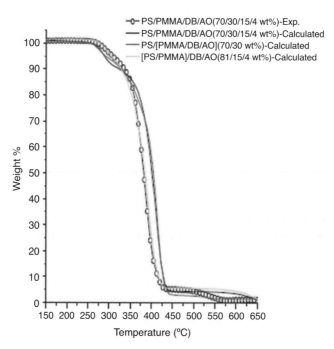

Figure 10.19 Experimental and calculated curves of TGA from PS/PMMA/DB/AO nanocomposites. Reproduced from Ref. [7] with permission from Elsevier.

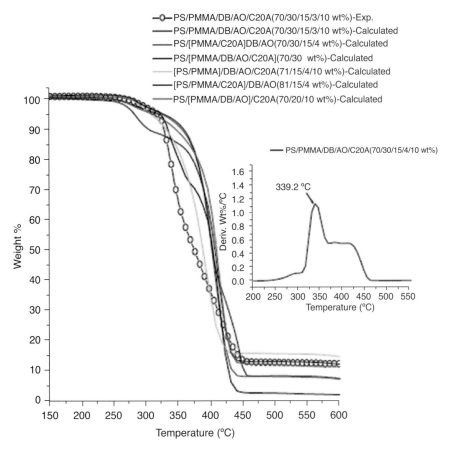

Figure 10.20 Experimental and calculated curves of TGA from PS/PMMA/DB/AO/C20A nanocomposites and DTG (an insert image). Reproduced from Ref. [7] with permission from Elsevier.

blend with FR components (solid line) is the best fit to the experimental trace of the PS/PMMA/DB/AO blend. This is consistent with the TEM images, where the FR particles exist as a third phase in the melt matrix. Therefore, the decomposition of the PS/PMMA blend is not affected by the addition of FR particles. However, in Figure 10.20, the TGA experimental trace of the blend containing 18 wt% FR particles and 10 wt% clays is plotted with several simulated TGA traces. From the figure we can see that the experimental trace is not a good fit for the expected TGA trace, which would be a superposition of the separate PMMA/DB/AO/clay trace and the PS trace, as observed in the absence of the FR. Hence, it was postulated that a synergistic effect may occur where the clay is catalyzing different reactions in the gas phase, which may be more efficient at consuming the decabrome. This synergy may be also observed in the derivatives shown in the insert, where a long plateau region appears after the large peak, mainly from the PMMA/DB/AO/C20A phase.

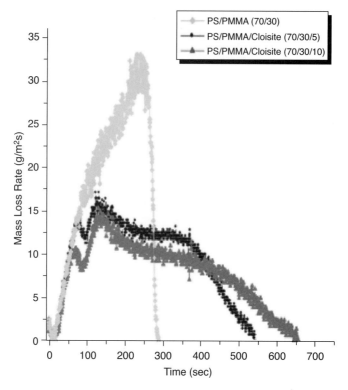

Figure 10.21 The effect of addition of Cloisite 20A on MLR of PS/PMMA at 50 kW/m² under nitrogen. Reproduced from Ref. [7] with permission from Elsevier.

10.3.5 Cone calorimetry

Combustion of polymeric materials involves a complex process, where both condensed and vapor-phase reactions occur at exposed surfaces that are sources of flame and/or thermal radiation of the most common parameters measuring the flammability of polymeric materials are heat release rate (HRR) and mass loss rate (MLR) from cone calorimetry. Recently, nanocomposites containing nanoparticles have been of great interest in the composite industries. In particular, polymer blends containing clays have not been comprehensively studied for their flammability, in spite of the fact that most plastic products are made out of blends of more than two polymer. Furthermore, because the dispersion of nanoparticles is a key factor in determining the HRR and MLR of nanocomposites [23–26], we investigated correlations between flammability and dispersion in air and under nitrogen, especially for polymer blends.

PS/PMMA/C20A polymer blends

In Figure 10.21 we plot the MLR of PS/PMMA blends with either 5% or 10% clays by weight under nitrogen. For comparison, we also plot the MLR of a pure PS/PMMA polymer blend in the same graph. From the figure we can see that the MLR trace of the PS/PMMA blend rapidly increases up to 250 s. After that, a decrease occurs between 250 and 300 s. This is a typical feature of a MLR in noncharring polymers, where all components vaporize without carbonaceous chars. However, when Cloisite 20A is added, the MLR trace of the blend significantly changes after 140 s, at which point the MLR is greatly affected by the addition of clays. The rate of decrease of the MLR seems to be proportional to the clay concentration: when 10% clays are added, the MLR trace becomes steeper. Additionally, the blend with 10% clays is consumed for longer times, up to 650 s. Thus, the addition of clays to the blend shows two characteristic regions in the mass loss rate: (1) a noncharring region, where the MLR is determined by the net heat flux; (2) a region of char growth, where the mass loss rate is adjusted by the amount of char yield in the melt matrix. This tendency of the MLR is seen in polymer–clay nanocomposites, where the clays are intercalated and/or exfoliated. The thermal decomposition of nanocomposites can be explained by a modified model of the relationship between MLR and net heat flux proposed by Quintiere and co-workers [27, 28],

$$m'' = \frac{Q_{\text{net}}}{L_{\text{g}}/(1-y)},$$ (10.1)

where Q_{net} is the net heat flux into the exposed surface, which is composed of an external heat flux and a reradiation from the surface in the gasification test, m is the rate of mass loss, L_{g} is heat of combustion, and $1-y$ is a char fraction. Because the pure blend undergoes a noncharring process, the MLR curve is controlled by the net heat flux, where there is no reradiation from the exposed surface. Therefore, all heat flux can go through the PS/PMMA blend and start vaporizing the melt matrix after ignition. For the MLR of the blends with the clays, small dips appear at around 100 s, which may be evidence of a weak protective layer on the surface. The trace gradually decreases after 140 s. It is believed that the net heat flux is reduced by additional reradiation from the solid char layers. Therefore, the mass burning rate of the blend with 10 wt% clays is much lower than that of the blend with 5 wt% clays because the addition of 10 wt% clays makes better char layers for reradiation, which could result in a lower MLR curve. In Figure 10.22, we also plot the HRR and the MLR of the blends with clays in air. From the figure we can see that the PS/PMMA blend has the highest peak heat release rate (PHRR) in this study, 1507.8 kW/m^2. However, when the 10% clays are added, the PHRR is reduced to 563.1 kW/m^2 at 75 s. The MLR of the PS/PMMA blend is also reduced, which is shown in Figure 10.22 (b). Therefore, the intercalated clays in the polymer blend can reduce the flammability of nanocomposites either in the inert gas or air condition, forming constant regions on the MLR and HRR.

PS/PMMA/FR and PS/PMMA/FR/Cloisite 20A

In order to investigate the effect on the HRR and MLR of PS/PMMA blend when either FR particles alone or combination FR particles withclays are added, we also plot the results of HRR and MLR from the PS/PMMA/FR and PS/PMMA/FR/C20A blends in Figure 10.22. Data are summarized in Table 10.3. From the figure we can see that the addition of 13% FR

Table 10.3 *Results of cone calorimetry of PS/PMMA or PS/PMMA/DB/AO without the clay and with the clay nanocomposites*

Sample (concentration wt%)	Peak HRR (KW/m^2)	Average HRR (KW/m^2)	Average MLR (g/s)	Lenition time (s)
PS/PMMA(70/30)	1507.8	792.9	0.123	25 ± 1
PS/PMMA/Cloisite 20A (70/30/10)	563.2	288.6	0.050	21 ± 2
PS/PMMA/DB/AO (70/30/10/3)	721.0	386.2	0.123	15 ± 1
PS/PMMA/DB/AO/Cloisite 20A (70/30/10/3/5)	320.2	228.7	0.072	18 ± 2
PS/PMMA/DB/AO (70/30/15/4)	570.0	287.5	0.116	12 ± 2
PS/PMMA/DB/AO/Cloisite 20A (70/30/15/4/3)	375.0	211.5	0.111	15 ± 2
PS/PMMA/DB/AO/Cloisite 20A (70/30/15/4/10)	219.0	157.8	0.057	18 ± 2

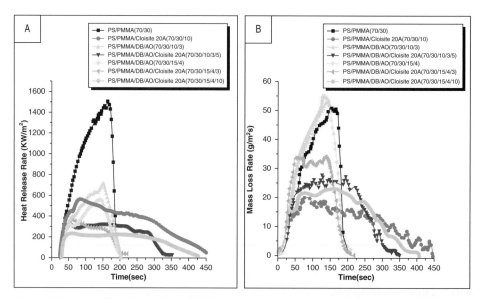

Figure 10.22 The effect of addition of DB, AO, and Cloisite 20A on HRR (A) and MLR (B) in PS/PMMA/DB/AO at 50 kW/m² in air. Reproduced from Ref. [7] with permission from Elsevier.

particles reduces the PHRR of the pure polymer blend by one-half, and even more when 19% FR particles are added. It is well known that halogenated FR particles mostly react in the gas phase, generating radical species that reduce the exothermic reactions with oxygen [29]. Furthermore, when incomplete volatile products exist in the gas phase, the efficiency of combustion processes is reduced, which can lead to a large reduction in the

HRR. Tewarson [30] proposed a relationship where the parameter (HRP), including the efficiency term χ, is proportional to HRR. Hence, the heat release rate (HRR) is described as

$$HRR = \chi(1 - y)\, H_c / L_g Q_{net}, \tag{10.2}$$

where H_c is heat of combustion and $\chi H_c / L_g$ is the parameter (HRP). This equation also has the term $(1 - y)$ for the fraction of char residues, because MLR is also directly proportional to HRR in noncharring polymers. As expected, the MLR of the PS/PMMA/FR blend is the same as that of the PS/PMMA blend, but slightly higher peaks are observed in the MLR, which may be explained by products with greater volatility being produced when the FR particles are added to the blend. Therefore, the HRR decreases with increase of halogen FR concentration because the addition of FR particles produces incomplete combustible products. From the figures we can see that when the clays are added to the PS/PMMA/DB/AO blend, there are two big changes in the HRR: (1) the elimination of PHRR, which is an important parameter for measuring the material flammability, and (2) the formation of plateau regions, where the volatile products containing the FR derivatives diffuse into the flame zone at a constant rate because solid char layers are uniformly formed. Hence, in the equation, the total HRR can be reduced by the combination of the fraction of char residues with the efficiency when the HRP is constant. From Figure 10.22 (A) we can confirm that after 75 s the HRR of PS/PMMA/C20A blends gradually decreases because of an increase of the fraction of char resides. Furthermore, because the addition of FR particles produces incomplete volatile products, the HRR of PS/PMMA/C20A is more reduced. On the other hand, the magnitude of the MLR in PS/PMMA/DB/AO/C20A blends is higher than that of the MLR in PS/PMMA/C20A until the MLR is affected by the efficiency in eq. (10.2).

PC/SAN24/FR and PC/SAN24/FR/C20A

In Figure 10.23, for a PC/SAN24/FR blend, we can see that the addition of clays to the blend has less effect on the HRR, and no effect on the MLR. From the figure we can see that there is still a PHRR at around 120 s, although the PHRR of PC/SAN24/FR is reduced from 516.90 to 390 kW/m^2. Furthermore, because PC is a good charring polymer and decomposes as an intumescing barrier to heat fronts, the HRR of the blend with FR particles moderately decreases after the PHRR appears at around 120 s. On the other hand, when clays are added to the PC/SAN24/FR blend, the HRR trace seems to be one of noncharring polymers, which is also confirmed by the MLR trace, where there is a big peak at the end of combustion [31]. This may be caused by the fact that the surfactants on the clays can facilitate the evolution of CO_2 and CH_4 from the carbonate groups in PC, which may hinder the rearrangement processes for producing intumescing chars [18]. Therefore, even though the addition of clays improves the compatibility of PC/SAN24 blends and increases the degree of dispersion of the FR particles in the blends, it is believed that the surfactants in the clays could greatly affect the combustion of the blend.

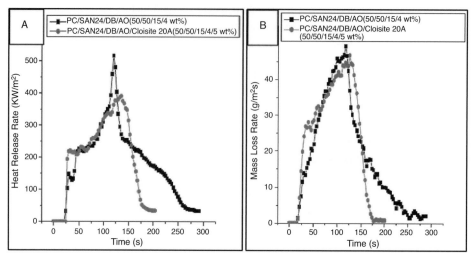

Figure 10.23 The effect of addition of DB, AO, and Cloisite 20A on HRR (A) and MLR (B) in PC/SAN24 at 50 kW/m^2. Reproduced from Ref. [7] with permission from Elsevier.

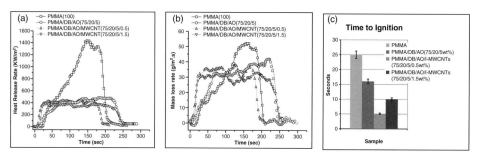

Figure 10.24 The effect of addition of *l*-MWCNT on (a) HRR, (b) MLR, and (c) time to ignition in PMMA/DB/AO(75/20/5) at 50 kW/m^2. Reproduced from Ref. [11] with permission from the American Chemical Society.

PMMA/FR/l-MWCNTs and PS/PMMA/FR/Cloisite 20A with either s-MWCNTs or l-MWCNTs

The combination of FR particles with MWCNTs can also affect both HRR and MLR traces. From Figure 10.24 (A, B) we can see that the PHRR of pure PMMA is decreased by nearly a factor of four to a small plateau, around 490 kW/m^2, when halogenated FR particles are added, whereas only small changes are observed in the MLR. This behavior is characteristic of a system where the flame retardant additive suppresses the combustion of hot gases. When 0.5 or 1.5 wt% MWCNTs are added to the PMMA/FR blend, only a small additional reduction in the HRR is observed. If we plot the early portion of the HRR curves, we can measure the changes in time to ignition induced by the nanoparticles. From the figure we can see that the time to ignition decreases slightly from 24 s for the pure

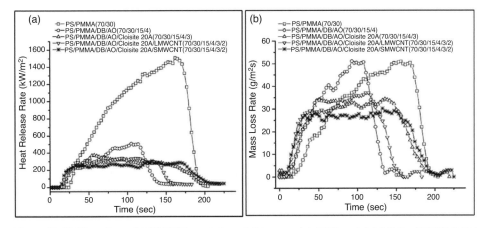

Figure 10.25 The effect of MWCNTs and clay additives on (a) HRR and (b) MLR of PS/PMMA/ DB/AO at 50 kW/m^2. Reproduced from Ref. [11] with permission from the American Chemical Society.

polymer to 19 s when the FR is added. When 0.5 wt% of the MWCNT is added, the time to ignition decreases even further to 5 s. If 1.5 wt% MWCNT is added, though, the time to ignition doubles to 10 s. A possible explanation for this phenomenon may be that all the added particles have lower heat capacities than the polymer. If the particles are dispersed within the matrix, the approaching heat front causes a sharper rise in temperature within the particles. The higher heat capacity of the matrix does not permit the excess heat to dissipate, leading to early combustion in the hotter area near the particles. As can be seen from the table, where we list the heat capacities of the components, the greatest decrease in time to ignition is caused by the particles with the lowest heat capacity. On the other hand, if the particles percolate and are in physical contact, the increased thermal conductivity should dissipate the heat from the approaching front from the interior to the exterior of the sample, thereby decreasing the time to ignition and subsequently the HRR and MLR. This can be seen from the figures, where this effect is seen to occur when the concentration of the MWCNT is increased. In this case, the MWCNT are long and can become entangled within the matrix, thereby dissipating hear.

In Figure 10.25, we plot the HRR and MLR of PS/PMMA blends. From the figure we can see the same effect as for the PMMA homopolymer, that when the FR particles are added the magnitude of HRR is greatly reduced, whereas the magnitude of MLR is not affected. The addition of Cloisite 20A clays (3 wt%) causes an even greater reduction. The change in thermal behavior of the PS/PMMA/FR blend with additional MWCNT is seen to depend on the length of the MWCNT. When long nanotubes are added the HRR and MLR increase, whereas when short nanotubes are added a significantly larger decrease is observed, which in fact allows the blend to qualify for a UL-94 V0 rating. In describing the added synergy between the MWCNT and the clays, we showed that the FR particles were more attracted to the MWCNT, thereby allowing the clays to localize at the polymer

Time to Ignition

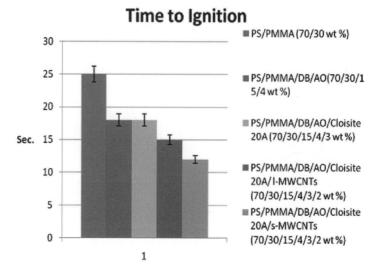

Figure 10.26 The effect of addition of MWCNTs and clays on time of ignition in HRR at 50 kW/m². Reproduced from Ref. [11] with permission from the American Chemical Society.

interfaces, whereas the MWCNT dispersed the nanoparticles. It is interesting to note that only the blend with short MWCNT was able to pass the UL-94 V0 diagnostic test. As before, the longer tubes are immobilized within the matrix, whereas the shorter tubes can diffuse between the phases, thereby providing more protection against a heat front.

In Figure 10.26 we plot the time to ignition for the blend. Here we see that the ignition time decreases from the bulk value regardless of the nanoparticle combination or concentration. In this case, the addition of the clays prevents intimate contact between the MWCNT, which in turns prevents the propagation of heat. From these data we conclude that to improve the time to ignition, conductive nanoparticles must be added, in order to increase the contact between the tubes and decrease the percolation threshold concentration.

This result may be explained by the model, in which the degree of heat dissipation is a key factor determining the self-extinction. Hence, we introduce a term corresponding to the degree of heat dissipation: thermal diffusivity α. The equation HRR is modified as follows:

$$\text{HRR} = \chi(1 - y)\alpha\frac{H_c}{L_g}m''. \tag{10.3}$$

From eq. (10.3), the thermal diffusivity is proportional to the HRR, where the thermal diffusivity would cause a small change in the HRR when no carbonaceous chars are formed during combustion. This can be confirmed by the results for HRR in PMMA/FR/MWCNTs. Because the addition of MWCNTs does not produce carbonaceous chars, the char friction can be neglected in the equation. Hence, the HRR is controlled by the term of thermal

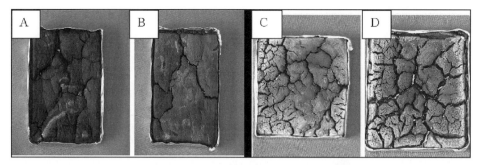

Figure 10.27 Images of residues after cone calorimetry under nitrogen: (A) PS/PMMA/C20A (70/30/5 wt%), (B) PS/PMMA/C20A(70/30/10 wt%). Images of residues after cone calorimetry in air: (C) PS/PMMA/DB/AO/C20A(70/30/10/3/5 wt%), (D) PS/PMMA/DB/AO/C20A(70/30/15/4/ 10 wt%). Reproduced from Ref. [7] with permission from Elsevier.

diffusivity. Therefore, good heat dissipation is a key parameter for designing self-extinguishing nanocomposites.

10.3.6 Analysis of chars

Morphology

The formation of chars is a process of thermal degradation, which can have an important effect on HRR and MLR. Char is a carbonaceous material, which remains after complete thermal decomposition of the polymer. Char is not flammable and can absorb heat from the incoming flux, thereby interrupting the diffusion of volatile products into a flame fed by the burning matrix. "Good char" is char that produces a uniform layer and an effective barrier, and good char–forming polymers are more flame retardant than those that do not produce chars when heated. In addition to affecting the MLR and HRR, addition of clays to polymer matrices has been shown to facilitate the formation of chars, even in cases, such as PMMA and PS, where no char production occurs [9]. Pack *et al.* therefore investigated the effects of clays on the mechanical and morphological properties of chars in polymer blend and homopolymer matrices [7, 10, 11].

Kashiwagi *et al.* [8] have compared the morphology of chars produced from homopolymer nanocomposites with that of single or multiwalled nanotubes and those with Cloisite clays. They concluded that, in the absence of additional FR formulations, the superior flame retardant response of the nanocomposites containing carbon nanotubes was in part because of the quality of the chars. The carbon nanotubes formed a dense network that was able to support a uniform char layer, whereas the clays did not pack as well and produced a layer with many cracks, through which gas bubbles from the interior could easily penetrate [9]. In Figure 10.27, we show the optical images of the residues after cone calorimetry of a PS and PMMA blend sample. From the figure we can see that, regardless of any other additives, robust char is only formed when clay is added. This explains the large reduction in MLR

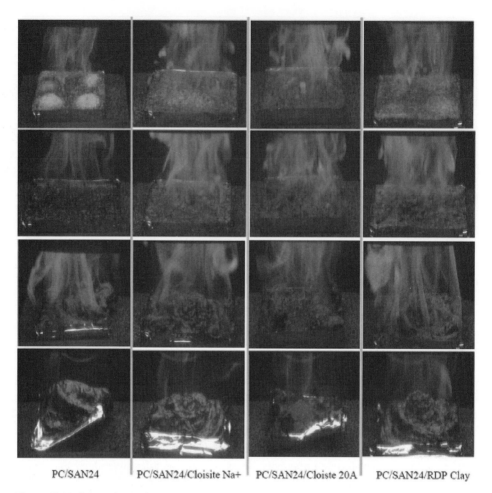

PC/SAN24   PC/SAN24/Cloisite Na+   PC/SAN24/Cloiste 20A   PC/SAN24/RDP Clay

Figure 10.28 Selected video images at 30, 100, 230, and 300 s from the gasification tests: PC/SAN24 (50/50 wt%) – the first column, PC/SAN24/Cloisite Na+ (50/50/5 wt%) – the second column, PC/SAN24/Cloisite 20A (50/50/5 wt%) – the third column, and PC/SAN24/RDP MMT-Na+ (50/50/5 wt%) – the fourth column. Reproduced from Ref. [10] with permission from the American Chemical Society.

and HRR that is always observed when clays are added. On the other hand, only the blends where the FR was added satisfied the UL-94 V0 criteria. Yet if we compare the appearance of the chars, we find that the ones formed with flame retardant appear to have more cracks. These results indicate that we must be careful in the interpretation of the cracks we observe in the cold chars. Video images of the samples as they are heated in the gasification test show that the chars form a shell, which becomes inflated by the gases in the interior of the sample, as shown in Figure 10.28. The best flame retardancy is obtained from the chars that are able to withstand the internal pressure of the expanding gases without breaking, or

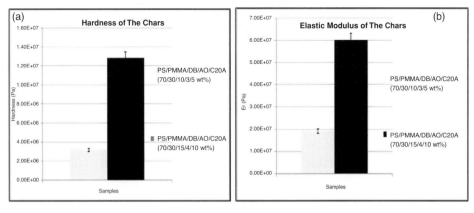

Figure 10.29 (a) The hardness and (b) the reduced modulus of the chars. Reproduced from Ref. [11] with permission from the American Chemical Society.

the chars that have the highest elasticity. Once the sample has cooled, the gases contract quickly and all chars collapse. The cracks we observe in the cold chars are formed during this collapse and do not necessarily correlate with cracks observed in the optical images after removal from the calorimeter.

Mechanical properties

In addition to the chemical reactions involved in the char, it is also important to measure the mechanical properties. Because chars are very brittle, it is not possible to mold or shape them and apply the standard mechanical analysis methods, such as tensile testing, DMA, or impact and hardness tests. Microscopically, though, chars have microscale intact regions that can be measured using new techniques for nanoscale measurements. A preferred method is nanoindentation, which can be used to scan localized sections. The moduli and hardness of two samples containing both clays and FR are shown in Figure 10.29. From the figure we can see that the modulus and the hardness increase nearly threefold when the clay concentration is doubled from 5% to 10%. The ability to self-extinguish is clearly due to the increased impact resistance of the char coating. Thus far, this is the only group that has applied this methodology to study chars. More microscopic measurements are needed to obtain a comprehensive understanding of char function at the molecular level.

10.4 Summary

Flame retardance is a process for disrupting burning, which has several stages involving physical and chemical reactions. The addition of flame retardant (FR) agents is a conventional method for achieving flame retardant materials and composites. However, for the FR to be effective, it must have maximum interaction with the host matrix, and hence, it must be properly dispersed. In addition, if the FR functions best in the vapor phase, it is also important that it enter the vapor phase at the same time as the matrix. In most cases, these conditions are difficult to meet. It is difficult to obtain good dispersion because most FR are immiscible with the polymer matrix, which can lead to formation of agglomerated FR agents. Most FR agents are more volatile than the polymer matrix, and hence large amounts are usually added in order to obtain even a small overlap of the vapor phases during heating. Polymer blends pose additional challenges that exacerbate these problems. In that case the FR will preferentially segregate in one of the phases. When the sample is heated even at a relatively low temperature (smoldering), phase separation progresses, increasing the rate of agglomeration. Adding large amounts of FR is not a practical solution, because it can degrade the mechanical properties. The addition of high–aspect ratio nanoparticles, such as organoclays and nanotubes, is known to improve the mechanical and thermal properties of polymeric materials when their interactions with the polymer matrix are favorable. It was also been previously established that when present in blends, these types of nanoparticles can be interfacially active, because they formed in situ grafts that were able to increase the degree of compatibilization of the polymers in host matrix. As a result, these particles were also combined with FR, and it was demonstrated that they had a synergistic effect. In homopolymers, they catalyzed the formation of chars, forming a penetration barrier to the release of gases from both halogenated FR and the polymer matrix, thereby greatly increasing the area of interaction, and increased the viscosity of the molten polymers, reducing dripping. The synergy was demonstrated by the large decrease in the HLR and MLR of the otherwise volatile polymer PMMA, which could now be rendered flame retardant according to the stringent UL-94 V0 criteria with the addition of either clays or MWCNT. Similar results were also obtained with other polymers, as long as favorable interactions existed between the host and nanoparticles, which allowed them to be well dispersed.

The results in blends were more complicated. Clays indeed prevented phase segregation, but to a lesser extent when the FR was added. TEM indicated that, even though the dispersion of the FR was improved by adsorption onto the clay surfaces, it nevertheless displaced the polymers, and competed with the in situ graft formation. As a result, clays or nanotubes alone improved the flame retardant properties of blends, but only to a V2 designation with concentrations below 10%. A new, additional synergy was obtained when nanotubes and clays were combined. In this case, the FR segregated preferentially to the nanotubes, where it was dispersed throughout the sample, whereas the clays retained their ability to form in situ grafts, which stabilized the entire systems against phase segregation. The combination of the two components was so favorable that a rating of UL-94 V0 could be obtained with smaller total amounts (less than 5%) of particle additives than with either of the nanoparticles.

The addition of nanoparticles also highlighted two other aspects. First, numerous sources have reported that despite the improved HRR and MLR characteristics imparted by clay, or carbon nanotubes, the ignition time of the matrix would invariably decrease in proportion to the quantity of particles added. This was postulated to be due to the significantly lower heat capacity of the particle inclusions, which allowed them to heat faster than the host, and reach the ignition temperature faster, igniting the matrix earlier. When only MWCNTs were added, the time to ignition was shown to decrease quickly with increasing concentration, till percolation was achieved. Afterward, the time to ignition began to increase with temperature. At low concentrations the particles were isolated within the low–thermal conductivity polymer matrix, and no effective mechanism existed for dissipating the accumulated heat. When percolation was reached, the low specific heat and high thermal conductivity of the tubes became assets, enabling them to conduct heat away from the core of the sample and increasing the time to ignition. This effect was not observed when mixed clay/MWCNT was used, and it was postulated, based on TEM images, that the clays in the matrix became intercalated within the MWCNT, preventing direct contact, and reducing the conductivity. This highlighted the need to develop more thermally conductive high–aspect ratio nanoparticles.

Finally, because the chars are central to flame retardancy, the role of clays and other nanoparticles in the char morphology was also investigated. SEM images indicated that the clays were a large component of the chars. Because the chars formed at the surfaces of the samples, these images indicated that some segregation of the clays to the free surface probably occurred during the burning process. The segregation was driven by the lower surface energy of the clays, and its degree could be controlled by tuning the free energy of the clays to that of the polymer matrix. In addition to thermal stability, the role of chars is also to prevent hot gases from the interior from escaping, and deflect the heat front. As a result, the mechanical properties of the hot chars become important. Previous work that analyzed optical images of the cracks formed was inconclusive, because cracks can form at any time when the char is cooled rapidly. Newer results were presented where microindentations were performed that showed that the additional requirement for good flame retardancy was chars of high impact modulus and strength, which could contain the high internal pressure generated by the hot gases without cracking.

References

1. A. C. Balazs, T. Emrick, and T. P. Russell, Nanoparticle–polymer composites: Where two small worlds meet. *Science*, 314 (2006), 1107–10.
2. M. Si, M. Goldman, G. Rudomen, M. Y. Gelfer, J. C. Sokolov, and M. H. Rafailovich, Effect of clay type on structure and properties of poly(methyl methacrylate)/clay nanocomposites. *Macromolecular Materials and Engineering*, 291 (2006), 602–11.
3. T. Kashiwagi, F. Du, J. Douglas, K. I. Winey, R. H. Harris, and J. R. Shields, Nanoparticle networks reduce the flammability of polymer nanocomposites. *Nature Materials*, 4 (2005), 928–33.
4. M. Gelfer, C. Burger, A. Fadeev, I. Sics, B. Chu, B. S. Hsiao, A. Heintz, K. Kojo, and S.-L. Hsu, M. Si, and M. Rafailovich, Thermally induced phase transitions and morphological changes in organoclays. *Langmuir*, 20 (2004), 3746–58.
5. M. Si, T. Araki, H. Ade, A. L. D. Kilcoyne, R. Fisher, J. C. Sokolov, and M. H. Rafailovich, Compatibilizing bulk polymer blends by using organoclays. *Macromolecules*, 39 (2006), 4793–801.

6. Y. Wang, Q. Zhang, and Q. Fu, Compatibilization of immiscible poly(propylene)/ polystyrene blends using clay. *Macromolecular Rapid Communications*, 24 (2003), 231–5.

7. S. Pack, M. Si, J. Koo, J. C. Sokolov, T. Koga, T. Kashiwagi, and M. H. Rafailovich, Mode-of-action of self-extinguishing polymer blends containing organoclays. *Polymer Degradation and Stability*, 93 (2009), 306–26.

8. T. Kashiwagi, M. Mu, K. I. Winey, B. Cipriano, S. Raghavan, S. Pack, M. Rafailovich, Y. Yang, E. Grulke, J. Shields, R. Harris, and J. Douglas, Relation between the viscoelastic and flammability properties of polymer nanocomposites. *Polymer*, 49 (2008), 4358–68.

9. M. Si, V. Zaitsev, M. Goldman, A. Frenkel, D. G. Peiffer, E. Weil, J. C. Sokolov, and M. H. Rafailovich, Self-extinguishing polymer/organoclay nanocomposites. *Polymer Degradation and Stability*, 92 (2007), 86–93.

10. S. Pack, T. Kashiwagi, C. Cao, C. Korach, M. Lewin, and M. Rafailovich, Role of surface interactions in the synergizing polymer/clay flame retardant properties. *Macromolecules*, 43 (2010), 5338–51.

11. S. Pack, T. Kashiwagi, D. Stemp, J. Koo, M. Si, J. C. Sokolov, and M. H. Rafailovich, Segregation of carbon nanotubes/organoclays rendering polymer blends self-extinguishing. *Macromolecules*, 42 (2009), 6698–709.

12. R. M. Aseeva and G. E. Zaikov, *Combustion of Polymer Materials* (Munich: Hanser, 1985).

13. W. C. Kuryla and A. J. Papa, eds., *Flame Retardancy of Polymeric Materials*, Vols. 1–5 (New York: Dekker, 1973–9).

14. M. Lewin, S. M. Atlas, and E. M. Pearce, eds., *Flame Retardant Polymeric Materials*, Vols. 1–3 (New York: Plenum, 1975–82).

15. Sergei Levchik and Charles A. Wilkie, Char formation. In: *Fire Retardancy of Polymeric Materials*, ed. A. E. Greand and C. A. Wilkie (New York: Dekker, 2000), pp. 171–215.

16. D. Wang, K. Echols, and C. A. Wilkie, Cone calorimetric and thermogravimetric analysis evaluation of halogen-containing polymer nanocomposites. *Fire and Materials*, 29 (2005), 283–94.

17. O. Segev, A. Kushmaro, and A. Brenner, Environmental impact of flame retardants (persistence and biodegradability). *International Journal of Environmental Research and Public Health*, 6 (2009), 478–91.

18. S. Levchik and E. Weil, Overview of recent developments in the flame retardancy of polycarbonates. *Polymer International*, 54 (2005), 981–98.

19. Joseph Green, Phosphorus-containing flame retardants. In: *Fire Retardancy of Polymeric Materials*, ed. A. E. Greand and C. A. Wilkie (New York: Dekker, 2000), pp. 147–70.

20. H. Ade, D. A. Winesett, A. P. Smith, S. Qu, S. Ge, J. Sokolov, and M. Rafailovich, Phase segregation in polymer thin film: Elucidations by X-ray and scanning force microscopy. *Europhysics Letters*, 45 (1999), 526–32.

21. H. L. Hampsch, J. Yang, G. Wong, and J. M. Torkelson, Thermal degradation of saturated poly(methyl methacrylate). *Macromolecules*, 21 (1988), 528–30.

22. T. Kashiwagi, A. Inaba, J. Brown. K. Hatada, T. Kitayama, and E. Masuda, Effects of weak linkages on the thermal and oxidative degradation of poly(methyl methacrylates). *Macromolecules*, 19 (1986), 2160–8.

23. J. W. Gilman, C. L. Jackson, A. Morgan, and R. Harris, Flammability properties of polymer-layered-silicate nanocomposites, polypropylene and polystrene nanocomposites, *Chemistry of Materials*, 12 (2000), 1866–73.

24. T. Kashiwagi, R. Harris, X. Zhang, R. M. Briber, B. H. Cipriano, S. R. Raghavan, W. H. Awad, and J. R. Shields, Flame retardant mechanism of polyamide 6–clay nanocomposites. *Polymer*, 45 (2004), 881–91.

25. J. Zhu, A. B. Morgan, F. J. Lamelas, and C. A. Wilkie, Fire properties of polystyrene–clay nanocomposites. *Chemistry of Materials*, 13 (2001), 3774–80.

26. J. W. Gilman, Flammability and thermal stability studies of polymer layered-silicate (clay) nanocomposites. *Applied Clay Science*, 15 (1999), 31–49.

27. B. T. Rhodes and J. G. Quintiere, Burning rate and flame heat flux for PMMA in a cone calorimeter. *Fire Safety Journal*, 26 (1996), 221–40.

28. D. Hopkins and J. G. Quintiere, Material fire properties and predictions for thermoplastics. *Fire Safety Journal*, 26 (1996), 241–68.

29. A. E. Greand and C. A. Wilkie, eds., *Fire Retardancy of Polymeric Materials* (New York: Dekker, 2000), pp. 245–84.

30. A. Tewarson, Heat release rate in diffusion flames. *Thermochimica Acta*, 278 (1996), 19–37.

31. B. Schartel and T. R. Hull, Development of fire-retarded materials-interpretation of cone calorimeter data. *Fire and Materials*, 31 (2007), 327–54.

11

Flame retardant polymer nanocomposites with fullerenes as filler

ZHENGPING FANG[a,b] AND PINGAN SONG[b,c]

[a]Ningbo Institute of Technology, Zhejiang University
[b]Institute of Polymer Composites, Zhejiang University
[c]Zhejiang Forestry University

11.1 Background

Organic polymers are rapidly and increasingly taking the place of traditional inorganic and metallic materials in various fields owing to their excellent properties, such as low density, resistance to erosion, and ease of processing [1, 2]. However, organic polymers are inherently flammable; their use can cause the occurrence of large fires and, consequently, loss of lives and properties. Thus, enhancing the flame retardancy of these organic polymers is becoming more and more imperative with their wider application, especially in fields such as electronics where high flame retardancy is required.[1]

For traditional flame retardants, on one hand, a very high loading is usually needed to meet flame retardancy demands, which can lead to the deterioration of mechanical properties; on the other hand, utilization of flame retardants can cause environmental problems. Fortunately, recent studies have suggested that nanoscale fillers such as clay minerals [3–7], carbon nanotubes (CNTs) [1, 8–12], and layered double hydroxides (LDHs) [13–16] can significantly improve the thermal stability and reduce the flammability of polymeric materials at very low loading levels, usually less than 3% by mass. Such nanoscale fillers are highly attractive because they are environmentally friendly and can simultaneously improve the physical properties and flame retardancy of polymeric materials. Many reports about clays, CNTs, and LDHs as flame retardants for polymer composites have been published in the past few years. Fullerenes have also recently been found to be another candidate as highly effective flame retardants for polymeric materials [17–20]. In this chapter, we will focus on the effects of fullerenes [C_{60}] on the thermal degradation behavior and flame retardancy of polypropylene (PP).

11.2 Flame retardancy of polymer/fullerene nanocomposites

11.2.1 Morphology

The diameter of the fullerene [C_{60}] molecule is 0.71 nm; its crystal size differs with different methods of fabrication [21], from 50 to 120 nm. PP/C_{60} nanocomposites were fabricated

[1] Financial support from the National Natural Science Foundation of China (No. 51073140) is acknowledged.

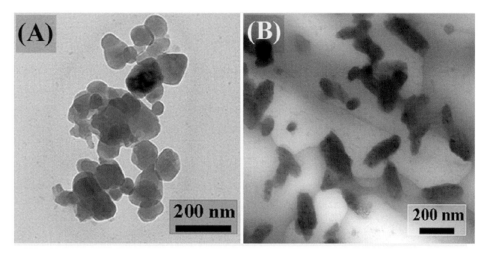

Figure 11.1 TEM images of (A) pristine C_{60} and (B) PP/1.0 wt% C_{60}. Reprinted with permission from Ref. [17]. © 2008 IOP Publishing Ltd.

via melt-compounding, and the composites containing 0.5, 1.0, and 2.0 wt% C_{60} were designated as PF1, PF2, and PF3. Figure 11.1 presents TEM images of pristine C_{60} and PP/1.0 wt% C_{60}. After being blended with PP, most of them are shaped in different forms such as spherical, ellipsoidal, and rodlike, caused by strong shear forces during the melt-compounding process. Shear force will destroy the stack state of C_{60} crystallites and they consequently are arranged to form different types. On the whole, the dispersion of C_{60} in the PP matrix is homogeneous.

11.2.2 Thermal degradation behavior

As shown in Figure 11.2, the thermogravimetric analysis of pristine C_{60} shows that it has very high thermal stability both in air and under nitrogen. The mass loss of C_{60} is only about 2 wt%, after it is heated to 600 °C in nitrogen, whereas in air a 66 wt% residue is obtained and it degrades completely until the temperature reaches 720 °C.

The PP/C_{60} nanocomposites were prepared via melt-blending and their thermal degradation of pure PP and composites was characterized by TG analysis, with curves in air and nitrogen presented in Figures 11.3 and 11.4, respectively. Pure PP experienced one-step degradation both under nitrogen and in air, starting to decomposes at around 263 and 418 °C, and the maximum weight loss occurs at around 338 and 482 °C in air and under nitrogen, respectively. PP is much easier to degrade in air than in nitrogen because the oxygen in air promotes the oxidative degradation of PP through oxidative dehydrogenization accompanied by hydrogen abstraction [22]. Under any conditions, no PP residue remains after it is heated to 600 °C.

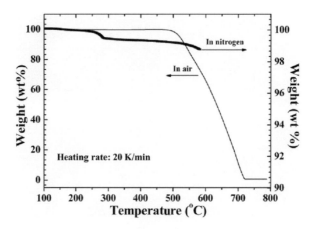

Figure 11.2 TG curves of pristine C_{60} in air and under nitrogen.

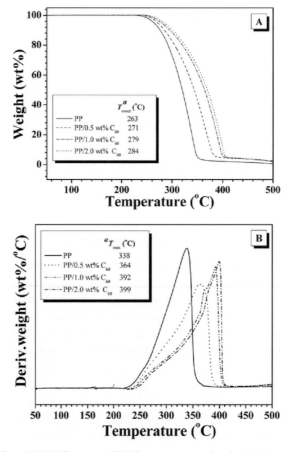

Figure 11.3 (A) TG and (B) DTG curves of PP/C_{60} nanocomposites in air. *Notes:* T_{onset}: the temperature where 5 wt% weight loss occurred; T_{max}: the temperature where maximum weight loss rate took place. Reprinted with permission from Ref. [17]. © 2008 IOP Publishing Ltd.

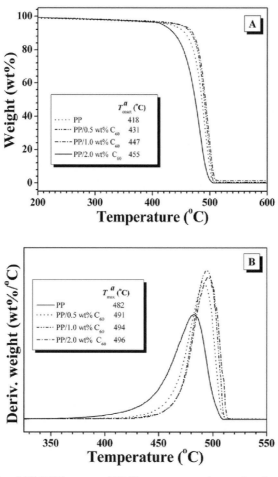

Figure 11.4 (A) TG and (B) DTG curves of PP/C$_{60}$ nanocomposites under nitrogen. Reprinted with permission from Ref. [19]. © The Royal Society of Chemistry 2009.

In air, both the T_{onset} and T_{max} values of PP are shifted to higher temperatures after incorporation of C$_{60}$ into the PP matrix, and both are monotonically and noticeably increased with increasing C$_{60}$ loading level. For instance, the T_{onset} and T_{max} values of PP/0.5 wt% are 271 and 364 °C, respectively, whereas those of PP/2.0 wt% are increased to 284 and 399 °C, respectively, 21 and 61 °C higher than the corresponding temperatures of pure PP. These data indicate that the presence of C$_{60}$ can considerably delay the thermal oxidative degradation of PP. Troitskii *et al.* [23, 24] also found that C$_{60}$ and C$_{70}$ were effective inhibitors for thermal and thermooxidative destruction of PMMA and polystyrene. Kashiwagi *et al.* [9, 10] have observed 1.0 vol% and 2.0 vol% carbon nanotubes (CNTs) can increase T_{max} (second step) of PP by 57 and 78 °C, respectively. However, two-step decomposition processes were also observed, such as for PP/1 wt% CNTs, with a lower first-step T_{max} around 260 °C, even lower than that (298 °C) for PP, and they attributed this complex thermal behavior to the presence of iron particles as a catalyst in MWNTs.

Though incorporating C_{60} can still increase the thermal stability of PP in nitrogen, as well as in air, the extent of improvement is slightly smaller. For example, the T_{onset} and T_{max} values of PP/0.5 wt% are 431 and 491 °C, 13 and 9 °C higher than those of pure PP, respectively. Even at 2.0 wt% loading level, T_{onset} and T_{max} are advanced to 455 and 496 °C, which are 37 and 14 °C higher than the corresponding temperatures for PP. Relative to the CNTs in the same fullerene family, at 2.0 vol% loading in PP, T_{max} is only 12 °C higher than for pristine PP [9].

On the whole, the presence of C_{60} can markedly raise the thermal stability and delay the thermal oxidative degradation of PP, even at very low loading levels. From scientific and application viewpoints, C_{60} is a highly effective antioxidant for PP, and can match CNTs. Thus, it can also be expected to be a highly effective thermal stability agent for other thermoplastics such as polyethylene and polystyrene.

11.2.3 Flame retardancy

Cone calorimetry is widely used to evaluate fire performance. Various parameters are collected from cone measurements, including the time to ignition (t_{ign}), heat release rate (HRR), peak heat release rate (PHRR), time to peak heat release rate (t_{PHRR}), total heat release (THR), average mass loss rate (AMLR), and average specific extinction area (ASEA, a measure of smoke). HRR and PHRR are very important terms in fire safety evaluation. HRR determines the speed of fire development during combustion tests and PHRR can characterize the dangerousness of a fire. It has been noted that microcomposites give minimal reductions in PHRR, whereas nanocomposite formation brings about larger reductions [25].

During the cone experiments, the pure PP resin first melts and subsequently starts to burn, with numerous bubbles bursting on the sample surface. At the end of the test, no residue is left, which is consistent with the thermal degradation process. The HRR curves and detailed data collected from cone experiments are plotted in Figure 11.5 and listed in Table 11.1.

The results show that the presence of C_{60} in the PP matrix not only prolonged the t_{ign} and t_{PHRR} of samples, but also considerably reduced the PHRR. PP/2.0 wt% C_{60} showed a t_{ign} of 36 s and a t_{PHRR} of 82 s, both of which were much longer than for pristine PP; this implied that C_{60} could delay the start of combustion. Furthermore, the PHRR of PP/C_{60} nanocomposites was around 60% (1.0 wt%) and 55% (2.0 wt%) of that of PP, demonstrating that the addition of C_{60} to the PP matrix improved the flame retardancy of PP. In comparison, CNTs and clay were reported to reduce the PHRR of PP by 70% for PP/1.0 wt% CNTs [9, 10] and around 50% for PP/clay nanocomposites [26]. Despite this, incorporating CNTs and clay would lead to a shorter time to ignition of nanocomposites, which suggested that nanocomposites containing CNTs and clay were easier to ignite than PP itself. Although

Table 11.1 *Cone calorimeter data for PP and its nanocomposites at a heat flux of 35 kW/m²*

Samples	t_{ign} (s)	t_{PHRR} (s)	PHRR (kW/m²)	AHRR (kW/m²)	AMLR (g/s)
PP	26	65	1345	282	0.048
PP/0.5 wt% C_{60}	37	76	930	234	0.037
PP/1.0 wt% C_{60}	39	78	810	212	0.033
PP/2.0 wt% C_{60}	38	82	750	186	0.028

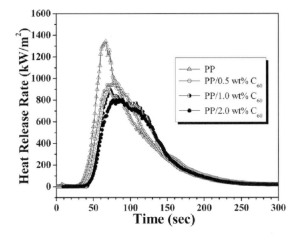

Figure 11.5 Heat release rate curves for PP and PP/C_{60} nanocomposites at a heat flux of 35 kW/m². Reprinted with permission from Ref. [17]. © 2008 IOP Publishing Ltd.

C_{60} was not as good as CNTs and clay in terms of reducing the PHRR, it at least conferred on PP a longer time to ignition, which made PP more difficult to burn. Therefore, it opens up a new strategy for reducing the flammability of polymers.

To obtain visual observations of the flammability of a PP/C_{60} sample, PP and its nanocomposites were heated from room temperature to 360 °C in a muffle and digital photos were taken. As shown in Figure 11.6, after heat treatment, pure PP degraded almost completely without leaving any char residue, whereas PP/C_{60} nanocomposites left a mass of black char residue on the outer surface, and no degradation took place on the inner surface, except for many bubble apertures from which gaseous degradation products evolve. Furthermore, with increased C_{60} loading level, not only was the outer surface of the char residue more compact, but also the number of bubble apertures in the inner surface was much less. All the visual photographs proved that the presence of C_{60} could improve the thermal stability and flame retardancy of PP, which agreed well with the results obtained from TGA and cone calorimetry.

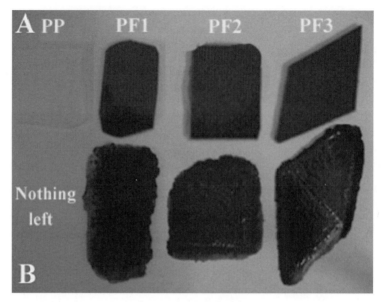

Figure 11.6 Digital photos for PP and PP/C_{60} nanocomposites before (A) and after (B) being heated up to 360 °C in a muffle. *Notes:* PF1, PF2, and PF3 represent PP composites containing 0.5, 1.0, and 2.0 wt% C_{60}. Reprinted with permission from Ref. [17]. © 2008 IOP Publishing Ltd.

11.3 Flame retardation mechanism of fullerenes

11.3.1 Char residue morphology

Generally, for condensed phase flame retardancy, the morphology of the char residue may help to clarify the combustion mechanism. Figure 11.7 presents TEM and SEM microphotographs of PP/1.0 wt% C_{60} nanocomposites after cone calorimetry. Interestingly, compared with the pristine C_{60} crystallites with a size of around 100 nm, after combustion the sizes were much smaller, 30–50 nm, though well proportioned. An explanation was that the combustion of polymers at elevated temperatures would destroy or disorder the stack state of C_{60} crystals, making the size of crystals tend to be the same.

The SEM image of the residue of the PP/C_{60} sample showed that C_{60} crystals only aggregated rather than forming a compact and continuous network. Thus, it was not the char residue that was responsible for the enhanced thermal stability and improved flame retardancy of PP/C_{60} nanocomposites. Because the char residue could not confer flame retardancy on polymers, possible reasons may be concealed in the primary state or in the heating or combustion of polymer materials.

11.3.2 Relationship between flammability properties and rheological behavior

To clarify the flame retardation mechanism, rheological measurements were introduced to investigate the viscoelastic behavior of the nanocomposites. First, the viscoelastic behavior

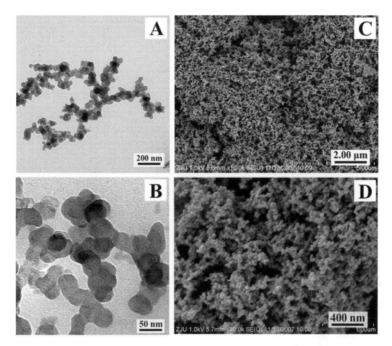

Figure 11.7 TEM images (A, B) and SEM images (C, D) for the char residue of PP/1.0 wt% after cone calorimeter tests. Reprinted with permission from Ref. [17]. © 2008 IOP Publishing Ltd.

of PP and its nanocomposites at 180 °C was studied. Figure 11.8 shows (A) the storage modulus (G') and (B) the complex viscosity (η^*) of PP and its nanocomposites as functions of frequency (ω). Some researchers have found that for nanocomposites containing CNTs [1, 10, 27] and clay [28, 29], both G' and η^* exhibit much larger values than those of parent polymers in the low-ω regime, usually accompanied by a so-called second plateau at a higher loading level. This suggests that the presence of CNTs and clay limits the relaxation and motion of polymer chains because of their spatial geometry. Most researchers attribute this viscoelastic behavior to the formation of CNTs and clay networks in the polymeric matrix. In the high-ω regime, their addition did not significantly affect the G' and η^* values of polymers. In contrast to CNTs and clay, whether in the low- or high-ω regime, not much difference in G' and η^* values was observed for PP/C$_{60}$ nanocomposites, implying that the presence of C$_{60}$ did not greatly affect the movement of polymer segments or the relaxation of molecular chains. Based on this analysis, it was concluded that C$_{60}$ did not form a network in the polymer matrix. Thus, the most likely reason for the improved flammability properties was the process of heating or combustion of polymeric materials.

To prove this inference, both dynamic temperature-scanning and dynamic time-scanning measurements were conducted to investigate the viscoelastic behavior of nanocomposites in the heating process. Figure 11.9 (A) presents the curves of temperature dependence of complex viscosity (η^*) for PP and its nanocomposites. η^* for four systems clearly first

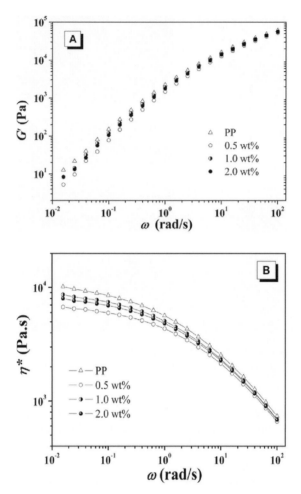

Figure 11.8 Frequency (ω) dependence of the dynamic viscoelastic properties for PP and PP/C_{60} nanocomposites: (A) G' and (B) η^*. Reprinted with permission from Ref. [17]. © 2008 IOP Publishing Ltd.

experienced a decrease, followed by a subsequent sharp increase from T_c (at which η^* starts to increase or a cross-linking reaction occurs) and a gradual reduction after T_d (where the destruction of the cross-linked network takes place). This behavior can be explained as follows. Upon heating, the easier movement of polymer chains results in a decrease in the complex viscosity of the sample or the degradation of polymer. The increase of η^* was due to the occurrence of oxidative cross-linking reactions, which overwhelmed the polymer decomposition. The decrease of η^* was because the degradation overpowered the oxidative cross-linking. In the case of pure PP, it gave a T_c of 320 °C and a T_d of 360 °C, and an S-shaped change of η^* was clearly observed over the whole temperature range. TGA results showed that the initial decomposition temperature (5.0 wt% mass loss occurs) for PP was

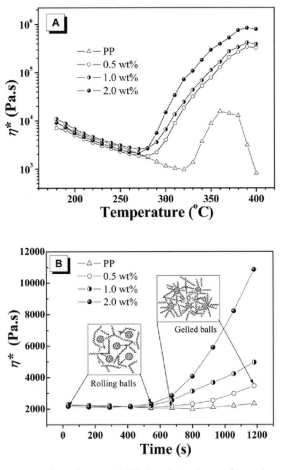

Figure 11.9 (A) Temperature dependence and (B) time dependence of complex viscosity (η^*) for PP and PP/C$_{60}$ nanocomposites. Reprinted with permission from Ref. [17]. © 2008 IOP Publishing Ltd.

about 264 °C; thus in the temperature range from 264 to 320 °C, oxidative cross-linking did not happen or was not strong enough to compensate for the reduction of η^* because of the degradation of the polymer, which led to a continuous decrease of complex viscosity. Above 320 °C, oxidative cross-linking predominated, resulting in a rapid increase of η^*. However, at a sufficiently high temperature, 360 °C for PP, the degradation started to overpower the oxidative cross-linking reaction, resulting in the destruction of the cross-linked network with a reduction of η^* as a signal. For PP/C$_{60}$ nanocomposites, T_c was around 40 °C lower than for pure PP, whereas T_d was 30 °C higher than for PP. These observations imply that another kind of cross-linking reaction occurs in the process of heating PP/C$_{60}$ systems, which speeded up the cross-linking reaction before and after oxidative cross-linking took place.

It is well known that the pyrolysis of polyolefin was a free radical chain reaction through β-scission of polymer chains, whereas the free radicals produced in turn speeded up the degradation of polymers. Many scientists have reported that the C_{60} fullerene molecule has high reactivity to free radicals: one C_{60} molecule can trap from 1 to 34 free radicals [30]. Based on these viscoelastic measurements, it was reasonable to conclude that C_{60} was likely to trap macromolecular free radicals (PP radicals) and other free radicals such as hydrogen and hydroxyl radicals created from the decomposition of PP at elevated temperatures and form cross-linked networks in situ. On the other hand, the radical-trapping effect of C_{60} also provided partial PP radicals more time to recombine, which resulted in a remarkable increase in both storage modulus and complex viscosity of nanocomposites.

To exclude the effect of the oxidation reaction as much as possible, Figure 11.9 (B) gives the time-dependent viscoelastic properties of nanocomposites determined in a dynamic time-scanning mode at 260 °C for 20 min, where oxidation reactions happened slightly or not at all, as evidenced by TGA and temperature scanning tests. Thus, this method was more favorable for investigating the reaction of C_{60} with radicals. Apparently, no change was observed for η^* of pure PP in the whole scanning process, suggesting that no oxidation reaction took place or the oxidative cross-linking reaction was not severe, whereas for PP/C_{60} systems, η^* gradually increased after 8 min. Moreover, the greater the content of C_{60} in PP, the higher the complex viscosity of nanocomposites at the same scanning time. For PP/2.0 wt%, η^* at 20 min was around five times that at the beginning of scanning. All the rheological properties confirmed this hypothesis: C_{60} in the PP matrix trapped the macromolecular free radicals to form a cross-linked network in situ, and other free radicals (H· and HOO·), on the other hand, provided partial macromolecular radicals a longer time to recombine, both of which may be responsible for the enhanced thermal properties and flame retardancy of PP.

11.3.3 Heat-treatment studies

To validate these conclusions, the gel content of different systems was also determined. Figure 11.10 presents the gel content percentages for PP and nanocomposites heated to 360 °C for 15 min. For pure PP, the gel content remained very small, whereas for nanocomposites, gel content gradually increased with increasing C_{60} loading. Thus, gel content measurements provided proof for the proposed flame retardation model. Besides gel content tests, IR spectra were also employed to investigate the change of chemical bonds. As shown in Figure 11.11, for PP, clearly, a new absorption peak appeared at 1,725 cm^{-1}, attributed to the stretching vibration of C=O because of the oxidation of polymer after heat treatment. For PP/1.0 wt% C_{60}, after heat treatment, in addition to a weak absorption peak at 1,725 cm^{-1}, a new strong absorption peak was observed at 725 cm^{-1}. This peak was not likely assigned to the stretching mode of C–O or the bending mode of C–OOH, because this assignment implied that the stretching vibration of O–H, at around 3,520 cm^{-1}, could also be observed; this peak did not appear in our IR spectrum (not given). On the other

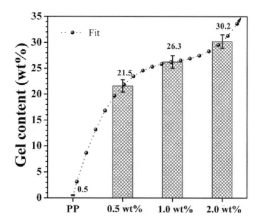

Figure 11.10 Gel content for PP and PP/C$_{60}$ samples after heat treatment at 360 °C for 15 min. Reprinted with permission from Ref. [17]. © 2008 IOP Publishing Ltd.

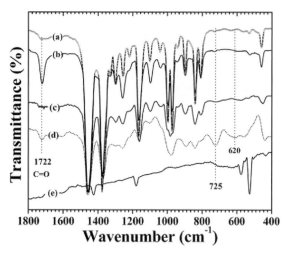

Figure 11.11 FTIR spectra of PP (a) before and (b) after, and PF2 (c) before and (d) after heat treatment at 260 °C for 15 min, with (e) FTIR spectra of pristine C$_{60}$ as a comparison. Reprinted with permission from Ref. [17]. © 2008 IOP Publishing Ltd.

hand, the peak at 725 cm^{-1} also did not belong to vibration modes of any sample, including PP, PP-t, and the C$_{60}$ molecule, which has four strong absorption peaks (1,429, 1,183, 577, and 528 cm^{-1}) reported by several researchers [30–33]. Taking into account that our systems after heat treatment, only involved PP, C$_{60}$, and the newly created compounds C$_{60}$(R)$_n$ (R refers to macromolecular free radicals and H· and HOO· free radicals), this new peak (725 cm^{-1}) suggested the formation of a new chemical bond in the C$_{60}$(R)$_n$ molecule. This new peak was not attributed to vibration of C$_{60}$−H, in that hydrogen atoms were small, distortion of chemical bonds did not occur, and thus vibration modes of C−H

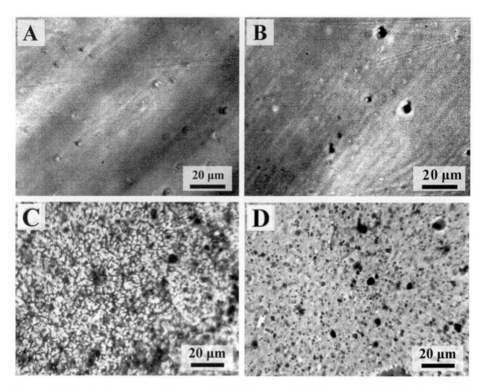

Figure 11.12 Optical micrograph images of PP (A) before and (B) after, and PF2 (C) before and (D) after heat-treatment at 260 °C for 15 min. Reprinted with permission from Ref. [17]. © 2008 IOP Publishing Ltd.

should be the same as those in PP. Furthermore, on the basis of previous reports and our observations [30–33], it was reasonable that this new absorption peak was assigned to the bending vibration mode of $C_{59}\sim C-C\sim R$ because of strong distortion of $C-C$ bonds. On the other hand, considering the increase in complex viscosity, this peak was more likely to belong to the bending mode of $C_{59}\sim C-C\sim R$.

Some visual images were also used to support these conclusions. Figure 11.12 shows optical microscope (OM) images for PP and its nanocomposites. Apparently, C_{60} was able to disperse well in the PP matrix, although there were also some aggregative domains, which was consistent with TEM observation. After heat treatment, the number of C_{60} aggregates in the PP matrix was much smaller, indicating a chemical reaction between C_{60} and polypropylene. On the other hand, we also observed that the morphology of the char residue of PP/1.0 wt% C_{60} after cone calorimetry was similar to that of pure C_{60}. Many researchers [34, 35] have utilized atomic force microscopy (AFM) to investigate the surface and/or interfacial roughness of various materials. We also used AFM to detect the change of surface roughness of a sample before and after heat treatment directly. Figure 11.13 shows 3D images of surface roughness for PP and PP/1.0 wt% C_{60}, as well as the corresponding

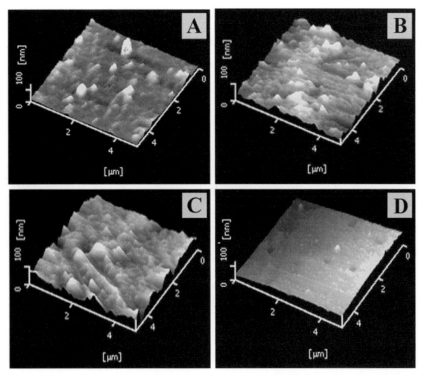

Figure 11.13 AFM images of PP (A) before and (B) after, and PF2 (C) before and (D) after heat treatment at 260 °C for 15 min. Reprinted with permission from Ref. [17]. © 2008 IOP Publishing Ltd.

heat-treatment samples. Clearly, the surface roughness increased to some extent with the incorporation of C_{60}. Comparing the AFM images of PP and PP/1.0 wt% C_{60} before and after heat treatment, it could be observed that there was no obvious change in roughness for PP, whereas the roughness of PF2 after heat treatment was much smaller than that before heat treatment, suggesting that a chemical reaction may occur between the PP matrix and C_{60}.

To elucidate the flame retardation mechanism of C_{60}, we put forward a simple model, presented in Figure 11.14. At temperatures below T_c (at which cross-linking reactions occurred), 280 °C for nanocomposites, C_{60} molecules could freely move or roll through Brownian motion, resulting in a plasticizing effect, with slightly small G' and η^*. In the temperature range from T_c to T_d (where the cross-linked network was destroyed), 390 °C for nanocomposites, aside from the partial oxidation cross-linking of PP itself, C_{60} trapped most of the macromolecular free radicals and other free radicals such as H· and HOO·. Consequently, PP radicals had more time to recombine, and macromolecular free radicals were cross-linked by C_{60}, forming a gelled-ball cross-link network, which could prevent the degradation of the polymer matrix by preventing heat flow and oxygen transfer from

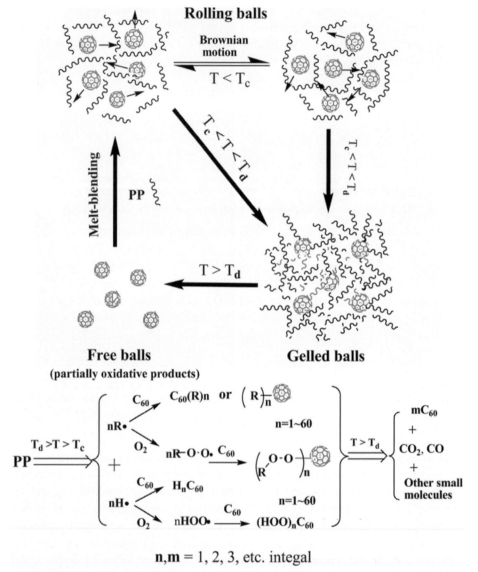

n,m = 1, 2, 3, etc. integal

R• = any macroradicals, any small free radicals and other free radicals

Figure 11.14 A schematic representation of a proposed model for C_{60} flame retardancy in PP. Reprinted with permission from Ref. [17]. © 2008 IOP Publishing Ltd.

the outer surface into the underlying polymer to sustain combustion. C_{60} itself, at the same time, changed into gelled balls. At elevated temperatures above T_d, the cross-linked network could also be destroyed by high heat flow and oxygen. Most of the cross-linking products ($C_{60}(R)_n$) changed into extremely stable free C_{60} molecules and other small molecules such as carbon dioxide, carbon monoxide, and other gas fuels. Of course, part of the C_{60} could also be oxidized or degraded. In summary, C_{60} molecules trapped the free radicals in the gas phase and were able to form cross-linked networks in situ and permit PP radicals more time to recombine. Although these cross-linked networks did not have sufficiently high thermal stability, they could still function as a barrier to both heat and mass transfer, like clay and CNTs, which made it more difficult for degradation products to escape to the gas phase and also prohibited the feedback of heat. In short, C_{60} fullerene indirectly reduced the flammability of polymer materials by trapping free radicals.

11.4 Synergism between fullerenes and intumescent flame retardants

11.4.1 Decoration of fullerenes with intumescent flame retardants

Generally, the degradation of polymers is thought to be a free radical chain reaction through β-scission of polymer chains. C_{60}, described as a radical sponge [30], has high reactivity toward free radicals, in that it has 30 carbon–carbon double bonds and can trap more than 34 free radicals [30, 36–38]. Thus, C_{60} can trap the macromolecular or any other radicals created by the pyrolysis of polymers and then form a gelled ball network in situ. The network can increase the melt viscosity and consequently slow the combustion of polymeric materials. Thus, in another sense, C_{60} may be considered to be a gas phase flame retardant.

Previous reports [16, 17, 39, 40] have shown that both the intumescent flame retardant, poly(4, 4-diaminodiphenyl methane-*o*-bicyclic pentaerythritol phosphate-phosphate) (PDBPP), and C_{60} can improve the thermal and flame retardancy properties of polypropylene. It is possible to achieve much better flame retardancy performance when gas and condensed phase flame retardants are employed simultaneously. The chemical structure of the PDBPP molecule and the schematic representation for a C_{60}-decorated oligomeric intumescent flame retardant (C_{60}-*d*-PDBPP) are presented in Figure 11.15. Covalently functionalizing C_{60} by grafting intumescent flame retardants with the goal of forming a new synergistic gas and condensed phase flame retardant will be expected to confer better thermal and flame retardancy performance on polymers.

Figure 11.16 shows that C_{60} exhibits four characteristic absorption peaks at about 1430, 1183, 576, and 521 cm^{-1} (intramolecular F_{1u} mode of C_{60}), which originate from stretching and bending modes of carbon within the highly distorted graphite structure and have been reported by many scientists [30, 41]. Absorption peaks at about 3,435, 3,368 cm^{-1}, and 1,611 cm^{-1}, attributed to stretching vibration and bending vibration of primary NH$_2$, respectively, as well as 1,387 cm^{-1} (P–N), 1,235 cm^{-1} (P=O), and 1,027 cm^{-1} (P–O) also appear in the spectrum of PDBPP [39, 40]. For C_{60}-*d*-PDBPP, in addition to the majority of the absorption peaks for both C_{60} and PDBPP, two new peaks appear, at about 2970 cm^{-1}, assigned to stretching vibration of C_{60}–H, as reported by Sahoo and Patnaik [41], and at about 1665 cm^{-1}, attributed to bending vibration of C_{60}–NH–C. The change of peaks proved that PDBPP reacted with C_{60}. ^{1}H NMR was performed on the resultant

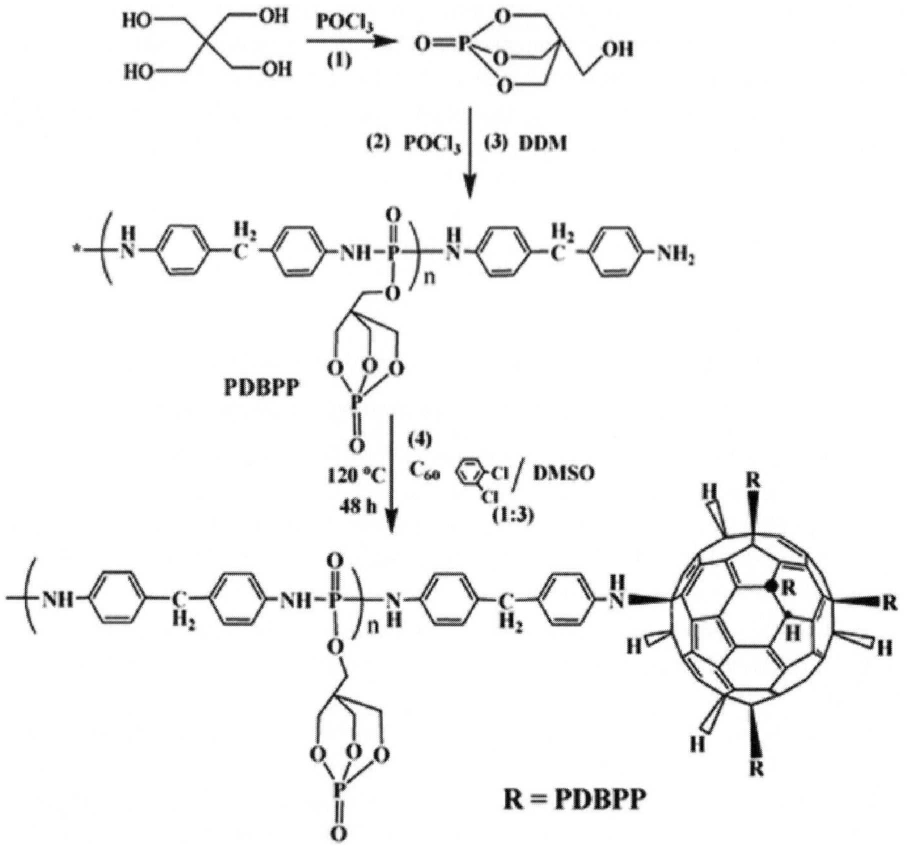

Figure 11.15 Schematic synthetic routes for C_{60}-decorated PDBPP, C_{60}-d-PDBPP. Reprinted with permission from Ref. [19]. © The Royal Society of Chemistry 2009.

product, shown in Figure 11.17. For PDBPP, the chemical shift values at lowest field $\delta =$ 6.7–7.4 ppm (H on benzene ring), 4.35–4.5 ppm (O–CH$_2$), 3.95–4.08 ppm (NH and NH$_2$), and about 3.8 ppm (–CH$_2$–) were found. For C_{60}-d-PDBPP, besides these chemical shifts, two new chemical shifts appeared at $\delta = 2.64$ ppm, assigned to the proton of C_{60}–H, and at $\delta = 5.23$ ppm, belonging to the proton in C_{60}–NH–C.

Figure 11.18 gives the XPS spectra of C_{60} and C_{60}-d-PDBPP in the binding energy range of 0–600 eV (A). C_{60} exhibits almost 100% carbon with a binding energy (BE) of about 284.5 eV (C–C), with a minor oxygen impurity. C_{60}-d-PDBPP spectra exhibited other elements BE peaks: O1s, N1s, P2p, suggesting that the C_{60}-d-PDBPP contained oxygen, nitrogen, phosphorus, and carbon. C1s scans of pure C_{60}, PDBPP, and C_{60}-d-PDBPP are also shown in Figure 11.18 (B, C, D). It is not difficult to observe BE peaks for C–C (284.58 eV) and C–O (286.67 eV), two satellite peaks in a C1s scan of C_{60} [42–44], and BE peaks for C–C, C–O, and C–N (285.68 eV) for PDBPP. For C_{60}-d-PDBPP, in addition

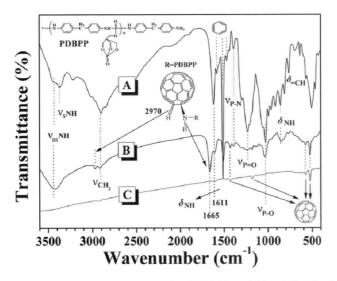

Figure 11.16 Infrared spectra of (A) PDBPP, (B) C_{60}-d-PDBPP, and (C) pristine C_{60}. Reprinted with permission from Ref. [19]. © The Royal Society of Chemistry 2009.

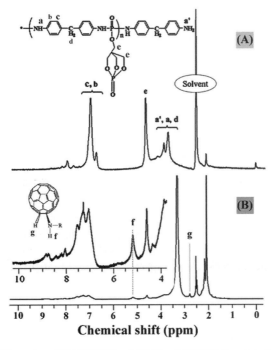

Figure 11.17 ^{1}H NMR spectra of (A) PDBPP and (B) C_{60}-d-PDBPP. Reprinted with permission from Ref. [19]. © The Royal Society of Chemistry 2009.

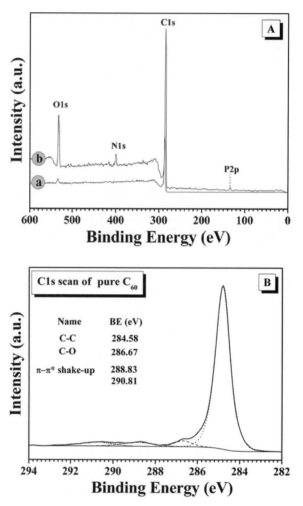

Figure 11.18 XPS spectra of (A-a) C_{60} and (A-b) C_{60}-d-PDBPP, and C1s scans of (B) C_{60}, (C) PDBPP, and (D) C_{60}-d-PDBPP. Reprinted with permission from Ref. [19]. © The Royal Society of Chemistry 2009.

to the BE at 284.64 eV (C–C, C–CH), 285.5 eV (C–N), and 286.83 eV(C–O), a satellite peak at 289.02 eV was also observed. This surface element analysis further supported the infrared spectra and ^{1}H NMR results.

To determine how many PDBPP molecules were grafted to one C_{60} molecule, we conducted TGA of C_{60}, PDBPP, and their reaction product under nitrogen (avoiding the oxidation effects due to O_2 in air), shown in Figure 11.19. The TGA curve of PDBPP showed a char residue of 39 wt% at 600 °C, but of about 50 wt% for PDBPP synthesized previously. Pristine C_{60} hardly degrades below 600 °C, with a char residue of about 93 wt%. The tendency of C_{60}-d-PDBPP was basically similar to that of PDBPP below 260 °C, and 58 wt% char was left at 600 °C. Because C_{60} itself can trap the free radicals, it will be

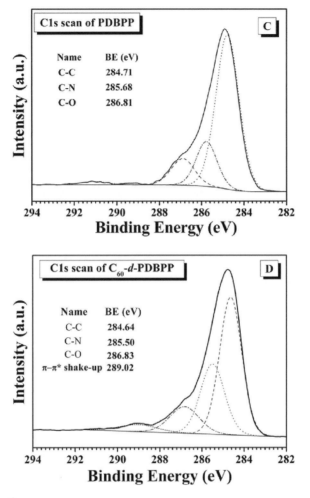

Figure 11.18 (*cont.*)

sure to affect the degradation of PDBPP; thus, theoretically, the char residue (58 wt%) was not accurate in a strict sense, and we could evaluate the grafting degree of PDBPP only approximately. According to the calculation methods in a previous report [12], the grafting degree of C_{60} was about 70 wt%. In view of the number average molecular weights (M_n) of pristine C_{60} (720 g/mol) and PDBPP (760 g/mol), the mole ratio of C_{60}/PDBPP was about 2:3: one C_{60} molecule was chemically bonded to 1.5 PDBPP molecules on the average.

11.4.2 Morphology

PP/C_{60}-*d*-PDBPP nanocomposites containing 0, 0.5, 1.0, and 2.0 wt% of C_{60}-*d*-PDBPP (designated as PP, PFP1, PFP2, and PFP3) were fabricated via melt-blending at 180 °C in a ThermoHaake Rheomix for 8 min with a rotor speed of 60 rpm. Figure 11.20 shows

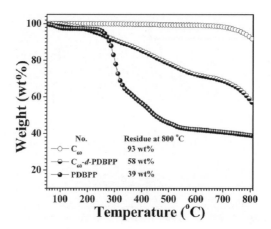

Figure 11.19 TGA curves of PDBPP, C_{60}, and C_{60}-d-PDBPP under nitrogen. Reprinted with permission from Ref. [19]. © The Royal Society of Chemistry 2009.

the TEM images of C_{60} (A), PDBPP (B), and C_{60}-d-PDBPP (C), as well as SEM images of C_{60}-d-PDBPP (D) and of PFP2 at low magnification (E) and high magnification (F). We could observe from both TEM and SEM images that a number of certain compound-coated nanoparticles instead of isolated spherical ones were present in C_{60}-d-PDBPP, which provided additional proof of the formation of C_{60}-d-PDBPP. Interestingly and expectedly, starlike nanoparticles (see C) could also be observed from the SEM image of C_{60}-d-PDBPP (see D), which may reflect the dendrimerlike molecular architecture of C_{60}-d-PDBPP from a macroscopic view. Pristine C_{60} nanoparticles were present in the PP matrix with a size of 200–300 nm, as previously reported by us [17], whereas TEM and SEM images of PFP2 clearly displayed that C_{60}-d-PDBPP with a size of about 100 nm was homogeneously dispersed in the PP matrix. This may be because the PDBPP prevented the aggregation of C_{60} particles, which favored the dispersion of C_{60}-d-PDBPP in the PP matrix. Good dispersion generally was more likely to maximize the properties conferred on polymer materials by the nanofillers.

11.4.3 Thermal oxidation degradation behavior

Because polymer materials are usually used in air environment, from a practical application viewpoint, it is important to evaluate the thermal properties of materials in air as well as under nitrogen. Figure 11.21 presents the TGA curves (A) and DTG curves (B) of PP and its nanocomposites. Oxidation degradation of pure PP starts at 264 °C (T_5, or T_{onset}), and is basically complete after 350 °C. Compared with pure PP, the initial degradation temperature or T_{onset} of nanocomposites was remarkably delayed, and the T_{onset} values shifted to higher temperature with increasing C_{60}-d-PDBPP loading level. For example, the T_{onset} of PFP2 was 332 °C, which was about 68 °C higher than that of pure PP. Besides the T_{onset}, the maximum weight loss rate temperature (T_{max}) was another important

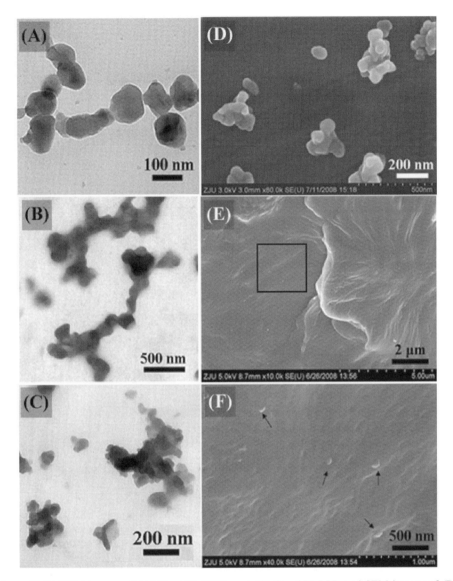

Figure 11.20 TEM images of (A) C_{60}, (B) PFP2, and (C) C_{60}-*d*-PDBPP, and SEM images of (D) C_{60}-*d*-PDBPP and (E, F) cross-section for PFP2. Reprinted with permission from Ref. [19]. © The Royal Society of Chemistry 2009.

parameter for evaluating thermal degradation of polymer materials. PP had a T_{max} of 338 °C, whereas PFP1, PFP2, and PFP3 exhibited T_{max} of 404, 409, and 418 °C, respectively. For PFP3, T_{max} was delayed about 80 °C, an enhancement in thermal oxidation degradation properties.

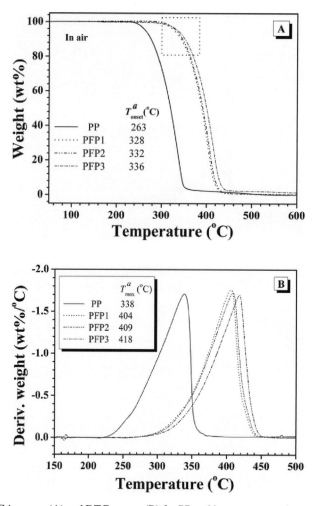

Figure 11.21 TGA curves (A) and DTG curves (B) for PP and its nanocomposites. T_{onset}: the temperature at which 5.0 wt% weight loss occurred; T_{max}: the temperature at which maximum weight loss rate occurred. Reprinted with permission from Ref. [19]. © The Royal Society of Chemistry 2009.

For the PP nanocomposite PF2, which contains 1.0 wt% pristine C_{60} [17], T_{onset} and T_{max} were 279 and 392 °C, respectively, both of which were far lower than those of PFP2, which contains 1.0 wt% C_{60}-d-PDBPP. The results suggest that C_{60}-d-PDBPP was much better than pristine C_{60} for delaying the thermal oxidation degradation of polypropylene. The delayed oxidation degradation of nanocomposites is not difficult to understand. Free radical trapping by C_{60}, known as a radical sponge, could be responsible, and has been evidenced by our previous report [17, 18]. C_{60}-d-PDBPP was much more effective than C_{60} in terms of oxidation degradation, which is attributed to the presence of PDBPP. Because the presence of PDBPP will prevent the aggregation of C_{60} molecules and makes

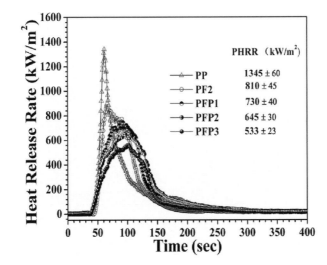

Figure 11.22 Heat release rate curves for PP and PP/C_{60}-*d*-PDBPP nanocomposites at a heat flux of 35 kW/m². Reprinted with permission from Ref. [19]. © The Royal Society of Chemistry 2009.

it disperse more uniformly in the polymer matrix, more C_{60} molecules will participate in the radical-trapping reaction.

11.4.4 Flame retardancy

Figure 11.22 presents the HRR curves for PP and its nanocomposites. The time to ignition (t_{ign}) of pure PP was about 35 s, and its HRR reached a maximum of 1,345 kW/m² after about 60 s (t_{PHRR}), whereas for PP nanocomposites, their t_{ign} was prolonged to different extents, and PHRR gradually decreased with increasing C_{60}-*d*-PDBPP content. For PFP2 containing 1 wt% C_{60}-*d*-PDBPP, PHRR exhibited a 52% decrease. When the concentration of C_{60}-*d*-PDBPP reached 2.0 wt% (PFP3), t_{ign} was delayed 8 s, and the time to PHRR was prolonged 102 s, which was very significant for improving the flame retardancy of materials. For nanocomposites containing the same C_{60} content as in [17], the PHRR values were higher than for those containing C_{60}-*d*-PDBPP; for example, the PHRR of PF2 containing 1.0 wt% pristine C_{60} was 810 kW/m², about 165 kW/m² higher than that of PFP2. Similarly to HRR curves, the mass exhibited slower loss after incorporation of C_{60}-*d*-PDBPP, indicating that adding C_{60}-*d*-PDBPP could slow the burning of nanocomposites. Other parameters such as average specific extinction area (ASEA) and average mass loss rate (AMLR) displayed reduction to some extent relative to pure PP.

Figure 11.23 displays digital photos of PP and its nanocomposites before and after heat treatment. From Figure 11.23, one can readily observe that PP basically left nothing only a small residue, whereas for nanocomposites, the residue was much greater, and increased with increasing C_{60}-*d*-PDBPP content. Furthermore, for PFP1 and PFP2, the

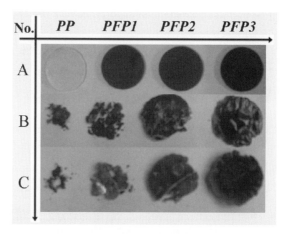

Figure 11.23 Digital photos for PP and PP/C_{60}-d-PDBPP nanocomposites before (A) and upper surface (B) and inner surface (C) after being heated up to 360 °C in a muffle. Reprinted with permission from Ref. [19]. © The Royal Society of Chemistry 2009.

inner surface of the residue had many small apertures because of vigorous bubbling during degradation, whereas PFP3 was much more compact without any small apertures. These visual photos directly suggest that C_{60}-d-PDBPP could significantly delay the thermal oxidation degradation and improve the fire retardancy of polypropylene.

On the basis of this cone analysis and digital photos, C_{60} functionalized via grafting flame retardant PDBPP exhibits better flame retardancy than pristine C_{60}. In [17, 18], we have found that C_{60} considerably enhanced the thermal oxidation degradation temperature and markedly reduced the flammability of polypropylene, and we assumed that the free radical–trapping effect of C_{60} was largely responsible for the improvement in thermal and flame retardancy properties. Because PP/C_{60}-d-PDBPP systems exhibited much more improvement in thermal and flame retardancy, their action mechanism should have some relationship to PP/C_{60} systems, and extra investigation of flame retardant PDBPP should be performed.

11.4.5 Rheological behavior and char residues

Figure 11.24 presents plots for the dependence of complex viscosity (η^*) on frequency (A), temperature (B, C), and time (D). Through the whole frequency-scanning regime, unlike nanocomposites containing C_{60}, whose complex viscosity was lower than that of PP [17], nanocomposites containing C_{60}-d-PDBPP had higher complex viscosity than that of PP. The dendrimerlike structure of C_{60}-d-PDBPP may contribute to form a network when its concentration reaches a critical value and, consequently, leads to higher complex viscosity. Because higher viscosity of composites in the melt state will limit the volatilization of degradation products during the burning process, higher melt viscosity may contribute to flame retardancy of the nanocomposites. On the other hand, temperature scanning plots

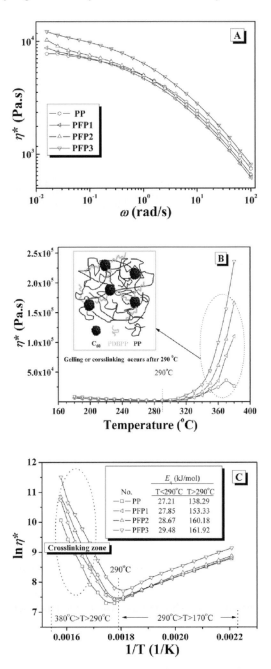

Figure 11.24 Plots for the dependence of complex viscosity (η^*) on (A) frequency (ω), (B) temperature, and (D) time at 260 °C. (C) plots of ln η^* versus $1/T$. Reprinted with permission from Ref. [19]. © The Royal Society of Chemistry 2009.

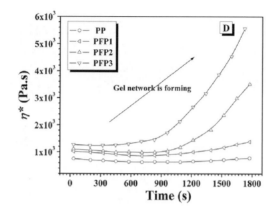

Figure 11.24 (*cont.*)

suggested that the complex viscosity of the nanocomposites were slightly higher than that of pure PP below 290 °C, whereas above that temperature, the viscosities were far higher than that of PP. Moreover, the η^* values of the nanocomposites dramatically increased with increasing temperature after 290 °C, suggesting the formation of a gel network or the occurrence of an additional cross-linking reaction. This increasing viscosity was due to the free radical–trapping effects of C_{60}. We could obtain the viscous-flow activation energy (E_η) of PP and the nanocomposites by plotting $\ln\eta^*$ versus $1/T$, as presented in Figure 11.24 (C). Apparently, whether the temperature was below or above 290 °C, the E_η of the nanocomposites was much higher than that of PP (27.21 kJ/mol vs. 138.29 kJ/mol due to slight oxidation cross-linking). Moreover, the E_η of the composites gradually increased with increasing C_{60}-d-PDBPP content, and PFP3 had the highest E_η, about 29.48 kJ/mol ($T < 290$ °C) and 161.92 kJ/mol ($T > 290$ °C). The significant increase in E_η indicated the formation of a gel network, which may be responsible for the improvement in the thermal oxidation degradation and flame retardancy of the nanocomposites.

As expectedly, time-scanning measurements 260 °C indicated that the melt viscosity of pure PP basically remained unchanged, demonstrating that no oxidative cross-linking reaction occurred, whereas the complex viscosities of nanocomposites rapidly increased after ca. 10 min with increasing time and increasing C_{60}-d-PDBPP content. Based on viscoelastic measurements, we could conclude that free radical–trapping effects of C_{60} increased the melt viscosity at elevated temperatures (above degradation temperature, even to burning), which limited the escape of degradation products during combustion, and consequently slowed the combustion process.

Because intumescent flame retardants, known as condensed phase flame retardants, form a compact char layer on the surface of the materials, useful information may be obtained through investigating the morphology and chemical composition of char residue. The SEM image, IR spectrum, and chemical composition of the collected char residue are shown in Figure 11.25. One can clearly see that the char residue of PP containing 1.0 wt% C_{60} (A) was an incompact network structure with lots of apertures, whereas for PP composite

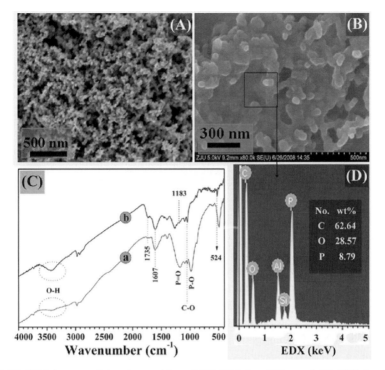

Figure 11.25 SEM images of the char residue of (A) PP/1.0 wt%C_{60} and (B) PP/1.0 wt%C_{60}-*d*-PDBPP; IR spectra of the char residue of (C, a) PDBPP and (C, b) PP/1.0 wt%C_{60}-*d*-PDBPP; and (D) EDX spectrum of char residue of PP/1.0 wt%C_{60}-*d*-PDBPP. Reprinted with permission from Ref. [19]. © The Royal Society of Chemistry 2009.

containing 1 wt% C_{60}-*d*-PDBPP, the char residue was a whole compact structure nearly without pores. The IR spectrum of the collected char residue for PFP2 (11.25C) shows that besides some absorption peaks at 980–1025 cm^{-1} (P−O), 1235 cm^{-1} (P=O), and 1607 cm^{-1} (benzene rings or polyaromatic compounds), which came from PDBPP [39, 40], 524 cm^{-1} and 572 cm^{-1}, which belonged to the characteristic absorption peaks of C_{60} [30, 41], also appeared. Interestingly, for the IR spectrum of char for PFP2, its peak for O−H (3400–3500 cm^{-1}) was much stronger than that of char from PDBPP, suggesting that some C_{60} molecules may be oxidized to $C_{60}(OH)_n$. EDX analysis also indicated that the residues contained 62.64 wt% C, 28.57 wt% O, and 8.79 wt% P. As expectedly, the percentage content of P was lower than that in char from pure PDBPP, whereas the percentage content of C was much higher. IR and EDX analysis indicated that the char residues were mainly residues derived from PDBPP and C_{60} particles.

This analysis of char residue may contribute to understanding the interaction between C_{60} and flame retardant PDBPP. On one hand, it should be pointed out that improvement in thermal and flame retardancy mainly resulted from C_{60}, which has been studied in our previous reports [17, 18], partially from the char layer produced by PDBPP, because the added amount of common intumescent flame retardant was generally above 25 wt% [45, 46]. On the other hand, the interaction between them was not negligible; from the morphology, chemical structure, and elemental composition, some complicated reactions must occur, such as free radical reactions, which contribute to forming a char layer more compact than that from pristine C_{60}. Therefore, it was the interaction between C_{60} and PDBPP, as well as the self-degradation of PDBPP and C_{60}, that led to the formation of a more compact char layer that also contributed to the improvement in thermal and flame retardancy. In brief, the improvements in flame retardancy and thermal oxidation degradation were attributed to the synergism of free radical trapping by C_{60} and a compact char layer from residues of PDBPP after burning.

11.5 Synergism between fullerenes and carbon nanotubes

11.5.1 Decoration of carbon nanotubes with fullerenes

In the past few years, we have witnessed that carbon nanotubes (CNTs) are being considered to be another candidate for fire retardant additives, because of their unique structure and properties [47–50]. At present, flammability properties of polymer/CNTs nanocomposites with various resins such as PP [8–10], PMMA [1, 51], PA6 [52], and EVA [53] have been investigated. Much research shows that carbon nanotubes can greatly reduce the heat release rates of polymers at very low loading levels. Because fullerene (C_{60}) was proved to be a new flame retardant in the previous section, is there a synergistic effect between CNTs and C_{60}? On the other hand, considering the good solubility of C_{60} in many organic solvents such as toluene and dichlorobenzene, grafting C_{60} onto the surface of CNTs will not only improve the solubility of carbon nanotubes and dispersion in polymer matrix, but also combine the unique fire performance of C_{60} and CNTs.

The synthesis strategy consists of a three-step reaction: first hydroxylation using KOH/ethanol treatment, then amino-functionalization using 3-aminopropyltriethyoxylsilane (APTES), and finally C_{60}-decoration of CNTs through the addition of amino groups to C_{60} molecules. The three-step resultant products were named as CNT-OH, CNT-NH$_2$, and C_{60}-*d*-CNTs, respectively, by a synthesis route presented in Figure 11.26.

XPS spectra for Pristine CNTs, CNT-OH, CNT-NH$_2$, and C_{60}-*d*-CNT, as well as N1s spectra for CNT-NH$_2$ and C_{60}-*d*-CNT are shown in Figure 11.27, with their chemical composition listed in Figure 11.27 (A). For the pristine CNTs, aside from the main C−C peak at 284.5 eV, another weak photoemission at higher BE of ca. 286.1 eV (O1s, atomic: 1.78) was attributed to atmospheric moisture or oxidation during purification [54], which was consistent with the IR results mentioned previously. As expected, CNTs-OH displayed much stronger photoemission at BE of 286.1 eV belonging to O1s. The concentration of oxygen increased up to 8.86 atomic% and the C1s spectrum showed a BE at 531.5 eV assigned as C−OH, which implied that the CNTs were successfully hydroxylated. In comparison with CNT-OH, two new photoemission peaks appeared in the XPS spectrum of CNT-NH$_2$, namely, at BE of 102.8 eV (Si2p, 1.99 atomic%) and about 399.3 eV (N1s,

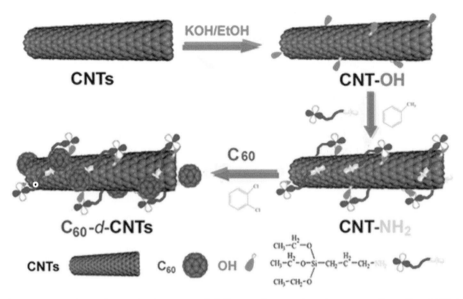

Figure 11.26 Schematic synthesis route of fullerene-decorated carbon nanotubes, C$_{60}$-*d*-CNTs. Reprinted with permission from Ref. [20]. © The Royal Society of Chemistry 2009.

2.02 atomic%), and the higher surface oxygen content, as high as 13.14 atomic%, was attributed to the hydrolysis of APTES, which provided the proof of amino-functionalization of CNTs. As for C$_{60}$-*d*-CNT, a very important point was the increase in surface content of carbon, increasing from 82.85 atomic% for CNT-NH$_2$ to 86.49 atomic% for C$_{60}$-*d*-CNT, which was due to the introduction of C$_{60}$ molecules onto the surface of CNT-NH$_2$, and also suggested that about 0.67% C$_{60}$ molecules were attached onto the surface of CNTs (ca. 2.71 wt% by mass obtained from TGA). Another direct proof was the splitting of the N1s peak (Figure 11.27B), namely, from one photoemission peak at BE ca. 399.6 eV for CNT-NH$_2$ to three peaks at BE 399.5 eV (C–N) and 400.3 eV and 401.4 eV for C$_{60}$-*d*-CNTs, respectively, indicating that the primary amino group (NH$_2$) was chemically bonded to other atoms. In our system, it was obvious that the amino groups were only likely to react with C$_{60}$ molecules via the amine-addition reactions reported by many organic chemists. The BE at 400.3 eV and 401.4 eV for C$_{60}$-*d*-CNTs were 0.7 eV and 2.0 eV higher than that for CNT-NH$_2$. This higher binding energy may be due to the strong electro-withdrawing effect of C$_{60}$ molecules, which severely lack electrons. The BE at 400.3 eV may be due to the photoemission peak of N1s in the biaddition products, CNT–N(C$_{60}$)$_2$, whereas 401.4 eV is the peak of N1s in the monoaddition products, CNT–NH–C$_{60}$.

As shown in Figure 11.28, compared with pristine CNTs (A), the diameter of CNT-OH (B) is still maintained at 20–50 nm, except that the surface of the latter was much smoother, which was because of the fact that the hydroxylation reaction got rid of the amorphous carbon on the surface of pristine CNTs. However, after CNT-OH was amino-functionalized using APTES, not only did the diameter of CNTs increase to some degree, but also its

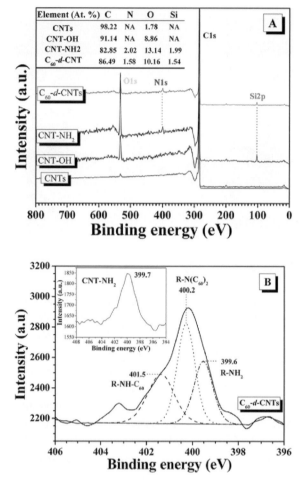

Figure 11.27 (A) XPS spectra of pristine CNTs, CNT-OH, CNT-NH$_2$, and C$_{60}$-d-CNTs, as well as (B) N1s spectra of CNT-NH$_2$ and C$_{60}$-d-CNTs. Reprinted with permission from Ref. [20]. © The Royal Society of Chemistry 2009.

surface became rough relative to that of CNT-OH, and a thin polymeric layer could be observed [see Figure 11.28 (C), marked by arrows]. Combining these findings with the XPS analysis, we could draw the conclusion that the layer is the self-condensation products of APTES. Spherical C$_{60}$ nanoparticles with a diameter of 80–130 nm can be observed readily in Figure 11.28 (D). For C$_{60}$-d-CNTs [Figure 11.28 (E)], in contrast to CNT-NH$_2$, besides further increased diameter, up to 90 nm, some salient parts could be observed (marked by arrows), especially on the ends of CNTs, because of high activation of the ends of tubes. The size of C$_{60}$ attached to the surfaces of CNTs was much larger than reported in the literature [55–58]. Because the raw C$_{60}$-d-CNTs product has been washed three times via ultrasonication using o-dichlorobenzene to remove the unreacted and physically

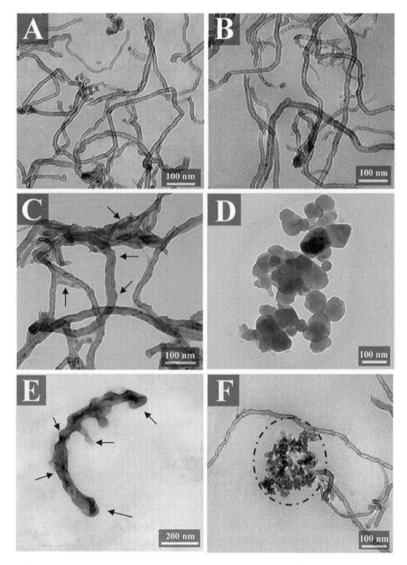

Figure 11.28 TEM images of (A) pristine CNTs, (B) CNT-OH, (C) CNT-NH$_2$, (D) C$_{60}$, (E) C$_{60}$-*d*-CNTs, and (F) physical mixture of C$_{60}$ and CNTs. Reprinted with permission from Ref. [20]. © The Royal Society of Chemistry 2009.

absorbed C$_{60}$ molecules, besides the large size of C$_{60}$ molecules, with diameter around 0.7 nm, the occurrence of condensation polymerization of APTES created a large number of amino groups, which further produced more multiadducts [CNT-N (C$_{60}$)$_2$] on reacting with excess C$_{60}$. To further verify this analysis, TEM images of a physical mixture of CNTs and C$_{60}$ are also presented as comparisons [Figure 11.28 (F)]. Obviously, they are separated for self-aggregated molecular crystalline C$_{60}$, which also supports the observation.

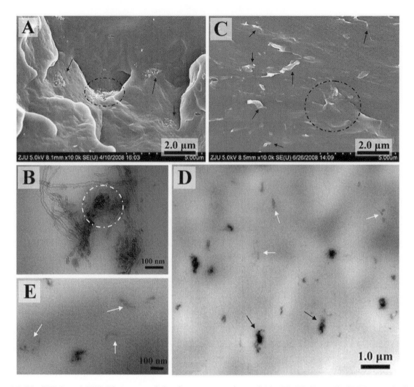

Figure 11.29 SEM and TEM images of the fracture section of (A, B) PP/1 wt% CNTs, and (C, D, E) PP/1 wt% C$_{60}$-*d*-CNTs nanocomposites.

11.5.2 *Morphology*

Figure 11.29 presents SEM and TEM images of the fracture section of (A, B) PP/1 wt% CNTs and (C, D, E) PP/1 wt% C$_{60}$-*d*-CNTs nanocomposites. Because the surface of C$_{60}$-*d*-CNTs contains some unreacted active groups such as –OH and –NH$_2$, maleic anhydride–grafted polypropylene (PP-g-MA) was introduced as the compatibilizer for this system. For pristine CNTs, many agglomerations can apparently be observed, whereas for C$_{60}$-*d*-CNTs, the diameter increased by reaction with PP-g-MA during melt-blending, and the dispersion was much more homogeneous than for pristine CNTs. TEM images showed that even single CNTs could be observed, indicating that C$_{60}$-*d*-CNTs could disperse well with the aid of the compatibilizer.

11.5.3 *Flame retardancy*

Figure 11.30 presents the heat release rate curves of polypropylene (PP) and its composites with 1 wt% CNTs or C$_{60}$-*d*-CNTs. The incorporation of CNTs considerably reduced the peak heat release rate (PHRR) of PP (reduction around 66). At the same loading level,

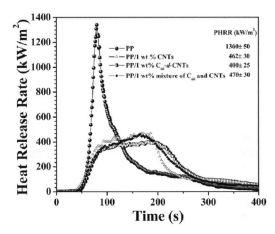

Figure 11.30 Heat release rate curves for pure PP, PP/1.0 wt% CNTs, PP/1.0 wt% C_{60}-d-CNTs, and 1.0 wt% of physical mixture C_{60} and CNTs at a heat flux of 35 kW/m². *Note:* PHRR is peak heat release rate. Reprinted with permission from Ref. [20]. © The Royal Society of Chemistry 2009.

C_{60}-d-CNTs could not only make the PHRR decrease further to some extent, but also slow down the combustion process, which indicates that C_{60}-d-CNTs could confer better flame retardancy on PP than pristine CNTs. To clarify the grafting effect or the effect of a simple combination of C_{60} and CNTs, we conducted a controlled experiment by physically mixing C_{60} and CNTs in PP. The result exhibited a PHRR value higher than that of C_{60}-d-CNTs at the same loading level, which indicated that it was a grafting effect, not the effect of a simple combination of the two. According to the investigations reported by Kashiwagi [1] and our group [17], the improvement in flame retardancy may be because of the free radical–trapping effect and the barrier effect of the CNT network. Because pristine C_{60} is too small to aggregate in a PP matrix, it was not easy to optimize its action. However, grafting will make it disperse better in PP and therefore endow PP with better flame retardancy.

11.6 Conclusions and future prospects

Polymer/fullerene [C_{60}] nanocomposites can be considered environmentally friendly alternatives to some traditional flame retardants. The presence of C_{60} can markedly delay thermal oxidative degradation and reduce the flammability of polypropylene at very low loadings. It can decrease the heat release rate of polymeric materials by trapping the free radicals created through thermal degradation and combustion, and subsequently forming three-dimensional gelled networks. This network can increase the melt viscosity and consequently slow down combustion. Furthermore, the incorporation of C_{60} does not affect the physical properties of the polymer.

Obviously, polymer/C_{60} nanocomposites offer novel strategies for developing flame retardant polymeric materials. However, they do not behave well in traditional fire retardation tests, such as LOI and the UL test. Thus, in the future, more work needs to be done on this problem and the synergistic effects of C_{60} and other traditional fire retardants.

References

1. T. Kashiwagi, F. Du, K. I. Winey, J. F. Doulas, K. I. Winey, R. H. Harris, and J. R. Shields, Nanoparticle networks reduce the flammability of polymer nanocomposites. *Nature Materials*, 4 (2005), 928–33.
2. S. Bourbigot, S. Duquesne, and J. M. Leroy, Modeling of heat transfer of a polypropylene-based intumescent system during combustion. *Journal of Fire Sciences*, 17 (1999), 42–56.
3. J. K. Pandey, K. R. Reddy, A. P. Kumar, and R. P. Singh, An overview on the degradability of polymer nanocomposites. *Polymer Degradation and Stability*, 88 (2005), 234–50.
4. E. Giannelis, Polymer layered silicate nanocomposites. *Advanced Materials*, 8 (1996), 29–35.
5. J. Zhu and C. A. Wilkie, Thermal and fire studies on polystyrene–clay nanocomposites. *Polymer International*, 49 (2000), 1158–63.
6. A. B. Morgan, Flame retarded polymer layered silicate nanocomposites: A review of commercial and open literature systems. *Polymers for Advanced Technologies*, 17 (2006), 206–17.
7. H. Y. Ma, L. Tong, Z. B. Xu, and Z. P. Fang, Clay network in ABS-graft-MAH nanocomposites: Rheology and flammability. *Polymer Degradation and Stability*, 92 (2007), 1439–45.
8. T. Kashiwagi, E. Gruke, J. Hilding, K. Groth, R. Harris, K. Butler, J. Shields, S. Kharchenko, and J. Douglas, Thermal and flammability properties of polypropylene/carbon nanotube nanocomposites. *Polymer*, 45 (2004), 4227–39.
9. T. Kashiwagi, E. Gruke, J. Hilding, K. Groth, R. Harris, W. Awad, and J. Douglas, Thermal degradation and flammability properties of poly (propylene)/carbon nanotube composites. *Macromolecular Rapid Communications*, 23 (2002), 761–5.
10. T. Kashiwagi, F. Du, K. I. Winey, K. M. Groth, J. Shields, S. P. Bellayer, H. Kim, and J. F. Douglas, Flammability properties of polymer nanocomposites with single-walled carbon nanotubes: Effects of nanotube dispersion and concentration. *Polymer* 45 (2005), 471–81.
11. H. Ma, L. Tong, Z. B. Xu, and Z. P. Fang, Synergistic effect of carbon nanotube and clay for improving the flame retardancy of ABS resin. *Nanotechnology*, 18 (2007), 1–8.
12. H. Y. Ma, L. F. Tong, Z. B. Xu, and Z. P. Fang, Functionalizing carbon nanotubes by grafting on intumescent flame retardant: Nanocomposite synthesis, morphology, rheology, and flammability. *Advanced Functional Materials*, 18 (2008), 414–21.
13. L. Ye, P. Ding, M. Zhang, and B. J. Qu, Synergistic effects of exfoliated LDH with some halogen-free flame retardants in LDPE/EVA/HFMH/LDH nanocomposites. *Journal of Applied Polymer Science*, 107 (2009), 3694–701.
14. M. Zhang, P. Ding, and B. J. Qu, Flammable, thermal, and mechanical properties of intumescent flame retardant PP/LDH nanocomposites with different divalent cations. *Polymer Composites*, 30 (2009), 1000–1006.

15. C. M. Jiao, Z. Z. Wang, X. L. Chen, and Y. Hu, Synthesis of a magnesium/aluminum/iron layered double hydroxide and its flammability characteristics in halogen-free, flame-retardant ethylene/vinyl acetate copolymer composites. *Journal of Applied Polymer Science*, 107 (2007), 2626–31.

16. B. Ramaraj and K.-R. Yoon, Thermal and physicomechanical properties of ethylene-vinyl acetate copolymer and layered double hydroxide composites. *Journal of Applied Polymer Science*, 108 (2008), 4090–95.

17. P. A. Song, Y. Zhu, L. F. Tong, and Z. P. Fang, C_{60} reduces the flammability of polypropylene nanocomposites by in situ forming gelled-ball network. *Nanotechnology*, 19 (2008), 1–10.

18. Z. P. Fang, P. A. Song, L. F. Tong, and Z. H. Guo, Thermal degradation and flame retardancy of polypropylene/C_{60} nanocomposites. *Thermochimica Acta*, 473 (2008), 106–8.

19. P. A. Song, H. Liu, Y. Shen, B. X. Du, and Z. P. Fang, Fabrication of dendrimer-like fullerene (C_{60}) decorated oligomeric intumescent flame retardant for reducing the thermal oxidation and flammability of polypropylene nanocomposites. *Journal of Materials Chemistry*, 19 (2009), 1305–13.

20. P. A. Song. Y. Shen, B. X. Du, Z. H. Guo, and Z. P. Fang, Fabrication of fullerence decorated carbon nanotubes and its application in flame retarding polypropylene. *Nanoscale*, 1 (2009), 118–21.

21. L. W. Zhu, D. H. Wu, and D. C. Xu, *Carbon Nanotube*, 2nd ed. (Beijing: China Machine Press, 2003).

22. M. Zanetti, P. Camino, P. Reichert, and R. Mülhaupt, Thermal behaviour of poly (propylene) layered silicate nanocomposites. *Macromolecular Rapid Communications*, 22 (2001), 176–80.

23. B. B. Troitskii, L. S. Troitskaja, A. S. Yakhnov, M. A. Lopatin, and M. A. Novikova, Retardation of thermal degradation of PMMA and PVC by C_{60}. *European Polymer Journal*, 33 (1997), 1587–90.

24. B. B. Troitskii, L. S. Troitskaja, A. A. Dmitriev, and A. S. Yakhnov, Inhibition of thermo-oxidative degradation of poly(methyl methacrylate) and polystyrene by C_{60}. *European Polymer Journal*, 36 (2000), 1073–84.

25. S. P. Su, D. D. Jiang, and C. A. Wilkie, Poly(methyl methacrylate), polypropylene and polyethylene nanocomposite formation by melt blending using novel polymerically-modified clays. *Polymer Degradation and Stability*, 83 (2004), 321–31.

26. J. G. Zhang, D. D. Jiang, and C. A. Wilkie, Polyethylene and polypropylene nanocomposites based on a three component oligomerically-modified clay. *Polymer Degradation and Stability*, 91 (2006), 641–8.

27. S. B. Kharchenko, J. F. Douglas, J. Obrzut, E. A. Grulke, and K. B. Migler, Flow-induced properties of nanotube-filled polymer materials. *Nature Materials*, 3, (2004) 564–8.

28. M. A. Treece and J. P. Oberhauser, Soft glassy dynamics in polypropylene − clay nanocomposites. *Macromolecules*, 40 (2007), 571–82.

29. C. O. Rohlmann, M. D. Failla, and L. M. Quinzani, Linear viscoelasticity and structure of polypropylene–montmorillonite nanocomposites. *Polymer*, 47 (2006), 7795–804.

30. P. J. Krusic, E. Wasserman, P. N. Keizer, J. R. Morton, and K. F. Preston, Radical reactions of C_{60}. *Science*, 22 (1991), 1183–5.

31. W. Krätschmer, K. Fostiropoulos, and D. R. Huffman, The infrared and ultraviolet absorption spectra of laboratory-produced carbon dust: Evidence for the presence of the C_{60} molecule. *Chemical Physics Letters*, 170 (1990), 167–70.

32. Z. C. Wu, D. A. Jelski, and T. F. George, Vibrational motions of buckminsterfullerene. *Chemical Physics Letters*, 137 (1987), 291–4.
33. D. E. Weeks and W. G. Harter, Rotation–vibration spectra of icosahedral molecules. II. Icosahedral symmetry, vibrational eigenfrequencies, and normal modes of buckminsterfullerene. *Journal of Chemical Physics*, 90 (1989), 4744–71.
34. J. B. Zhang, T. P. Lodge, and C. W. Macosko, Interfacial morphology development during PS/PMMA reactive coupling. *Macromolecules*, 38 (2005), 6586–91.
35. C. Neto and V. S. Craig, Colloid probe characterization: Radius and roughness determination. *Langmuir*, 17 (2001), 2097–9.
36. L. B. Gan, S. Huang, H. X. Zhang, A. X. Zhang, B. C. Cheng, H. Cheng, X. L. Li, and G. Shang, Fullerenes as a *tert*-butylperoxy radical trap, metal catalyzed reaction of *tert*-butyl hydroperoxide with fullerenes, and formation of the first fullerene mixed peroxides $C_{60}(O)(OOtBu)4$ and $C_{70}(OOtBu)10$. *Journal of the American Chemistry Society*, 124 (2002), 13384–5.
37. B. Z. Tang, S. M. Leung, H. Peng, and N. T. Yu, Direct fullerenation of polycarbonate via simple polymer reactions. *Macromolecules*, 30 (1997), 2848–52.
38. T. Cao and S. E. Webber, Free radical copolymerization of styrene and C_{60}. *Macromolecules*, 29 (1996), 3826–30.
39. P. A. Song, Z. P. Fang, L. F. Tong, and Z. B. Xu, Synthesis of a novel oligomeric intumescent flame retardant and its application in polypropylene. *Polymer Engineering and Science*, 49 (2009), 1326–31.
40. P. A. Song, Z. P. Fang, L. F. Tong, Y. M. Jin, and F. Z. Lu, Effects of metal chelates on a novel oligomeric intumescent flame retardant system for polypropylene. *Journal of Analytical and Applied Pyrolysis*, 82 (2008), 286–91.
41. R. R. Sahoo and A. Patnaik, Surface confined self-assembled fullerene nanoclusters: A microscopic study. *Applied Surface Science*, 245 (2005), 26–38.
42. Y. F. Zhu, T. Yi, B. Zheng, and L. L. Cao, The interaction of C_{60} fullerene and carbon nanotube with Ar ion beam. *Applied Surface Science*, 137 (1999), 83–90.
43. J. Onoe, A. Nakao, and K. Takeuchi, XPS study of a photopolymerized C_{60} film. *Physical Review B*, 55 (1997), 10051–5.
44. L. Qian, L. Norin, J.-H. Guo, C. Såthe, A. Agui, U. Jansson, and J. Nordgren, Formation of titanium fulleride studied by X-ray spectroscopies. *Physical Review B*, 59 (1999), 12667–71.
45. R. Delobel, M. Le Bras, N. Ouassou, and F. Alistiqsa, Thermal behaviours of ammonium polyphosphate–pentaerythritol and ammonium pyrophosphate–pentaerythritol intumescent additives in polypropylene formulations. *Journal of Fire Sciences*, 8 (1990), 85–108.
46. B. K. Kandola, A. R. Horrocks, P. Myler, and D. Blair, The effect of intumescents on the burning behaviour of polyester-resin-containing composites. *Composites, Part A*, 33 (2002), 805–17.
47. G. Beyer, Short communication: Carbon nanotubes as flame retardants for polymers. *Fire and Materials*, 26 (2002), 291–3.
48. G. Beyer, Organoclays as flame retardants for PVC. *Fire and Materials*, 29 (2005), 61–9.
49. S. Bourbigot, S. Duquesne, and C. Jama, Polymer nanocomposites: How to reach low flammability? *Macromolecular Symposia*, 233 (2006), 180–90.
50. M. Moniruzzaman and K. I. Winey, Carbon nanotubes reinforced nylon-6 composite prepared by simple melt-compounding. *Macromolecules*, 39 (2006), 5194–205.

51. M. C. Costache, D. Y. Wang, M. J. Heidecker, E. Manias, and C. A. Wilkie, The thermal degradation of poly(methyl methacrylate) nanocomposites with montmorillonite, layered double hydroxides and carbon nanotubes. *Polymers for Advanced Technologies*, 17 (2006), 272–80.

52. B. Schartel, P. Potschke, U. Knoll, and M. Abdel-Goad, Fire behaviour of polyamide 6/multiwall carbon nanotube nanocomposites. *European Polymer Journal*, 41 (2005), 1061–70.

53. B. B. Marosfoi, G. J. Marosi, A. Szep, P. Anna, S. Keszei, B. J. Nagy, H. Martvona, and I. E. Sajo, Complex activity of clay and CNT particles in flame retarded EVA copolymer. *Polymers for Advanced Technologies*, 17 (2006), 255–62.

54. P. C. Ma, J. K. Kim, and B. Z. Tang, Functionalization of carbon nanotubes using a silane coupling agent. *Carbon*, 44 (2006), 3232–8.

55. W. Wu, H. R. Zhu, L. Z. Fan, and S. H. Yang, Synthesis and characterization of a grapevine nanostructure consisting of single-walled carbon nanotubes with covalently attached [60]fullerene balls. *Chemistry – A European Journal*, 14 (2008), 5981–7.

56. D. M. Guldi, E. Menna, M. Maggini, M. Marcaccio, D. Paolucci, F. Paolucci, S. Campidelli, M. Prato, G. M. Aminur Rahman, and S. Schergna, Supramolecular hybrids of [60]fullerene and single-wall carbon nanotubes. *Chemistry – A European Journal*, 12 (2006), 3975–83.

57. J. L. Delgado, P. D. L. Cruz, A. Urbina, J. T. L. Navarrete, J. Casado, and F. Langa, The first synthesis of a conjugated hybrid of C_{60}–fullerene and a single-wall carbon nanotube. *Carbon*, 45 (2007), 2250–52.

58. T. Kashiwagi, M. F. Mu, K. Winey, B. Cipriano, S. R. Raghavan, S. Pack, M. Rafailovich, E. Grulke, J. Shields, R. Harris, and J. Douglas, Relation between the viscoelastic and flammability properties of polymer nanocomposites. *Polymer*, 49 (2008), 4358–68.

12

Flame retardant polymer nanocomposites with alumina as filler

ABDELGHANI LAACHACHI[a] AND JOSÉ-MARIE LOPEZ CUESTA[b]

[a]AMS – Centre de Recherche Public Henri Tudor
[b]CMGD – Ecole des Mines d'Alès

12.1 Introduction

It has been reported that approximately 12 persons are killed and 120 are severely injured because of fire every day in Europe. Fire has considerable impact on the environment in terms of destruction of substructures and production of toxic and/or corrosive compounds such as CO, dioxins, HCN, and polycyclic aromatic compounds. Consequently, it is necessary to limit this kind of risk by designing new materials with improved flammability properties. Nowadays, many companies (building and civil engineering, transportation, cable-making and electrotechnical material, etc.) are directly concerned with this topic.

Buildings contain increasing calorific value in the form of highly combustible polymeric materials replacing more traditional materials (wood, alloys, metals, etc.) with the aim of improving the comfort of occupants (pieces of furniture, carpets, toys, household and leisure electric components, and data processing equipment, etc.). Potential sources of fire tend to grow with the multiplication of electric and electronic devices. The increasing sophistication and miniaturization of electronics (with increasingly powerful and fast microprocessors) have as a consequence a stronger concentration of energy, leading to an increased risk of localized overheating and thus of fire.

Under these conditions, regulations impose the use of materials possessing thermal stability as well as efficient fire retardant properties. In parallel, emissions of smoke must be low, not very opaque, not very toxic, and not very corrosive. This evolution toward greater safety seriously limits the use of many materials and involves the rejection of solutions largely used so far and, in particular, halogen-based flame retardants, on account of environmental concerns. Moreover, analysis of various statistics on plastic consumption (180 MT/yr, with a global annual growth rate of approximately 8%) shows the economic importance of this field and illustrates the world's industrial stake in it.

Because of their chemical structure, mainly composed of carbon and hydrogen, polymers are highly combustible [1]. To understand the mode of actions of flame retardants (FR), it is essential to have a clear vision of the sequences involved in the polymer combustion process [2]. Combustion results from two factors: combustibles (reducing agents) and combustives (oxidizing agents (O_2)). The whole process usually starts when the polymeric material is degraded in the presence of a source of heat, inducing polymer bond scission and releasing the flammable species. The volatile fraction of the resulting polymer fragments diffuses into the air and creates a combustible gaseous mixture (also called fuel). This gaseous mixture ignites when the autoignition temperature (defined as the temperature at which the activation energy of the combustion reaction is attained) is reached, liberating heat. Alternatively, the fuel can ignite at a lower temperature (called the flash point) upon reaction with an external source of intense energy (spark, flame, etc.) The life span of the combustion cycle depends on the quantity of heat liberated during the combustion of the fuel. When the amount of heat liberated reaches a certain level, new decomposition reactions are induced in the solid phase, and more combustibles are produced. The combustion cycle is thus maintained, as described by the fire triangle model.

Flame retardants are added to different polymers to reduce the risk of fire. They save lives, prevent injuries and property losses, and protect the environment by helping to prevent fires from starting and to limit fire damage. Currently, it is possible to treat most potentially flammable materials with special additives to make them more difficult to ignite and to significantly reduce fire spread. FR can thus make a decisive contribution to the fire safety of buildings, furniture, electric and electronic apparatus, textiles, and public transport and cars. Depending on their nature, FR systems can act either physically (by cooling, forming a protective layer, or diluting fuel) or chemically (by reaction in the condensed or gas phase). They can interfere with the various processes involved in polymer combustion (heating, pyrolysis, ignition, propagation of thermal degradation). Many different FR, reactive or not, have been developed with different modes of action. The main FR materials are halogen, phosphorus, and inorganic and nitrogen compounds.

In this chapter we review the effects of hydrated and anhydrous alumina as filler on FR properties of polymer nanocomposites. Nanocomposites are particle-filled polymers for which at least one dimension of the dispersed particles is in the nanometer range. In the following sections, the mechanism and mode of action of nanometric alumina trihydrate (ATH), alumina monohydrate (AlOOH), and alumina (Al_2O_3) will be discussed.

12.2 Alumina trihydrate flame retardant

ATH is an inorganic FR widely used because of its low cost and nontoxicity and easy to incorporate into various thermoplastics and thermosets. It decomposes 180 and 200 °C (12.1). However, it is very difficult to dehydrate it completely. It is generally used in plastic formulations with processing temperatures around 190 °C, such as polyolefins for electric cables, PVC, PU, EVA, thermosets such as unsaturated polyesters, phenolics, epoxy, and rubber.

$$2Al(OH)_3 \rightarrow Al_2O_3 + 3H_2O(\Delta H = -298 \text{ kJ/mol}) \qquad (12.1)$$

As a FR, ATH has several modes of action. When it degrades thermally, the polymer is cooled by endothermic decomposition, which implies less release of volatile products. The aluminum oxide resulting from the dehydration of ATH forms a barrier layer on the surface of the material and is more or less mixed with the carbonaceous remaining structure (char). The layer forms an insulating barrier to further decomposition of the underlying material. Finally, the water released as vapor dilutes the gas phase, which reduces the amount of oxygen at the surface of the material [3, 4]. The aluminum oxide (Al_2O_3) formed during decomposition can also act as a barrier layer in the condensed phase [5–7]. Although ATH acts as a FR by several mechanisms, its effectiveness is limited and high loadings (more than 50 wt%) are generally needed to achieve high levels of fire retardancy, leading in some cases to processing difficulties and marked deterioration in the mechanical properties of the polymer [8, 9].

Despite the great interest of polymer nanocomposites, few papers illustrate the effect of ATH nanoparticles on fire resistance of polymers, except for the work reported by Zhang *et al.* [10]. ATH particles in the micrometric range are mainly used in conventional composites. Zhang *et al.* have shown that it is necessary to add 60 wt% of nano-ATH as loading into EVA to increase the limiting oxygen index (LOI) of EVA from 17 to 30. It has also been shown that the use of a titanate coupling agent between the EVA and the nano-ATH was advantageous to increase the fire resistance of the EVA. This improvement was mostly attributed to better dispersion of surface-modified nano-ATH and strong adhesion between the nanofiller and matrix [10]. Similarly, Daimatsu *et al.* [11] have grafted Phosmer PP (a phosphonated compound) onto the nano-ATH surface, followed by radical polymerization in mass with methyl methacrylate (MMA). The addition of 3% of grafted nano-ATH contributes to significant slowing of the horizontal burning rate. In comparison with the burn rate of pure PMMA, that of the prepared PMMA nanocomposite was almost halved. This result reflects the synergy between the nano-ATH and phosphorus compounds acting as FR. An improvement of thermal stability is also observed and is attributed to the presence of covalent bonds between nano-ATH particles and PMMA.

In addition, several studies have reported the use of conventional ATH in combination with other types of nanofillers, particularly organoclays, to improve the FR properties of polymer nanocomposites [12, 13]. Organoclays are not considered as FR despite their ability to decrease peak heat release rates of several polymers under firelike conditions [14]. For instance, compared with unfilled EVA/PE blend, the incorporation of organoclays (5 phr) into polymer blend results in a small decrease of LOI [15]. It is believed that, under LOI tests, the layered clays in the nanocomposites cannot form a rigid insulation layer on the burning surface and may flow away with dripping molten flaming material. Thus, the addition of organoclay at a low level is insufficient to provide effective flame retardancy. However, the addition of organoclay (5 phr) to EVA/PE blends containing ATH FR results in significant improvement in flammability properties, and an increase in the LOI value by more than 15% is observed for the case of EVA/PE–organoclay–ATH nanocomposite [15]. This result suggests that a low level of added organoclay may act as a synergist for conventional ATH FR. To explain this synergistic effect, the authors suggested that thermal decomposition of ATH, at the burning surface of the nanocomposite, forms a protective layer of aluminum oxide, which is reinforced by reassembling silicate layers to give an expanded and cohesive insulated surface layer. This insulated layer is suggested to be responsible for the synergistic effect on flame retardancy, as well as for smoke suppression observed in the flame-retarded nanocomposite. However, the results obtained by Zhang *et al.* for the same polymer blend have shown no clear indication of a synergistic effect between organoclay and ATH [16].

Better fire retardant properties have been also obtained by Morgan in EVA–organoclay–ATH nanocomposites than in EVA–ATH [17]. As the EVA–organoclay–ATH nanocomposite is heated, the modifier in the organoclay begins to degrade first, forming acid sites on the clay surface that catalyze cross-linking and aromatization reactions during the EVA decomposition, resulting in a carbon char that resists combustion. This carbon char formation and the decomposition mechanism of the ATH improve the fire retardant properties of the EVA matrix [17]. The effects of particle size and surface treatment of ATH fillers, in combination with organoclays, on FR properties have been studied by Cardenas *et al.* [18]. Synergistic effects on FR properties were observed for some EVA–organoclay–ATH nanocomposites through their peak heat release rate (HRR) values being lower than in the EVA–ATH composite. However, it was found that FR properties were affected by the particle size and the surface treatment of ATH fillers in different ways. High LOI and reduced time-to-ignition values were achieved with small particle size and silane-coated ATH filler. Silane-coated ATH fillers also led to the best char stability after a cone calorimeter test.

Hull *et al.* suggested that the incorporation of organoclay into EVA–ATH composites appears to reinforce the protective layer formed by ATH and its resulting oxide [19]. They observed a large reduction in the rate of thermal decomposition because of the formation of a protective layer and its reinforcement by both fillers, ATH and organoclay. Indeed, the catalytic deacetylation of EVA by clay, which occurred slightly earlier than for EVA alone, could have increased the time available for cross linking, strengthening the protective layer. The influence of clay, which forms a protective layer, seems to have a negligible effect on the yield of carbon monoxide, under both fuel-lean and fuel-rich conditions. The influence of the residual Al_2O_3 from ATH in catalyzing the conversion of more organic material to CO also appears to be unaffected by the presence of the clay [19].

Beyer clearly observed a delay in thermal degradation using TGA (in air) with the addition of a small amount of organoclay to EVA–ATH composite [20]. The char of the EVA–ATH–organoclay compound formed in a cone calorimeter was rigid, with only a few small cracks, whereas the char of the EVA–ATH composite was much less rigid (reduced mechanical strength) and had many big cracks. These observations allowed the author to explain the reduction of the peak heat release rate to 100 kW/m^2 for the nanocomposite, compared to 200 kW/m^2 for the EVA–ATH composite [21].

The synergistic effect between ATH and organoclay has been also observed in other polymers such as polypropylene (PP) [22]. An increase in the LOI value and no dripping in the UL-94 vertical burning test were shown for PP–organoclay–ATH nanocomposite compared with PP–ATH or PP–organoclay systems. Zhang and Wilkie have demonstrated that the combination of 20% ATH with 5% inorganic clay in PP gives an 80% reduction of peak HRR, which is the same reduction that is obtained when 40% ATH is used [23].

In epoxy resin, the combination of ATH and phosphonium-modified clay additives showed superposition or even synergetic behavior for nearly all fire retardancy properties. Schartel *et al.* suggested that the presence of ATH resulted in an increase in residues and a small decrease in effective heat of combustion because of dilution of the pyrolysis products [24]. Both fire retardancy mechanisms have their primary source in the conversion of ATH into aluminum oxide, which increased the residues, and water, which diluted and cooled the flame zone. In addition, the presence of organophosphorus decreased the effective heat of combustion through a gas phase. Most of the phosphorus was liberated during polymer pyrolysis and influenced the fire behavior through flame inhibition.

Finally, it has been shown in all these works that the combination of layered silicates and ATH is a promising approach to sufficiently fire retardant materials.

12.3 Alumina monohydrate flame retardant

Boehmite is aluminum oxyhydroxide (AlOOH) powder inorganic FR (also called alumina monohydrate). AlOOH is composed of Al–O double layers connected by hydrogen bonds between the hydroxyl groups that can react with organic sulfates, for example, to improve compatibility with hydrophobic polymers [25]. It is used whenever the polymer processing temperature exceeds 200 °C, as for PMMA ($\approx$ 230 °C), which excludes the use of ATH because of its initial decomposition. Despite several decades of ATH use, it is only recently that AlOOH started to be studied as a FR. Moreover, there are still few papers devoted to the effect of this mineral on the fire reactions of polymers. However, it will be highlighted in this section that AlOOH can be promising in terms of effectiveness as a FR. The mode of action of AlOOH as a FR is similar to that of ATH and is based on physical and chemical processes. In the presence of an ignition source (a flame or a heat source), the thermal decomposition of the metal hydroxide into metal oxide and water takes place according to the reaction

$$2\text{AlOOH} \rightarrow \text{Al}_2\text{O}_3 + \text{H}_2\text{O} \qquad (12.2)$$

During this process, the energy of the ignition source is depleted, as the decomposition is an endothermic reaction. At the same time, the released water vapor cools the surface of the polymer and particularly dilutes the burnable gases in the surrounding area. The remaining Al_2O_3 residue has a high internal surface where sooty particles, respectively, polycyclic aromatic hydrocarbons, are absorbed. Additionally, the oxide residue acts as a barrier, disabling the further release of low–molecular weight decomposition products, and also as a heat barrier protecting the polymer against further decomposition. The maximum processing temperature range of AlOOH is higher than 340 °C, compared with 200 °C for ATH. In contrast to ATH, AlOOH has a layered structure. Analogously to other layered structures such as montmorillonite, it offers a promising potential for nanocomposites that build up surface layers during burning to work as barriers to mass and heat transfer [26, 27].

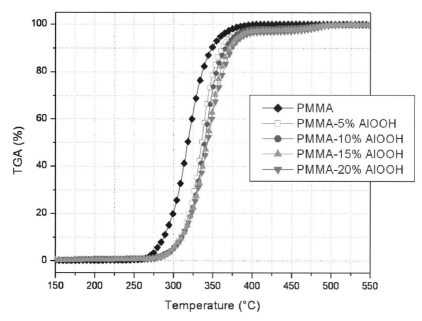

Figure 12.1 TGA curves for pure PMMA and PMMA–AlOOH composites at 5, 10, 15, and 20 wt% AlOOH under air (heating rate 10 °C/min).

Camino *et al.* [28] have examined the effect of AlOOH in comparison to those of ATH and $Mg(OH)_2$ on the fire retardant effectiveness of an EVA matrix. They concluded that AlOOH improves the char-forming properties but is not as effective as $Al(OH)_3$ with respect to flame retardancy [28]. Further investigations are needed to optimize the system and to understand the mechanism of AlOOH as a FR. Laachachi *et al.* [29, 30] have studied the effect of AlOOH on the thermal stability and FR properties of PMMA nanocomposite. The nanocomposite specimens were prepared by mixing molten PMMA pellets and AlOOH in appropriate ratios in a Haake PolyLab 60 cm^3 mixer rheometer at 225 °C and 50 rpm. Figure 12.1 presents the TGA curves in air for pure PMMA and PMMA–AlOOH composites with different contents of the inorganic phase (5, 10, 15, and 20 wt%). The run was carried out under dynamic conditions at a constant heating rate of 10 °C.min^{-1}. In addition to the mass loss corresponding to the degradation of PMMA, additional mass loss is observed on the TGA curves toward 480 °C. It corresponds to AlOOH decomposition, which releases water as shown by the TGA curve of pure AlOOH according to the reaction in eq. (2). The onset temperature (defined as the temperature corresponding to a 2% mass loss ($T_{2\%}$)) for all PMMA–AlOOH composites is increased by 17 °C from that for pure PMMA. The midpoint temperature (defined as the temperature corresponding to a 50% mass loss ($T_{50\%}$)) is also increased from that for pure PMMA by 19 °C for PMMA–5%AlOOH, 23 °C for PMMA–10%AlOOH, 29 °C for PMMA–15%AlOOH, and 31 °C for PMMA–20%AlOOH. It appears from these data that the AlOOH fillers improve the thermal stability of PMMA. It was proposed that this improvement of the thermal stability was the result of two main

Table 12.1 *Cone calorimetry data for PMMA and its composites with AlOOH at 35 kW.m^{-2} [43]*

Sample	PMMA	PMMA– 5% AlOOH	PMMA– 10% AlOOH	PMMA– 15% AlOOH	PMMA– 20% AlOOH
TTI (s)	69	80	74	88	82
TOF (s)	318	372	382	573	879
Peak HRR (kW.m^{-2})	624	503	489	424	348
THR (MJ.m^{-2})	112	109	109	103	99
Residual weight (%)	0	5	8	13	18
TCOR (g/kg)	6.7	7.4	6.8	8.7	14.4
TSR (m^2/m^2)	430	540	527	497	323

Note. TTI: time to ignition, TOF: time of flame out, peak HRR: peak heat release rate, THR: total heat release, TSR: total smoke release, and TCOR: total CO release.

factors: (i) restriction of mobility of the polymer chains caused by steric hindrance due to the presence of particles and (ii) bonds due to the adsorption of polymer on the oxide surface via methoxycarbonyl groups (–C(O)OCH$_3$) [29, 30].

The evaluation of the flammability properties of PMMA and its composites with AlOOH was done using a cone calorimeter (Fire Testing Technology). A 100 × 100 × 4 mm sheet was exposed to a radiant cone (35 kW.m^{-2}). The HRR was calculated from the oxygen consumption as measured with an oxygen analyzer. Table 12.1 gathers the parameters obtained in comparison with the pure PMMA sample.

All results obtained with the cone calorimeter showed a significant improvement of the behavior of PMMA in fire in the presence of AlOOH:

1. The time to ignition increases by more than 10 s for all PMMA–AlOOH samples.
2. The peak HRR decreases with the concentration of AlOOH in PMMA: from a reduction of 20% by only 5% of filler, to a 45% reduction for 20 wt% of filler.
3. The total heat released (THR) is slightly lower (99 MJ.m^{-2} for PMMA–20AlOOH compared with 112 MJ.m^{-2} for PMMA).
4. At 20 wt% of filler, the time of flameout is significantly increased (879 s) in comparison with that for all other studied systems. This indicates a real modification of the combustion kinetics of the polymer.
5. The total smoke released (TSR) is minimal for the composite at 20 wt% AlOOH. However, the total CO released (TCOR) increases during combustion when the concentration of AlOOH in PMMA also increases.

To conclude, the PMMA–20%AlOOH composite shows interesting behavior in fire. Its FR effectiveness can be attributed to the following factors:

- The endothermic decomposition of AlOOH to Al_2O_3 withdraws heat from the material and hence retards the thermal degradation.
- The accompanying release of water dilutes the fuel supply present in the gas phase.
- AlOOH and Al_2O_3 function as a barrier to insulate the polymer from the flame and a barrier to mass transfer of the polymer.
- The presence of 20 wt% of mineral filler in polymer matrix acts as a solid phase diluent.
- AlOOH and Al_2O_3 have a catalytic effect to form a char during combustion (black color for the residues). The char formation was previously reported when other metallic oxides [30, 31] and other nanofillers such as organomodified clay were used [32].

Concerning investigations of the use of AlOOH as a synergistic agent, Pawlowski and Schartel [33] showed that the combination of AlOOH and phosphorus-based additives in polycarbonate/acrylonitrile–butadiene–styrene copolymers (PC/ABS) resulted in the formation of $AlPO_4$. A synergistic effect was observed, leading to enhancement of the barrier layer of the fire residue during combustion.

12.4 Alumina flame retardant

Previous papers of the authors of this chapter showed that polymer–oxide nanocomposites generally exhibit higher thermal stability than pure polymers [29, 30]. Moreover, it has been found that the degradation mechanisms for polymers are very different in the presence of metal oxide compared to pure polymer. For example, the effect of alumina (Al_2O_3) nanoparticles (median particle size 13 nm and BET specific surface area 100 $m^2.g^{-1}$) on the thermal stability and flammability properties of PMMA nanocomposite has been studied. The PMMA–Al_2O_3 nanocomposite was prepared by mixing, in an appropriate ratio, molten PMMA pellets and Al_2O_3 in a Haake PolyLab 60-cm^3 mixer rheometer at 225 °C and 50 rpm. TEM analyses of the PMMA–Al_2O_3 were performed in order to investigate the distribution and the dispersion of oxide nanoparticles. Figure 12.2 shows a typical photograph obtained. It has been found that the oxide is well distributed in the material but with some tendency to aggregation, the size of the aggregates being much less than 200 nm.

Figure 12.3 shows the TGA curves for pure PMMA and its nanocomposites containing 5, 10, and 15 wt% of Al_2O_3 nanoparticles. The onset temperature is increased for all PMMA–Al_2O_3 nanocomposites by about 17 °C compared with that of pure PMMA. The midpoint temperature is increased compared with that of pure PMMA by about 35 °C for all PMMA-Al_2O_3 nanocomposites. These TGA data illustrated an improvement in the thermal stability of PMMA in the presence of Al_2O_3. However, this increase in temperature decomposition of PMMA with Al_2O_3 is lower than the results obtained previously with other metal oxide nanoparticles, such as TiO_2 and Fe_2O_3 [31]. But the comparison between the effects of AlOOH and Al_2O_3 on the thermal stability of PMMA shows that the improvement of thermal stability seems to be greater for Al_2O_3 than for AlOOH. This improvement of the thermal stability of PMMA in the presence of Al_2O_3 can be explained by the following:

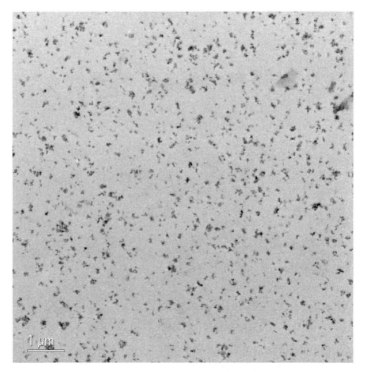

Figure 12.2 TEM image of PMMA–Al$_2$O$_3$ nanocomposite at 5 wt% Al$_2$O$_3$.

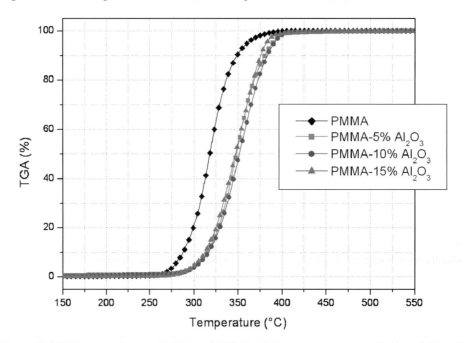

Figure 12.3 TGA curves for pure PMMA and PMMA-Al$_2$O$_3$ nanocomposites at 5, 10, and 15 wt% Al$_2$O$_3$ under air (heating rate 10 °C/min).

Table 12.2 *Cone calorimetry data for PMMA and its nanocomposites with different alumina at 35 kW.m^{-2} [38]*

Sample	Size (nm)	S_{BET} (m^2/g)	TTI (s)	Peak HHR (KW/m^2)	THR (MJ/m^2)	TSR (m^2/m^2)
PMMA	–	–	59±2	638±19	126±2	483±92
PMMA–5%AluA	13	100	55±4	589±1	123±2	490±8
PMMA–5%Inf	150	3	47±1	535±6	120±2	470±35
PMMA–5%Alu	470	6	57±2	601±15	124±1	456±39
PMMA–5%NF	3 × 100	280	42±2	467±10	119±3	609±28

Note. TTI: time to ignition, peak HRR: peak heat release rate, THR: total heat release, TSR: total smoke release, and S_{BET}: BET specific surface area of alumina.

i. The restriction of the mobility of polymer chains. It is well known that the filler surface has a marked effect on molecular mobility in a filled polymer [34].
ii. The radical-trapping effect of these mineral fillers.
iii. The adsorption of polymer onto filler surfaces via the methoxycarbonyl group.

Several degradation mechanisms involving acid–base or ionic interactions between surface groups of alumina and the polar groups of PMMA have been reported in the literature [35–37]. Hydrogen bonds between the methoxy groups and the hydroxyl groups of oxides promote reactions with the Lewis base sites of alumina. This would lead to strong bonds (ionic or covalent) to help in the stabilization of the nanocomposite system during thermal degradation. Interactions between carbonyl groups and the Lewis acid groups of the oxide may also intervene and delay the depolymerization. To conclude, the increase of the thermal stability of PMMA in the presence of alumina observed in TGA is ascribed to a restriction of the mobility of polymer chains and limitation of the release of gaseous products due to blocking of depolymerization.

Concerning the effect of particle size and specific surface of alumina on FR properties, Cinausero [38] has studied four types of alumina with different morphologies. Table 12.2 gathers the parameters obtained for PMMA filled with 5 wt% of each alumina type in comparison with the pure PMMA sample. Note that all different aluminas studied have a spherical shape, except the NF type, which has a fibrous shape.

All composites exhibit times to ignition shorter than for the pure polymer. The values of peak HRR depend on the effectiveness of the protective barrier formed during combustion, which acts as a barrier to insulate the polymer from the flame and a barrier to mass transfer of the polymer. Therefore, the total recovery of the sample surface during combustion by NF alumina contributed to a greater decrease of peak HRR (−27%) in comparison with the samples that were partially recovered by the alumina (Figure 12.4). The HRR results of the other nanocomposites exhibit small or negligible decrease of peak HRR (PMMA−5%Alu (−6%), PMMA−5%AluA (−8%), and PMMA−5%Inf (−16%)). In contrast, the combustion of composite PMMA−5%NF produces a greater quantity of smoke.

Figure 12.4 Photos of the residues obtained after calorimeter test of PMMA–Al$_2$O$_3$ at 35 KW/m^2.

Yang *et al.* have reported that the shape of alumina has a limited influence on the thermal stability of polymer nanocomposites. However, the particle size of alumina may greatly affect the degradation temperature of polycarbonate, because interactions between polycarbonate molecular chains and nanoparticles seem to occur principally when the particle size matches the segmental chain length of polycarbonate [39].

The effect on flammability of varying Al$_2$O$_3$ nanoparticle amounts in PMMA–Al$_2$O$_3$ nanocomposites was studied using a cone calorimeter (Table 12.3). It was shown that peak HRR was significantly reduced in comparison to that of pure PMMA. The reduction of peak HRR increased as the mass fraction of Al$_2$O$_3$ increased. At 5, 10, and 15 wt%, peak HRR was, respectively, lowered by 18, 34, and 44%. Thus, compared with AlOOH, the same peak HRR reduction could be obtained with a slightly lower loading rate (15 wt% instead of 20 wt%). The THR was also lowered for PMMA–15Al$_2$O$_3$ (81 MJ.m^{-2} instead of 112 MJ.m^{-2} for PMMA). However, TTI was significantly increased only by 19 s for PMMA–Al$_2$O$_3$ nanocomposite at 15 wt% of Al$_2$O$_3$ in comparison to pure PMMA. At 15 wt% of Al$_2$O$_3$, the duration of combustion was increased. This accounted for a modification of the degradation pathway of PMMA, in which other processes than depolymerization seemed to have occurred. Moreover, the amount of smoke released doubled in the presence of Al$_2$O$_3$ (Table 12.3).

Table 12.3 *Cone calorimeter data for PMMA and its nanocomposites*
with Al$_2$O$_3$ at 35 kW.m^{-2}

Sample	PMMA	PMMA–5% Al$_2$O$_3$	PMMA–10% Al$_2$O$_3$	PMMA–15% Al$_2$O$_{3-}$
TTI (s)	69	70	70	88
TOF (s)	318	350	390	500
Peak HRR (kW.m^{-2})	624	552	414	350
THR (MJ.m^{-2})	112	105	106	81
Residual weight (%)	0	4	10	14
TCOR (g/kg)	6.7	7	8	14
TSR (m^2/m^2)	430	566	660	800

Note. TTI: time to ignition, TOF: time of flame out, peak HRR: peak heat release rate, THR: total heat release, TSR: total smoke release, and TCOR: total CO release.

Figure 12.5 shows photographs of the residues collected after cone calorimeter tests for PMMA–Al$_2$O$_3$ nanocomposites. The morphology of the residues helps to understand the significant improvement of FR property of PMMA by using Al$_2$O$_3$ nanoparticles. First, polymers such as PMMA, which is only depolymerizing (does not form a char) as a pure polymer, degraded with char formation as a nanocomposite (black residues were obtained). The char formation was previously reported when other metallic oxides [30, 31] and other nanofillers such as organoclays were used [32]. Second, Figure 12.4 shows the formation of a particulate layer presenting a certain compactness, which is expected to create a barrier effect between the degradation products of the polymer and the flame. Thermogravimetric analysis of the residue after a cone calorimeter test of the PMMA–15%Al$_2$O$_3$ sample showed the presence of a slight mass loss at about 500 °C, attributed to the amount of carbonaceous compounds in the residue. This catalytic effect of Al$_2$O$_3$ nanoparticles in the formation of the char layer is probably ascribable to the presence of OH functions at the surface of the oxide.

Yang *et al.* have also observed an enhancement of the residues of polymer–alumina nanocomposites after calorimeter tests on several polymers (PS, PMMA, and PC). In contrast, they have also found that the reduction in peak HHR happened at a very low loading level of nanometric alumina. This phenomenon has been explained by a catalyzing effect of alumina. As a matter of fact, the presence of alumina may accelerate the decomposition of polymers by a catalyzing effect, as well as enhancing the thermal stability by radical trapping and chain mobility restriction [39].

The effect of varying Al$_2$O$_3$ nanoparticle amounts (median particle size 13 nm and BET specific surface area 100 m^2.g^{-1}) in PS–Al$_2$O$_3$ nanocomposites on flammability was studied by Cinausero using a cone calorimeter (Table 12.4) [38]. The PS–Al$_2$O$_3$ nanocomposites were prepared by mixing, in an appropriate ratio, molten PS pellets and Al$_2$O$_3$ in a Haake PolyLab 60 cm^3 mixer rheometer at 200 °C and 50 rpm. It was shown that peak HRR

Table 12.4 *Cone calorimeter data for PS and its nanocomposites with Al$_2$O$_3$ at 35 kW.m^{-2} [38]*

Sample	PS	PS–1.5% Al$_2$O$_3$	PS–3% Al$_2$O$_3$	PS–5% Al$_2$O$_3$	PS–10% Al$_2$O$_3$	PS–15% Al$_2$O$_3$
TTI (s)	83±0	90±3	82±2	79±5	81±1	96±3
TOF (s)	401±15	401±30	404±7	443±0	646±5	774±45
Peak HRR (kW.m^{-2})	752±10	742±23	749±14	547±14	422±6	387±8
THR(MJ.m^{-2})	131±0	133±2	135±1	132±2	128±4	122±1
Residual weight (%)	3.1±0.9	2.8±0.1	3.2±0	4±0.8	15.2±1	19.3±0.3
TCOR (g/kg)	2.29	2.29	2.31	2.25	2.39	2.37
TSR (m^2/m^2)	5163	5225	5236	5222	4685	4480

Note. TTI: time to ignition, TOF: time of flameout, peak HRR: peak of heat release rate, THR: total heat release, TSR: total smoke release, and TCOR: total CO release.

Figure 12.5 Photos of the residues collected after cone calorimetry tests at 35 kW/m^2 for PMMA–Al$_2$O$_3$ for different amounts.

was significantly reduced in comparison to that for pure PS. The reduction of peak HRR increased as the mass fraction of Al$_2$O$_3$ increased. At 5, 10, and 15 wt%, peak HRR was, respectively, lowered by 28, 44, and 49%. Thus, compared with PMMA–Al$_2$O$_3$, the same result could be obtained in terms of peak HHR reduction with a loading rate. The total heat released (THR) was also lowered for PS–15%Al_2O_3 (122 MJ.m^{-2} instead of 131 MJ.m^{-2} for PS). However, TTI was significantly increased only by about 10 s for PS–Al_2O_3 nanocomposite at 1.5 and 15 wt% of Al_2O_3 in comparison to pure PS. At 15 wt% of Al_2O_3, the duration of combustion was increased. This accounted for a modification of the degradation pathway of PS, in which processes other than depolymerization seem to have occurred. In contrast to the PMMA, the amount of smoke released by PS decreased in the presence of Al_2O_3 (Table 12.4).

In order to improve the efficiency of nanoalumina as FR, Cinausero *et al.* modified the surface groups of nanometric alumina with bis-phosphonic acid-based oligomers with the formation of covalent bonds between the oxide and the oligomers [40]. Thermal stability and fire behavior of PMMA were improved because of physical and physicochemical processes involving the presence of the nanometric alumina. Significant improvements in these properties in relation to the grafting of the mineral were only noticed for the PDMS phosphonic acid-based formulation, despite the very small amount of oligomer present in PMMA. It is suggested that this compound could act in the condensed phase because of its important thermal stability and could also promote modifications of the degradation pathway of PMMA in the interphase region surrounding the alumina particles [40].

The action of nanoalumina alone, in improving the fire retardancy of polymers, proved to be insufficient for ensuring adequate fire resistance to meet the required standards. However, their association with other FR systems such as phosphorated compounds could potentially be a very interesting approach. Several recent works have focused on such methods. For example, Laachachi *et al.* [41, 42] combined the FR action of nano-Al_2O_3 with the char formation induced by phosphorated FR systems (ammonium polyphosphates and phosphinates) in PMMA. In the case of aluminum phosphinate supplied by Clariant under the trade name Exolit OP930 (hereafter denoted phosphinate), cone calorimetry results showed that partial replacement of phosphinate by alumina nanoparticles promoted synergistic effects, with a marked decrease in peak HRR. However, no significant effect could be achieved with TiO_2 nanoparticles. Although Al_2O_3 and TiO_2 promote positive FR effects in PMMA, their combination with phosphinate did not automatically lead to a synergistic effect. Observation of residues involving alumina nanoparticles essentially showed a continuous solid layer, as typically illustrated by the Al_2O_3 9 wt%–phosphinate 6 wt% sample. However, with TiO_2–phosphinate combinations, the char residues did not cover the entire sample surface, leading to a poor barrier effect, which can explain the limited performance of these compositions. It has been shown that there was no chemical reaction between Al_2O_3 (or TiO_2) and phosphinate, but only the formation of a vitreous layer, promoted by the phosphorated compound and reinforced by alumina particles. It appeared that, in addition to their role in char reinforcement, alumina particles also had a positive catalytic effect on the formation of the protective layer with phosphinate (not provided by titanium oxide particles).

With the same objective, Gallo *et al.* have also studied the synergistic effect between aluminum diethlyphosphinate (AlPi) and nanometric metal oxides such as TiO_2 or Al_2O_3 on flame retardancy of poly(butylene terephthalate) (PBT) [43]. In particular, different active flame retardancy mechanisms were proposed. It was disclosed that AlPi acted mainly in the gas phase through the release of diethylphosphic acid, which provides flame inhibition. Part of the AlPi remains in the solid phase, reacting with the PBT to form phosphinate terephthalate salts that decompose to aluminum phosphate at higher temperatures. The metal oxides interact with the PBT decomposition and promote the formation of additional stable carbonaceous char in the condensed phase. A combination of metal oxides and AlPi gains a better classification in the UL-94 test thanks to the combination of the different mechanisms [43].

12.5 Conclusion

The use of nanoparticles is of prime interest for improving the flame retardancy of polymers. Organomodified layered silicates or clays are the main category of nanoparticles leading to significant improvements of thermal stability and flame retardancy of nanocomposites. Among the other categories of nanoparticles able to improve the flame retardancy of polymers, metallic oxides and hydroxides play a particular role because of their specific modes of action based on restriction of polymer chain mobility, catalytic effects on polymer degradation, and limitation of combustible volatiles and oxygen transfer (barrier effects), as well as endothermic water vapor release for hydroxides.

   Apart from micrometric ATH, which is one of the most used FR, other kinds of alumina at nanometric scale have been investigated as FR agents or components of FR systems. Alumina monohydrate (AlOOH) also called boehmite, despite its lower water content, in comparison with ATH, is able to create barrier effects, regarding its aspect ratio, and also able to promote the formation of a charred structure by catalytic action of its surface. Moreover, synergistic effects have also been noticed in combination with phosphorus-based additives, leading to aluminum phosphate. In the case of anhydrous alumina, the absence of water release can be overcome by a much higher specific surface area because of the very small size of the commercial alumina particles used, leading to stronger catalytic activity and better reactivity with phosphorus compounds. Particularly for PMMA nanocomposites, it has been shown that anhydrous alumina nanoparticles of around 13 nm could react with either ammonium polyphosphate or phosphinates. Synergistic effects on heat release rate were noticed and have been ascribed to the formation of charred and cohesive layers containing aluminum phosphates.

References

1. G. Pal and H. Macskasy, *Plastics: Their Behavior in Fires* (New York: Elsevier, 1991).
2. R. G. Gann, R. A. Dipert, and M. J. Drews, Flammability. In: *Encyclopaedia of Polymer Science and Engineering*, 2nd ed., ed. H. F. Mark and J. I. Kroschwitz (New York: Wiley, 1987).
3. R. N. Rothon and P. R. Hornsby, Flame retardant effects of magnesium hydroxide. *Polymer Degradation and Stability*, 54 (1996), 383–5.

4. D. Price, F. Gao, G. J. Milnes, E. B., C. I. Lindsay, and P. T. McGrail, Laser pyrolysis/time-of-flight mass spectrometry studies pertinent to the behaviour of flame-retarded polymers in real fire situations. *Polymer Degradation and Stability*, 64 (1999), 403–10.

5. J. T. Yeh, H. M. Yang, and S. S. Huang, Combustion of polyethylene filled with metallic hydroxides and crosslinkable polyethylene. *Polymer Degradation and Stability*, 50 (1995), 229–34.

6. F. Carpentier, Procédés de formulations de polymers hautement chargés. Application à un copolymère éthylène–acétate de vinyle. PhD thesis, Université des sciences et technologie de Lille, France, 2000.

7. A. Durin-France, L. Ferry, J.-M. Lopez Cuesta, and A. Crespy, Magnesium hydroxide/zinc borate/talc compositions as flame-retardants in EVA copolymer. *Polymer International*, 49 (2000), 1101–5.

8. C. M. Liauw, G. C. Lees, and S. J. Hurst, The effect of filler surface modification on the mechanical properties of aluminium hydroxide filled polypropylene. *Plastics Additives and Compounding*, 24 (1995), 249–60.

9. U. Hippi, J. Mattila, M. Korhonen, and J. Seppala, Compatibilization of polyethylene/aluminum hydroxide (PE/ATH) and polyethylene/magnesium hydroxide (PE/MH) composites with functionalized polyethylenes. *Polymer*, 74 (2003), 1193–1201.

10. X. Zhang, F. Guo, J. Chen, G. Wang, and H. Liu, Investigation of interfacial modification for flame retardant ethylene vinyl acetate copolymer/alumina trihydrate nanocomposites. *Polymer Degradation and Stability*, 87 (2005), 411–18.

11. K. Daimatsu, H. Sugimoto, Y. Kato, E. Nakanishi, K. Inomata, Y. Amekawa, and K. Takemura, Preparation and physical properties of flame retardant acrylic resin containing nano-sized aluminum hydroxide. *Polymer Degradation and Stability*, 92 (2007), 1433–8.

12. *Flameproof Polymer Composition*, World Intellectual Property Organization Patent WO0068312A1 (November 2000).

13. J. W. Gilman and T. Kashiwagi, Polymer-layered silicate nanocomposites with conventional flame retardants. In: *Polymer-Clay Nanocomposites*, ed. T. J. Pinnavaia and G. W. Beall (New York: Wiley, 2002).

14. M. C. Costache, M. J. Heidecker, M. J. Manias, and C. A. Wilkie, Preparation and characterization of poly(ethylene terephthalate)/clay nanocomposites by melt blending using thermally stable surfactants. *Polymers for Advanced Technologies*, 17 (2006), 764–71.

15. T. H. Chuang, W. Guo, K.-C. Cheng, S.-W. Chen, H.-T. Wang, and Y.-Y. Yen, Thermal properties and flammability of ethylene–vinyl acetate copolymer/montmorillonite/polyethylene nanocomposites with flame retardants. *Journal of Polymer Research*, 11 (2004), 169–74.

16. J. Zhang, J. Hereid, M. Hagen, D. Bakirtzis, M. A. Delichatsios, A. Fina, A. Castrovinci, G. Camino, F. Samyn, and S. Bourbigot, Effects of nanoclay and fire retardants on fire retardancy of a polymer blend of EVA and LDPE. *Fire Safety Journal*, 44 (2009), 504–13.

17. B. Morgan, Flame retarded polymer layered silicate nanocomposites: A review of commercial and open literature systems. *Polymers for Advanced Technologies*, 17 (2006), 206–17.

18. M. A. Cardenas, D. Garcia-Lopez, I. Gobernado-Mitre, J. C. Merino, J. M. Pastor, J. Martinez, J. Barbeta, and D. Calveras, Mechanical and fire retardant properties of

EVA/clay/ATH nanocomposites – Effect of particle size and surface treatment of ATH filler. *Polymer Degradation and Stability*, 93 (2008), 2032–7.

19. T. R. Hull, D. Price, Y. Liu, C. L. Wills, and J. Brady, An investigation into the decomposition and burning behaviour of ethylene–vinyl acetate copolymer nanocomposite materials. *Polymer Degradation and Stability*, 82 (2003), 365–71.

20. G. Beyer, Nanocomposites – A new class of flame retardants. *Plastics Additives and Compounding* (March/April 2009), 16–17, 19–21.

21. G. Beyer, Flame retardant properties of EVA-nanocomposites and improvements by combination of nanofillers with aluminium trihydrate. *Fire and Materials*, 25 (2001), 193–7.

22. Q. Kong, Y. Hua, L. Songa, and C. Yi, Synergistic flammability and thermal stability of polypropylene/aluminum trihydroxide/Fe-montmorillonite nanocomposites. *Polymers for Advanced Technologies*, 20 (2009), 404–9.

23. J. Zhang and C. A. Wilkie, Fire retardancy of polypropylene–metal hydroxide nanocomposite. In: *Fire and Polymers*, A.C.S. Symposium Series 922, ed. C. Wilkie and G. Nelson (Washington, DC: American Chemical Society, 2006), pp. 61–74.

24. B. Schartel, U. Knoll, A. Hartwig, and D. Pütz, Phosphonium-modified layered silicate epoxy resins nanocomposites and their combinations with ATH and organo-phosphorus fire retardants. *Polymers for Advanced Technologies*, 17 (2006), 281–93.

25. W. Lertwimolnun and B. Vergnes, Influence of compatibilizer and processing conditions on the dispersion of nanoclay in a polypropylene matrix. *Polymer*, 46 (2005), 3462–71.

26. S. J. Wilson, The dehydration of boehmite, γ-AlOOH, to γ-Al$_2$O$_3$. *Journal of Solid State Chemistry*, 30 (1979), 24–557.

27. J. W. Gilman, Flammability and thermal stability studies of polymer layered-silicate (clay) nanocomposites. *Applied Clay Science*, 15 (1999), 31–49.

28. G. Camino, A. Maffezzoli, M. Braglia, M. De Lazzaro, and M. Zammarano, Effect of hydroxides and hydroxycarbonate structure on fire retardant effectiveness and mechanical properties in ethylene–vinyl acetate copolymer. *Polymer Degradation and Stability*, 74 (2001), 457–64.

29. A. Laachachi, M. Cochez, M. Ferriol, J. M. Lopez-Cuesta, and E. Leroy, Influence of TiO$_2$ and Fe$_2$O$_3$ fillers on the thermal properties of poly(methyl methacrylate) (PMMA). *Materials Letters*, 59 (2005), 36–9.

30. A. Laachachi, E. Leroy, M. Cochez, M. Ferriol, and J. M. Lopez-Cuesta, Use of oxide nanoparticles and organoclays to improve thermal stability and fire retardancy of poly(methyl methacrylate). *Polymer Degradation and Stability*, 89 (2005), 344–52.

31. A. Laachachi, E. Leroy, M. Cochez, M. Ferriol, and J. M. Lopez-Cuesta, Thermal degradation and flammability of poly(methyl methacrylate) containing TiO$_2$ nanoparticles and modified montmorillonite. In: *Fire and Polymers*, A.C.S. Symposium Series 922 (2005), 36–47.

32. X. Zheng and C. A. Wilkie, Flame retardancy of polystyrene nanocomposites based on an oligomeric organically-modified clay containing phosphate. *Polymer Degradation and Stability*, 81 (2003), 539–50.

33. K. H. Pawlowski and B. Schartel, Flame retardancy mechanisms of aryl phosphates in combination with boehmite in bisphenol A polycarbonate/acrylonitrile–butadiene–styrene blends. *Polymer Degradation and Stability*, 93 (2008), 657–67.

34. Yu. S. Lipatov, V. F. Rosovitskii, and V. F. Babich, Effect of filler on relaxation-time spectra of filled polymers. *Mechanics of Composite Materials*, 11 (1975), 933–6.

35. D. Laachachi, M. Ferriol, M. Cochez, D. Ruch, and J.-M. Lopez-Cuesta, The catalytic role of oxide in the thermooxidative degradation of poly(methyl methacrylate)–TiO_2 nanocomposites. *Polymer Degradation and Stability*, 93 (2008), 1131–7.

36. Y. Grohens, M. Auger, R. Prud'homme, and J. Schultz, Adsorption of stereoregular poly(methyl methacrylates) on gamma-alumina: Spectroscopic analysis. *Journal of Polymer Science: Part B: Polymer Physics*, 37 (1999), 2985–95.

37. S.-C. Liufu, H.-N. Xiao, and Y.-P. Li, Thermal analysis and degradation mechanism of polyacrylate/ZnO nanocomposites. *Polymer Degradation and Stability*, 87 (2005), 103–10.

38. N. Cinausero, Etude de la dégradation thermique et de la réaction au feu de nanocomposites à matrice PMMA et PS. PhD thesis, Université de Montpellier, France (2008).

39. F. Yang, I. Bogdanova, and G. L. Nelson. Flammability of polymer–inorganic nanocomposites. In: *Fire and Polymers*, ed. C. A. Wilkie, A. B. Morgan, and G. L. Nelson, A.C.S. Symposium Series 1013 (2009).

40. N. Cinausero, N. Azema, M. Cochez, M. Ferriol, M. Essahli, F. Ganachaud, and J. M. Lopez Cuesta, Influence of the surface modification of alumina nanoparticles on the thermal stability and fire reaction of PMMA composites. *Polymers for Advanced Technologies*, 19 (2008), 701–9.

41. A. Laachachi, M. Cochez, E. Leroy, P. Gaudon, M. Ferriol, and J. M. Lopez Cuesta, Effect of Al_2O_3 and TiO_2 nanoparticles and APP on thermal stability and flame retardance of PMMA. *Polymers for Advanced Technologies*, 17 (2006), 327–34.

42. A. Laachachi, M. Cochez, E. Leroy, M. Ferriol, and J. M. Lopez-Cuesta, Fire retardant systems in poly(methyl methacrylate): Interactions between metal oxide nanoparticles and phosphinates. *Polymer Degradation and Stability*, 92 (2007), 61–9.

43. E. Gallo, U. Braun, B. Schartel, P. Russo, and D. Acierno, Halogen-free flame retarded poly(butylene terephthalate) (PBT) using metal oxides/PBT nanocomposites in combination with aluminium phosphinate. *Polymer Degradation and Stability*, 94 (2009), 1245–53.

13

Polymer/layered double hydroxide flame retardant nanocomposites

LONGZHEN QIU[a] AND BAOJUN QU[b]

[a]Hefei University of Technology
[b]University of Science and Technology of China

13.1 Introduction

With their ease of processing and high performance, polymeric materials have become a common and important part of modern life.* However, because almost all polymers are composed predominately of hydrocarbons, these materials are flammable and thus greatly increase fire hazard to human life and property. As estimated for the United States, there are approximately 400,000 residential fires each year, 20% involving electrical distribution and appliances, and 10% concerning upholstered furniture and mattresses. These fires kill about 4,000 people, injure 20,000 people, and result in property losses amounting to about US$4.5 billion. Flame retardants are additives that can make flammable materials more difficult to ignite and significantly reduce the spread of fire. Use of flame retardants plays a major role in fire safety, saving lives, and preventing injuries and property damage. For example, in 1974, the number of recorded television set fires in the United Kingdom was more than 2,300, whereas this number had decreased to 470 in 1989, despite the number of television sets in use increasing many times. This is because effective flame retardants were developed for television sets.

The main flame retardant systems currently in use are halogen-containing. However, in recent years, there has been increasing concern over their use, owing to their ability to emit toxic gases and smoke that can choke the people and damage the costly equipment exposed to them. Consequently, a growing number of restrictions and recommendations from government have promoted the development of halogen-free flame retardant polymers [1, 2].

There are many kinds of halogen-free flame retardants, such as metal hydroxides, phosphoros-containing compounds, and nitrogen-containing compounds. Because of their low toxicity, anticorrosion properties, low cost, and low emission of smoke during burning, metal hydroxides, mainly magnesium hydroxide and aluminum hydroxide, have drawn comprehensive interest for flame retardant systems. Generally, the flame-retarding mechanism of metal hydroxides is endothermic decomposition into the respective oxides and water upon heating; the released water vapor isolates the flame and dilutes the flammable

---

* The authors wish to acknowledge the contributions of all the references and figures used in this chapter.

gases in the gas phase. However, one of the major drawbacks of this class of additives is their relatively low flame retardant efficiency. Therefore, a high loading (typically >50 wt%) is needed to achieve flame retardancy, which will no doubt lead to deterioration in the physical/mechanical properties of the matrices [3]. Making nanosized metal hydroxide additives is an effective way to increase the flame retardant efficiency.

Polymer/layered silicate nanocomposites (PLSNs) present a new flame retardant approach and offer significant advantages over conventional formulations, where high loadings are often required. In the pioneer work, Gilman, Kashiwagi, and Lichtenhan [4] reported that the presence of nanodispersed montmorillonite (MMT) clay in nylon-6 matrices produces a substantial improvement in fire performance. Cone calorimetry data show that the peak heat release rate (PHRR), the most important parameter for predicting fire hazard, is reduced by 63% for the nanocomposite containing 5% MMT relative to virgin nylon-6. The main mechanism for this enhanced flame retardancy has been identified as carbonaceous-silicate barrier formation during combustion, which prevents the heat from reaching the underlying polymer and the mass transport of material from degrading the polymer. In spite of encouraging results obtained from cone calorimetry, the Underwriters Laboratory (UL-94) and limiting oxygen index (LOI) results of PLSNs are still very poor. This is because the layered silicates simply act as a physical barrier and remain chemically inert during burning, which makes the PLSNs burn slowly but remains the total heat release unchanged.

Layered double hydroxides (LDHs) are a different kind of layered crystalline filler for nanocomposite formation. Because they combine the flame retardant features of conventional metal hydroxide fillers (magnesium hydroxide and aluminum hydroxide) with those of layered silicate nanofillers (montmorillonite), LDHs are considered to be a new emerging class of nanofillers favorable for the preparation of flame retardant nanocomposites. In the present chapter, recent progress in the study of polymer/LDH flame retardant nanocomposites is reviewed.

13.2 Structure and properties of layered double hydroxides

LDHs, also known as hydrotalcite-like materials, are a class of host–guest materials consisting of positively charged metal oxide or hydroxide sheets with intercalated anions and water molecules. Generally, their chemical structure is represented by the formula $[M^{2+}_{1-x}M^{3+}_x(OH)_2]^{x+}$ $[A^{m-}]_{x/m}^{x-}$ $\bullet 2H_2O]$, where M^{2+} is a divalent metal ion (such as Mg^{2+} or Zn^{2+}), M^{3+} is a trivalent metal ion (such as Al^{3+} or Cr^{3+}), A^{m-} is an anion with valency m (such as CO_3^{2-}, Cl^-, or NO_3^-), and the value of x is equal to the molar ratio of $M^{2+}/(M^{2+} + M^{3+})$ and is generally in the range 0.2–0.33 [5].

The typical structure of LDH materials is presented in Figure 13.1. It can be seen that the structure of LDHs is based on that of magnesium hydroxide, brucite, in which Mg^{2+} ions are arranged in sheets, each magnesium ion being octahedrally surrounded by six hydroxide groups, whereas each hydroxide spans three magnesium ions. Isomorphous substitution

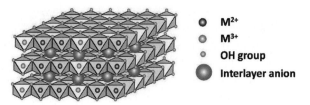

Figure 13.1 Layered crystal structure of layered double hydroxide.

for some fraction of the divalent ions of trivalent ions of comparable size (e.g., Al^{3+}, Fe^{3+}) forms mixed metal layers, that is, $[M^{2+}_{1-x}M^{3+}_x(OH)_2]^{x+}$, with a net positive charge. Electrical neutrality is maintained by the anions located in interlayer domains containing water molecules. These water molecules are connected to both the metal hydroxide layers and the interlayer anions through extensive hydrogen bonding. The presence of anions and water molecules leads to enlargement of the basal spacing from 0.48 nm in brucite to about 0.78 nm in Mg–Al LDH. Because the interlayer ions are confined to the interlayer space by a relatively weak electrostatic force, they can be removed without destroying the layered structures of LDH. This anionic exchange capacity is a key feature of LDHs, which makes them unique as far as inorganic materials are concerned.

Because of their unique layered structure and highly tunable chemical composition based on different metal species and interlayer anions, LDHs have many interesting properties, such as unique anion-exchanging ability, easy synthesis, high bond water content, "memory effect," nontoxicity, and biocompatibility. Based on these properties, LDHs are considered as very important layered crystals with potential applications in catalysis [6], controlled drugs release [7], gene therapy [8], improvement of heat stability and flame retardancy of polymer composites [9], controlled release or adsorption of pesticides [10], and preparation of novel hybrid materials for specific applications, such as visible luminescence [11], UV/photo stabilization [12], magnetic nanoparticle synthesis [13], or wastewater treatment [14].

13.3 Polymer/layered double hydroxide nanocomposites

The incorporation of polymers into LDHs to form polymer/LDH nanocomposites has been a subject of academic interest for more than 20 years. In their pioneer work, Sugahar *et al.* [15] prepared polyacrylonitrile (PAN)/MgAl LDH nanocomposite via in situ polymerization. In their work, acrylonitrile monomers were intercalated into MgAl LDH modified with dodecyl sulfate. The subsequent polymerization resulted in nanocomposites with well-ordered multilayer morphology built up from alternating polymeric and inorganic layers. Because the polymer content is low, this nanocomposite mainly presents the characteristics of inorganic fillers, and is preferably named organoceramic [16]. In the following decade, most efforts on polymer/LDH nanocomposites have concentrated on organoceramic [17–27]. However, there were few reports at that time on the exploration of polymer/LDH nanocomposites with polymeric matrices and LDH fillers.

Until 2003, Chen's [28], Qu's [29–31], and Hu's [32] groups independently reported nanocomposites with polymeric matrices for the first time the. In Hsueh and Chen's work, exfoliated polyimide/LDH was prepared by in situ polymerization of a mixture of aminobenzoate-modified Mg–Al LDH and polyamic acid (polyimide precursor) in *N,N*-dimethylactamide [28]. In other work, Chen and Qu successfully synthesized exfoliated polyethylene-g-maleic anhydride (PE-g-MA)/LDH nanocomposites by refluxing in a non-polar xylene solution of PE-g-MA [29, 30]. Then, Li *et al.* prepared poly(methyl methacrylate) (PMMA)/MgAl LDH by exfoliation/adsorption with acetone as cosolvent [32]. Since then, polymer/LDH nanocomposites have attracted extensive interest. The wide variety of polymers used for nanocomposite preparation include polyethylene (PE) [29, 30, 33–49], polystyrene (PS) [48, 50–58], poly(propylene carbonate) [59], poly(3-hydroxybutyrate) [60–62], poly(vinyl chloride) [63], syndiotactic polystyrene [64], polyurethane [65], poly[(3-hydroxybutyrate)-co-(3-hydroxyvalerate)] [66], polypropylene (PP) [48, 67–70], nylon 6 [9, 71, 72], ethylene vinyl acetate copolymer (EVA) [73–77], poly(L-lactide) [78], poly(ethylene terephthalate) [79, 80], poly(caprolactone) [81], poly(*p*-dioxanone) [82], poly(vinyl alcohol) [83], PMMA [32, 47, 48, 57, 84–93], poly(2-hydroxyethyl methacrylate) [94], poly(styrene-co-methyl methacrylate) [95], polyimide [28], and epoxy [96–98]. These nanocomposites often exhibit enhanced mechanical, thermal, optical, and electrical properties and flame retardancy. Among them, the thermal properties and flame retardancy are the most interesting and will be discussed in the following sections.

13.4 Thermal stability of polymer/layered double hydroxide nanocomposites

The thermal stability of a material is characterized mainly by thermogravimetric analysis (TGA), in which the sample mass loss due to volatilization of degraded by-products is monitored as a function of temperature. Qu *et al.* first reported the improved thermal stability of polymer/LDH nanocomposites that combined PE-g-MA and MgAl LDH [29] (Figure 13.2). TGA profiles of PE-g-MA/MgAl LDH nanocomposites show faster charring from 210 to 360 °C and greater thermal stability above 370 °C than PE-g-MA. The decomposition temperature of tnanocomposites, defined as 50% weight loss, can be 60 °C higher with 5 wt% MgAl LDH than that of PE-g-MA.

The dynamic Fourier transfer infrared (FTIR) spectrum is a relative new method that can be used to characterize the thermooxidative behavior of flame-retardant materials [99, 100]. FTIR can provide the structural changes and identify the products formed during thermooxidative degradation by recording spectra in situ during the degradation process. Figure 13.3 shows the changes of dynamic FTIR spectra obtained from the thermooxidative degradation of pure PE-g-MA [Figure 13.3 (A)] and PE-g-MA/5% MgAl LDH nanocomposite [Figure 13.3 (B)] samples in the condensed phase with increasing pyrolysis temperature from room temperature (RT) to 320 °C. There are two interesting regions in Figure 13.3. The first is located in the range 2800–3000 cm^{-1}, and the intensities of peaks at 2925 and 2854 cm^{-1} assigned to the CH_2 or CH_3 asymmetric and symmetric vibration,

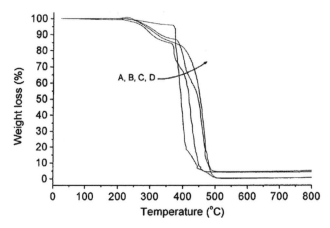

Figure 13.2 TGA profiles for (A) pure PE-g-MA and PE-g-MA/MgAl LDH exfoliation nanocomposites: (B) 5% LDH refluxed in air, (C) 2% LDH refluxed under nitrogen, and (D) 5% LDH refluxed under nitrogen. Reproduced with permission from Ref. [29].

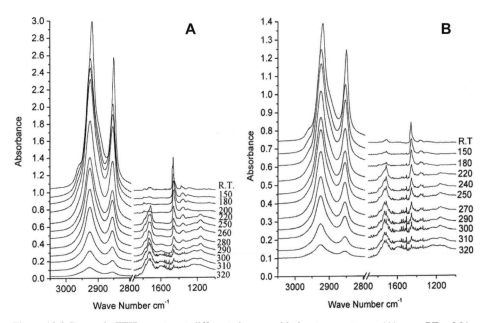

Figure 13.3 Dynamic FTIR spectra at different thermooxidative temperatures: (A) pure PE-g-MA; (B) PE-g-MA/5% MgAl LDH exfoliation nanocomposite. Reproduced with permission from Ref. [29].

respectively, decrease rapidly with the increase of pyrolysis temperature because of the thermooxidative degradation of PE-g-MA main chains. The second region of interest lies between 1700 and 1800 cm^{-1}, and the peaks at 1716 and 1780 cm^{-1} are assigned to the various C=O thermooxidative products. The peak intensities of these oxidative products

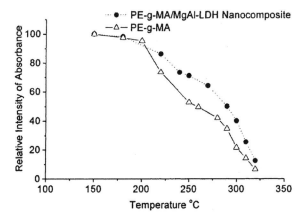

Figure 13.4 Relative peak intensities of absorbance at 2925 cm^{-1} assigned to CH_2 or CH_3 asymmetric vibration. Reproduced with permission from Ref. [29].

increase significantly for both PE-g-MA and PE-g-MA/MgAl LDH nanocomposite samples when the pyrolysis temperature reaches 220 °C. Although the dynamic FTIR spectra in Figure 13.3 (A, B) show very similar features, the rates of change of their peak intensities with increasing pyrolysis temperature are completely different (as shown in Figure 13.4). The PE-g-MA/MgAl LDH nanocomposite has a lower thermooxidative rate than pure PE-g-MA in the range from 200 to 320 °C.

There have been many reports of the improved thermal stability of nanocomposites prepared with various LDHs and polymer matrices. With the development of more and more nanocomposites, it is being found that improvement in thermal stability is a common feature of polymer/LDH nanocomposites. The barrier effect observed in polymer/layered silicate nanocomposites is generally used to explain this improvement. However, in a comparative experiment, Qiu, Chen, and Qu [35] found that PE/LDH nanocomposites show much higher thermal degradation temperatures than PE/MMT nanocomposites when they have the same filler content and similar structures (Figure 13.5). Isoconversional kinetic analysis is a common method of studying the kinetics of polymer degradation, which may provide information on the change of activation energy during thermooxidative degradation as well as offering mechanistic clues. Figure 13.6 shows the relationship of the activation energy (E_a) values calculated by the Ozawa–Flynn–Wall method with the extent of conversion (a) for the thermooxidative degradation of virgin LLDPE and its nanocomposites. It is clear that the mechanisms of enhanced thermal stability in the PE/MMT and PE/LDH nanocomposites are very different. The former is based mainly on protective charred layers formed by the MMT catalytic dehydrogenation of PE molecules, whereas the latter is based on the barrier effect of LDH layers with very high activation energy, which prevents the diffusion of oxygen from the gas phase into the polymer nanocomposites and thus not only protects the C–C main chain from thermal degradation but also hinders the dehydrogenation of PE molecules.

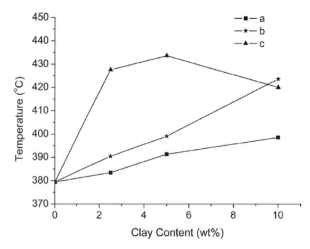

Figure 13.5 The variation of the temperature of 20 wt% mass loss ($T_{0.2}$) with the content of MMT (a), MgAl LDH (b), and ZnAl LDH (c) in the LLDPE nanocomposites. Reproduced with permission from Ref. [35].

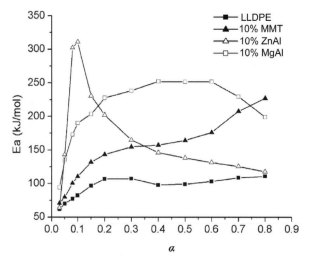

Figure 13.6 Dependence of the effective activation energy on the extent of conversion for the thermooxidative degradation of virgin LLDPE and its nanocomposites. Reproduced with permission from Ref. [35].

13.5 Fire retardant properties of polymer/layered double hydroxide nanocomposites

Because of the similar structure of metal hydroxides in chemistry, LDHs have been used as flame retardants since the 1970s [101]. Conventionally, a high loading of LDH (>50 wt%) is needed to achieve an acceptable industrial flammability rating (such as UL-94 V0), which drastically worsens the mechanical properties of the final products. The use of LDHs as

Table 13.1 *UL-94 V ratings*

UL-94 V rating	Flame duration	Total duration (five samples)	Dripping
V0 (best)	Less than 10 s	Less than 50 s	No dripping
V1 (good)	Less than 30 s	Less than 250 s	No dripping
V2 (drips)	Less than 30 s	Less than 250 s	Dripping allowed

flame-retardant nanofillers is a relatively new concept and offers significant advantages over conventional formulations.

Several test methods have been developed for evaluating flame retardancy:

- Cone calorimetry: Cone calorimetry is identified as the best laboratory instrument for evaluating the fire performance of polymers. During a cone calorimetry investigation, a constant external heat flux is maintained to sustain the combustion of the test sample; that is, the test method creates forced flaming combustion. The parameters that can be evaluated by cone calorimetry include the heat release rate (HRR), peak heat release rate (PHRR), time to ignition (t_{ign}), volume of smoke (VOS), total heat released (THR), a measure of the extent to which the entire polymer burns, and the average mass loss rate (AMLR). HRR, and in particular PHRR, has been found to be the most important parameter for evaluating fire retardants.

- LOI: The value of LOI is defined as the minimal oxygen concentration $[O_2]$ in the oxygen/nitrogen mixture $[O_2/N_2]$ that either maintains flame combustion of the material for 3 min or consumes a length of 5 cm of the sample, with the sample placed in a vertical position (the top of the test sample is lit with a burner). It is expressed as follows:

$$LOI = \frac{[O_2]}{[O_2] + [N_2]} \times 100.$$

A high LOI value indicates that the test sample cannot be ignited easily and is less flammable.

- UL-94: UL tests were originally developed by Underwriters Laboratory. They include a range of flammability tests, such as vertical burning tests (UL-94 V), horizontal burning tests (UL-94 HB), 500 w (125 mm) vertical burning tests (UL-94 5V), and radiant panel flame-spread tests (UL-94 RP). Among them, the most commonly used is UL-94 V, which measures the ignitability and flame spread of vertical bulk materials exposed to a small flame. The time taken for the sample to self-extinguish and the dripping effects are the main results obtained from UL-94 V test. A specimen is classified as V0, V1, or V2 according to the criteria listed in Table 13.1.

In 2005, Costantino *et al.* [41] first examined the flame performance of PE/LDH nanocomposites by cone calorimetry. They found that the nanocomposites containing 5 wt% LDH show a 55% reduction in PHRR. At almost the same time, Zammarano

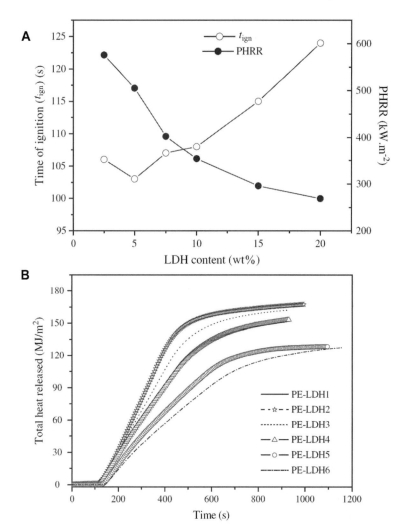

Figure 13.7 Cone calorimetry investigation results showing (A) variation of time of ignition (t_{ign}) and peak heat release rate (PHRR) with LDH content and (B) total heat released (THR) with time in LDPE/LDH nanocomposites. Reproduced with permission from Ref. [38].

et al. [98] noticed self-extinguishing behavior of epoxy/LDH nanocomposites during the UL-94 test and also greater reduction of the PHRR than in the montmorillonite-based nanocomposites. In a subsequent work, Costa, Wagenknecht, and Heinrich [38] reported the flammability and thermal properties of nanocomposites based on low-density polyethylene (LDPE) and MgAl LDH with exfoliated/intercalated structures. The cone calorimetry investigation revealed that the nanocomposites exhibit not only reduced THR (a measure of the propensity to produce long-duration fires), but also a smaller tendency to fast fire growth (measured by the ratio of PHRR and t_{ign}), as shown in Figure 13.7. The LOI and the dripping behavior are also improved by increasing LDH concentration. It has also been found that LDH materials facilitate the formation of carbonaceous char during combustion, making them more effective flame retardants than metal hydroxides. These results are very attractive because no relevant flame behavior enhancement in LOI and UL-94 tests except for dripping can be observed in polymer/layered silicate nanocomposites.

Because of the highly tunable properties of LDHs, it is possible to optimize their flame-retardant properties by adjusting many parameters, such as dispersibility, divalent metals, trivalent metals, and anions. Our aim in this section is to discuss recently published work on how these factors affect the flame retardancy of polymer/LDH nanocomposites.

13.5.1 Dispersibility

It is well known that good dispersion of layered silicates in the polymer matrix is necessary to obtain improved flame retardancy for polymer/layered silicate nanocomposites [102, 103]. The first question is then, what is the role of dispersibility in LDH-based nanocomposites? To answer this question, Wilkie and co-workers [89] prepared PMMA/LDH composites using unmodified hydrotalcite ($MgAlNO_3$ or $MgAlCO_3$), organo-modified MgAl LDHs (MgAlC16), and calcined oxides. The PMMA/MgAl-C16 exhibits an intercalated structure determined by both X-ray diffraction (XRD) and transmission electronic microscopy (TEM), whereas both $PMMA/MgAlNO_3$ and $PMMA/MgAlCO_3$ samples are found as microcomposites. Cone calorimetry experiments reveal that the PHRR of composites with 10% weight loading of filler is decreased by 51% for MgAl-C16, 20% for $MgAlNO_3$, 30% for $MgAlCO_3$, and 31% for calcined LDH compared with virgin PMMA. These results suggest that the dispersion state of LDHs strongly affects the flammability of composites.

Similar results are also observed in PS nanocomposites [56], which were prepared by free radical polymerization of styrene monomers in the presence of ZnAl and MgAl LDHs intercalated with 4,4′-azobis(4-cyanopentanoate) anions (LDH-ACPA). An intercalated–exfoliated morphology is observed for the composites of ZnAl-ACPA, whereas MgAl-ACPA shows microcomposite formation. The cone calorimetry results show good correlation between the reduction in PHRR and dispersion, in which the reduction in the peak heat release rate for 10% ZnAl-ACPA is 35% relative to the pristine polymer, whereas a 24% reduction is recorded for MgAl-ACPA at a similar loading.

In another work [48], a preliminary study to determine the compatibility of undecenoate-modified MgAl (MgAl-C11) with PE, PP, PS, and PMMA was carried out by melt-blending of the polymers with LDH filler. Morphological characterizations of all four systems using XRD and TEM reveal that good dispersion and nanocomposite formation were obtained in PMMA but not in the nonpolar polymers. As a result, the best reduction in PHRR at 10% loading of MgAl-C11 was observed from PMMA (52% reduction) and PS (20% reduction), whereas PE (7% reduction) and PP (11% reduction) showed no significant reductions. The surprise is that the maximum reduction for PMMA/layered silicate nanocomposites is about 30%. Therefore, the flame retardancy of LDHs is superior to that for layered silicates in PMMA nanocomposites. Moreover, it should be noted that a rather substantial reduction

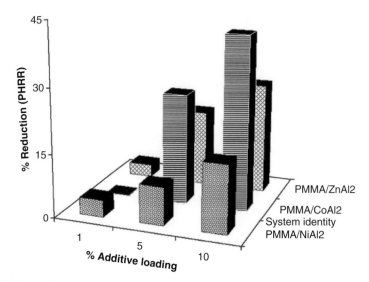

Figure 13.8 Comparing the fire properties of different PMMA–LDH systems: % reduction in PHRR versus % LDH loading versus system identity. Reproduced by permission of The Royal Society of Chemistry, Ref. [86].

in PHRR (20%) is still obtained, although the LDH is not well dispersed in the PS matrix. This is very different from MMT composites, for which a microcomposite will cause no change in the PHRR [104]. These observations suggest that MMT and LDH may affect flame retardancy by different pathways. More work is necessary to elucidate these pathways.

13.5.2 *Divalent metal cations*

To examine the role of divalent metal cations in the flame-retarding properties of nanocomposites, zinc–aluminum (Zn_2Al), cobalt–aluminum (Co_2Al), nickel–aluminum (Ni_2Al), and copper–aluminum (Cu_2Al) LDHs have been prepared by the coprecipitation method and used to prepare nanocomposites with PMMA [86]. XRD and TEM results suggest that an intercalated structure is obtained for Co_2Al, whereas the other LDHs lead to immiscible systems. In Figure 13.8, the reduction of PHRR is plotted against the additive loading and the system identity. The best reduction in PHRR (41%) was found with the PMMA/10% Co_2Al system, and the PMMA/10% Zn_2Al system gave a 26% reduction, whereas the smallest reduction, 16%, was obtained from the PMMA/10% Ni_2Al system. Obviously, the PHRR reduction in the polymer/LDH composite is highly dependent on the divalent metal cation.

Zn–Al and Mg–Al cation pairs are most commonly used to prepare LDHs. The effects of these two kinds of LDHs on dispersion, thermal, mechanical, and fire performance in various polymers have been systematically studied [47]. After being modified by oleate,

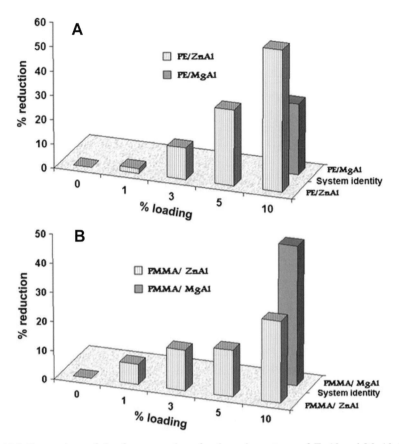

Figure 13.9 Comparison of the fire properties of polymeric systems of ZnAl and MgAl. The % reduction in PHRR is plotted versus the LDH loading (wt%): (A) PE systems; (B) PMMA systems. Reproduced with permission from Ref. [47].

Zn_2Al and Mg_2Al were melt-blended with PE, poly(ethylene-co-butyl acrylate) (PEBuA), and PMMA to prepare nanocomposites. Figure 13.9 shows the relationship between the reduction of PHRR and the LDH loading in PE and PMMA. A reduction in PHRR by 58% is recorded for PE/10% Zn_2Al, whereas a modest reduction is recorded for PE/10% Mg_2Al. In contrast, with the more polar polymer PMMA, a large reduction in PHRR (48%) is noted for 10% Mg_2Al, whereas Zn_2Al gives 29% at the same loading. However, either Zn_2Al or Mg_2Al LDHs can improve the fire properties of PEBuA. It is interesting that there is selective interaction between the LDHs and the polymers.

In a further work [70], a series of oleate-modified LDHs with varied ratios of zinc and magnesium were synthesized and used to prepare nanocomposites of PP. The reductions in the PHRR of the PP/oleate–LDHs relative to the pristine polymer during the combustion tests are shown in Figure 13.10 (A). At a 1% LDH loading, all five LDHs are ineffective in flame retardancy, as shown by the nonreduction in the PHRR for all five systems (that is,

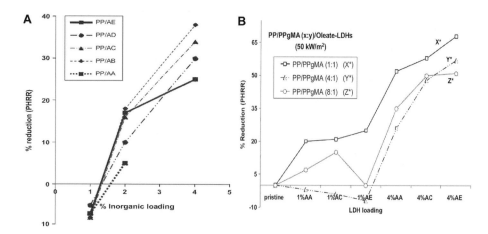

Figure 13.10 (A) Comparison of fire behavior of PP/oleate LDHs. % reduction in PHRR is plotted versus % LDH inorganic loading. (B) Effect of PP/PP-g-MA ratio and LDH loading on the reductions in PHRR. AA=Mg$_2$Al LDH, AB=Zn$_{0.5}$Mg$_{1.5}$Al LDH, AC= ZnMgAl LDH, AD=Zn$_{1.5}$Mg$_{0.5}$Al LDH, and AE= Zn$_2$Al LDH. Reproduced with permission from Ref. [70].

considering the ±10% error bars associated with cone calorimetry, there is no appreciable change in the PHRR at 1% LDH loading). Increasing the LDH loading results in steep increase of the reduction in PHRR for all LDHs. Systems with more magnesium show better fire retardancy. The Zn0.5Mg1.5Al LDH shows the best reduction in PHRR (38%). To optimize the dispersion of LDH in PP, polypropylene maleic anhydride copolymer (PP-MA) as a compatibilizer was incorporated into the PP matrix at different ratios. Figure 13.10 (B) provides a comparison between the reductions in PHRR recorded at different ratios of PP/PP-g-MA. Compared with PP results in Fig. 13.10 (A), it can be seen that the addition of PP-g-MA compatibilizer leads to greater reductions in PHRR (up to 68% for PP/PP-g-MA (1:1) containing 4% Zn$_2$Al LDH). Furthermore, the variation in the LDH composition also affects the reduction in PHRR. But in contrast to PP nanocomposites, the nanocomposites of PP/PP-g-MA have larger reductions in PHRR with increasing the content of zinc in LDHs.

Not only the divalent metal species but also the metal ratios may influence the fire retardant properties of LDHs. By comparing the cone calorimetry results for PMMA nanocomposites based on ZnAl LDH with Zn:Al ratios of 2:1 and 3:1, Wilkie and co-workers found that Zn$_3$Al gives a reduction of 35% in PHRR, whereas Zn$_2$Al gives 26% when both are at 10% LDH loading [86].

13.5.3 Trivalent metal cations

The effect of trivalent metal cations on fire properties was first investigated in microcomposites. Jiao *et al.* [105] studied the fire properties of EVA/MgAlFe ternary LDHs composites at 50% LDH loading. They found that the Fe^{3+} cation is important in improving the fire retardant properties of these systems. The EVA/MgAlFe–CO_3 composites containing a suitable amount of Fe^{3+} ion can reach the V0 rating in the UL-94 test, whereas the composites with an iron-free LDH (MgAlCO$_3$) cannot pass the UL-94 V0 rating.

In a recent work [87], undecenoate-modified Ca$_3$Al and Ca$_3$Fe LDHs were synthesized and used to prepare nanocomposites of PMMA. Although XRD results revealed that the diffraction peaks for both Ca$_3$Al and Ca$_3$Fe composites disappeared at all loading (1, 5, 10%), TEM images showed that dispersion of Ca$_3$Al LDHs was better than Ca$_3$Fe LDHs. Both LDH materials gave promising fire retardancy when combined with PMMA. PMMA/10% Ca$_3$Al sample gave greater reduction in PHRR (54%) relative to PMMA/10% Ca$_3$Fe (34%). However, it is still hard to conclude whether this improvement arises from the species of trivalent metal or from the difference in the dispersion of LDHs.

13.5.4 Intercalated anions

The organic modification of LDHs is an inevitable step in the process of polymer nanocomposites preparation. Therefore, the intercalated anions are important to the dispersion of LDHs and the resulting fire properties. Zammarano *et al.* [98] examined the effect of two organic modifiers, 3-aminobenzenesulfonic acid (ABS) and 4-toluenesulfonic acid monohydrate (TS), on the structure and properties of epoxy/LDH composites. Because of their reactivity with epoxy, ABS-modified LDHs (LDH-ABS) had better dispersion than LDH-TS. As a result, epoxy/LDH-ABS showed a greater reduction in PHRR (51%) than epoxy/LDH-TS at the same LDH loading.

Wilkie and co-workers [57] utilized the anions 2-ethylhexyl sulfate, bis(2-ethylhexyl) phosphate, and dodecyl benzenesulfonate as intercalated anions to synthesize organo-LDHs. Nanocomposites of PMMA and PS with organo-LDHs were prepared both by melt-blending and by bulk polymerization. XRD and TEM results revealed that the phosphate and sulfonate LDHs in PMMA show fairly good dispersion at the nanometer scale, whereas sulfa LDH is poorly dispersed. For PS, the LDHs are poorly dispersed and agglomerated LDH particles are observed. The reductions in PHRR for nanocomposites containing sulfate, phosphate, and sulfonate LDH are 27, 37, and 45% in PMMA and 32, 33, and 49% in PS, respectively. Both PMMA and PS samples obtained from bulk polymerization show poorer dispersion and less reduction in PHRR than samples obtained from melt-blending.

In another work, series of carboxylates from C-10 to C-22 have been used to examine the effect of anions on fire retardancy of PMMA/MgAl and PS/MgAl nanocomposites [90]. All the carboxylate-modified LDHs are well dispersed in PMMA, whereas none of them is well dispersed in PS. Figure 13.11 shows the plot of PHRR versus number of carbons in the modifiers for both PMMA and PS nanocomposites. There is a little dependence of PHRR on the length of the carboxylate for PMMA. The reduction in PHRR is between 49 and 58% for all of the carboxylate-modified LDHs at 10% LDH loading. However, the PHRR reduction of the PS system falls off as the carboxylate chain length increases: 56%

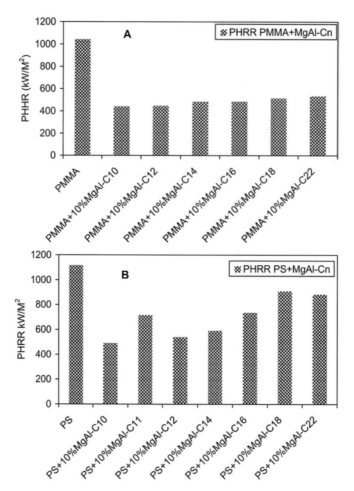

Figure 13.11 PHRR plots for (A) PMMA + MgAl–C$_n$ LDHs and (B) PS + MgAl–C$_n$ LDHs. Reproduced by permission of The Royal Society of Chemistry, Ref. [90].

for C-10 and 21% for C-22. It is surprising that a reduction of 56% in PHRR is achieved for a poorly dispersed PS/LDH.

Therefore, the effects of anions in LDHs on the fire properties of microcomposites were also studied [88]. MgAl-LDHs intercalated with various benzyl anions [i.e., benzoic (BA), benzenesulfonic (BS), 4-aminobenzoic (ABA), or benzenephosphonic acid (BP)] were prepared by different methods and then melt-blended with PMMA. For LDHs prepared using the coprecipitation method, the best reductions, 46% and 35%, are obtained with 10% MgAl-BA and 10% MgAl-ABA, whereas the systems with MgAl-BP and MgAl-BS give reductions that are much lower, 20% and 26%, respectively. Based on the HRR values, there are no significant differences in the fire retardant effectiveness of the LDHs prepared using different methods (i.e., coprecipitation, ion exchange, and rehydration methods).

13.6 Synergistic layered double hydroxide flame retardant systems

Although nanoscale dispersed LDHs can significantly improve the flammability properties of the polymer matrix, the utility of these alone is insufficient for ensuring adequate fire resistance to meet the required standards, such as LOI values and UL-94 ratings, especially at low LDH concentrations. Their association with other flame retardant systems could potentially be a very interesting approach to resolving this problem. Several recent works have focused on such methods. For example, Zammarano *et al.* [98] combined MgAl LDH with ammonium polyphosphate (APP) in epoxy-based composites. They found that the concentration of APP can be reduced from 30 to 16–20 wt% to obtain the UL-94 V0 rating when 4 wt% organo-LDH was added. Apparently, MgAl LDH and APP have a synergistic effect on the flammability of epoxy.

Qu *et al.* [73] reported the synergistic effects of exfoliated LDH with some halogen-free flame retardant (HFFR) additives, such as hyperfine magnesium hydroxide (HFMH), microencapsulated red phosphorus (MRP), and expandable graphite (EG), in the LDPE/EVA matrix. The sample with 100 phr HFMH showed a LOI value of 31 and failed to pass V0 rating in the UL-94 V test. Equivalent replacement of 5 phr HFMH with MRP or EG increased the LOI values to 34 and passed the V0 rating, but dripping still occurred. However, samples containing 90 phr HFMH, 5 phr MRP, and 5 phr LDH and samples containing 90 phr HFMH, 5 phr EG, and 5 phr LDH not only show higher LOI values of 36 and 38, but also can pass the V0 rating without dripping.

Recently, Wilkie's group studied the combination of LDHs with other flame retardants in PS, PMMA, PE, and EVA using cone calorimetry [46, 58, 85]. PS composites containing MgAl LDH and APP were prepared via melt-blending [58]. Composites with 5% and 10% MgAl LDH show reductions in PHRR of 17% and 27%, respectively. When APP is added to PS in the same weight fractions, smaller PHRR reductions were observed, 11% and 22%, respectively. However, the observed reduction of 42% in PHRR for PS containing 5% MgAl LDH and 5% APP is significantly higher than the linear additive result of 28% (as shown in Figure 13.12). Synergistic interactions between MgAl LDH and APP are thus implied. A combination of melamine and undecenoate-modified ZnAl LDH in PMMA has also been investigated [85]. Both melamine and LDH were found to be effective alone with PMMA, but a sample containing both melamine (10%) and LDH (5%) showed the best performance when reduction in PHRR, fire performance indices (FPI), and fire growth rate (FIGRA) were used as the indicators (Figure 13.13). The results showed that the combination of these two additives is beneficial for enhancing the flame retardant properties. However, in another work [46], the combination of ZnAl LDH with some conventional fire retardants, such as melamine polyphosphate (MPP), APP, triphenol phosphate (TPP), resorcinol diphosphate (RDP), decabromophenyl oxide (DECA), and antimony oxide (AO), in PE does not offer any advantage in PHRR reduction, and their effects on the time to ignition and/or the time to PHRR are also very limited. Therefore, synergistic effects 143 also related to the structure of the polymer matrix.

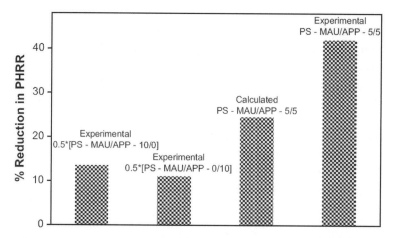

Figure 13.12 Experimental and calculated additive % reduction in PHRR for PS – MAU/APP – 5/0, PS – MAU/APP – 0/5, and PS – MAU/APP – 5/5. Reproduced with permission from Ref. [58].

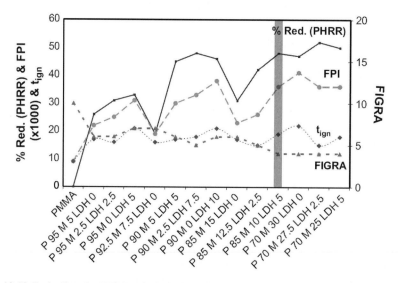

Figure 13.13 Reduction in PHRR, time to ignition (t_{ign}), fire performance indexes (FPI), and fire growth rate (FIGRA) of different PMMA, d ZnAl, and melamine compositions (P=PMMA; M=melamine; LDH=Zn_3Al undecenoate). FPI = t_{ign}/PHRR and FIGRA = PHRR/time to PHRR. Reproduced with permission from Ref. [85].

In a recent work, Zhang, Ding, and Qu examined the effects of LDHs with different divalent metal cations on the thermal and flammable properties of PP using intumescent flame retardant (IFR) and LDHs as flame retardants [68]. They found the addition of 1 wt% ZnAl, MgAl, CuAl, and CaAl LDHs can decrease the PHRR of the PP/IFR systems considerably. The lowest PHRR was observed in a PP/IFR/ZnAl-LDH sample.

Recently, a new method has been developed for preparing multicomponent flame retardant polymer composites. In this method the common flame retardants are intercalated into the gallery of LDHs instead of simply being blended. The intumescent flame retardants of ammonium polyphosphate (APP) and pentaerythritol (PER) were first intercalated into the galleries of ZnAl LDH, as reported by Qu and co-workers [69]. Although its flame performance did not change much, the new composite showed much better optical and mechanical properties than conventional PP/APP/PER/ZnAl LDH. The flammability characteristics and flame retardant mechanism of phosphate-intercalated LDH (MgAl-PO_4) in EVA have also been studied by the same group [74]. The cone calorimetry tests indicate that the HRR and MLR values of the EVA/MgAl-PO_4 samples are much lower than those of the EVA/carbonate-intercalated LDH (MgAl-CO_3) samples. The LOI values of EVA/MgAl-PO_4 samples are 2% higher than those of the corresponding EVA/MgAl-CO_3 samples in the range of 40–60 wt% loadings. The dynamic FTIR spectra reveal that the flame retardant mechanism of MgAl-PO_4 can be ascribed to its catalytic degradation of the EVA resin, which promotes the formation of charred layers with the P–O–P and P–O–C complexes in the condensed phase. The SEM observations give further evidence of this mechanism, in which the compact charred layers formed from the EVA/MgAl-PO_4 sample effectively protect the underlying polymer from burning. Nyambo and Wilkie also prepared LDH intercalated with borate anions and used it to prepare EVA compounds [77]. The fire properties of the composites, obtained by cone calorimetry, show that the addition of LDHs reduces the PHRR and AMLR significantly. The reductions in PHRR and AMLR increase with increasing loading of LDHs. At 40% loading, the reduction in PHHR observed from composites containing LDHs was significantly higher than for those consisting of zinc hydroxide, magnesium hydroxide, and their combinations, but similar to that for a combination of magnesium hydroxide and aluminum trihydroxide.

13.7 Possible mechanisms

Three main mechanisms for the effects of polymer/LDH nanocomposites on fire properties involve barrier formation during combustion, melt viscosity increase, and nanodispersed LDH layers.

Barrier formation during combustion is the most widely accepted mechanism for flame retardation in layered silicate nanocomposites. When the nanocomposites burn, a carbonaceous-silicate char layer forms by pyrolysis of the polymer and collapse of the nanocomposite's structure, which prevents the heat from reaching the underlying polymer and blocks the mass transport of material from the degrading polymer [106]. Zammarano *et al.* [98] first examined the mechanism of flame retardancy in epoxy/LDH nanocomposites. They found that the residue of the nanocomposite obtained after a cone calorimeter test has a thin shell structure. The thickness of this shielding layer is about 1 mm; its maximum height can reach 5 cm. This structure is brittle, but has good mechanical strength, integrity, coherence, and adherence to the substrate. As shown in Figure 13.14, the shell has a bilayered structure: The white and porous internal face is formed by metal oxides, whereas the

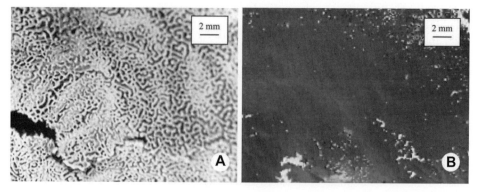

Figure 13.14 Residue of epoxy/LDH after cone calorimetry test: (A) white porous structure of mixed oxides on the internal face; (B) black compact carbonaceous residue on the external face. Reproduced with permission from Ref. [98].

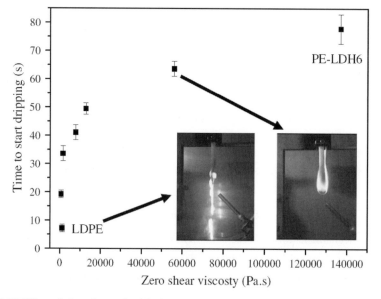

Figure 13.15 Effect of viscosity on the dripping tendency of the LDPE/LDH nanocomposites. Reproduced with permission from Ref. [38].

black and compact external face is formed by carbonaceous residues, some of which may be from soot formed in the gas phase. The char layer is a physical barrier that prolongs burning times without decreasing the total amount of combustible material. Similar results are also observed in other polymer/LDH nanocomposites [38, 47, 57, 86, 88, 89].

Increasing melt viscosity, which affects the dripping behavior, is reported for all kinds of nanocomposites. An increase in melt viscosity indicates efficiency in prevention of dripping. Drip behavior cannot be observed from cone calorimetry, as dripping is prevented in this test, but it can be observed from LOI tests and UL-94. Figure 13.15 presents the effect of viscosity on the dripping tendency of the LDPE/LDH nanocomposites. It can be seen that the time to dripping increases steadily with increasing viscosity of the nanocomposites. The LDPE/LDH nanocomposites pass the UL-94 HB rating and show much better performance in terms of burning rate than the unfilled polyethylene.

Nanodispersed metal hydroxides have been proved as efficient flame retardants for polymeric materials. It has been shown [107] that the LOI obtained from EVA containing 50 wt% $Mg(OH)_2$ increases from 24% to 38.3% when micrometric $Mg(OH)_2$ (2–5 µm) is replaced with nanometric $Mg(OH)_2$. The enhancement of EVA flame retardancy by nanosized $Mg(OH)_2$ was attributed to the good dispersion of the nanoparticles, which leads to the formation of more compact and cohesive charred layers during the combustion test. Therefore, the nanodispersed LDH layers may also contribute to the flame retardancy of polymer/LDH nanocomposites.

13.8 Conclusion and future trends

In this chapter, we discussed recent progress in the use of LDH as a new kind of inorganic filler for flame retardant polymer/LDH nanocomposites. As nanofillers, LDH materials combine the features of conventional metal hydroxide fillers such as magnesium hydroxide and aluminum hydroxide with those of layered silicate nanofillers such as montmorillonite. The former confer excellent flame retardant properties on LDHs, because they can undergo endothermic decomposition in the fire and thereby release the bound water and produce metal oxide layered residues to protect the underlayer polymer from further decomposition. The latter make it possible to achieve good dispersion in polymer matrices through intercalation, and thus enhance mechanical properties and gas barrier properties. Therefore, LDH materials are considered as flame retardants with great prospects for polymers. Much attention has been attracted to polymer/LDH nanocomposites in the past decade. TGA and dynamic FTIR revealed that most of these nanocomposites show improved thermal stability. Cone calorimetry revealed that the nanocomposites exhibit not only reduced PHRR, but also lower THR and longer t_{ign}. LOI and UL-94 ratings are also improved in some polymer/LDH nanocomposites. This flame performance is very different from that observed in polymer/layered silicate nanocomposites, where the improvement in flame retardancy can only be observed in cone calorimetry tests, not in UL-94 tests and LOI.

However, investigation of LDHs as nanofillers in polymers for fire retardancy is in its infancy. Many issues are still not resolved. First, it is still a challenge to obtain a high degree of exfoliation of inorganic clay materials in a polymer matrix through melt-compounding processes. Second, there is no clear mechanism for the improvement in thermal and flame properties. Third, the relationship between flame retardancy and the dispersion of the LDH in the polymer matrix is still unclear, because the characterization and quantification of the nanomorphology by the usual techniques (e.g., XRD and TEM) is not an easy task. Fourth, although some effects of chemical components (e.g., divalent metals, trivalent metals, and intercalated ions) on flame retardancy have been examined, the results are still very ambiguous, and some are even in conflict. To resolve these issues, much more work is necessary.

In spite of the encouraging results obtained in polymer/LDH flame retardant nanocomposites, the use of LDHs alone is insufficient for ensuring adequate fire resistance to meet the required standards, such as LOI values and UL-94 test ratings, especially at low LDH concentrations. The combination of LDH with conventional flame retardants is an effective way to avoid this limitation. By this means, it is possible to reach the flame retardancy required by the market with a halogen-free, nontoxic flame retardant system and improved mechanical properties. There are also many issues concerning the synergy between LDH and conventional flame retardants.

Clearly, LDHs offer a novel means of developing flame retardant polymeric materials. Increased research in this area continues to point to their potential. Once the present issues are resolved, polymer/LDH nanocomposites may satisfy the requirement for high-performance flame retardant systems.

References

1. R. C. Xie and B. J. Qu, Synergistic effects of expandable graphite with some halogen-free flame retardants in polyolefin blends. *Polymer Degradation and Stability*, 71 (2001), 375–80.
2. A. K. Sen, B. Mukherjee, A. S. Bhattacharya, L. K. Sanghi, P. P. De, and A. K. Bhowmick, Preparation and characterization of low-halogen and non-halogen fire-resistant low-smoke (FRLS) cable sheathing compound from blends of functionalized polyolefins and PVC. *Journal of Applied Polymer Science*, 43 (1991), 1673–84.
3. Z. Z. Wang, B. J. Qu, W. C. Fan, and P. Huang, Combustion characteristics of halogen-free flame-retarded polyethylene containing magnesium hydroxide and some synergists. *Journal of Applied Polymer Science*, 81 (2001), 206–14.
4. J. W. Gilman, T. Kashiwagi, and J. D. Lichtenhan, Nanocomposites: A revolutionary new flame retardant approach. *SAMPE Journal*, 33 (1997), 40–46.
5. D. G. Evans and R. C. T. Slade, Structural aspects of layered double hydroxides. *Structure and Bonding*, 119 (2006), 1–87.
6. B. M. Choudary, B. Bharathi, C. V. Reddy, M. L. Kantam, and K. V. Raghavan, The first example of catalytic N-oxidation of tertiary amines by tungstate-exchanged Mg–Al layered double hydroxide in water: A green protocol. *Chemical Communications* (2001), 1736–7.
7. J. H. Choy, S. Y. Kwak, Y. J. Jeong, and J. S. Park, Inorganic layered double hydroxides as nonviral vectors. *Angewandte Chemie International Edition*, 39 (2000), 4042–5.
8. L. Desigaux, M. Ben Belkacem, P. Richard, J. Cellier, P. Leone, L. Cario, F. Leroux, C. Taviot-Gueho, and B. Pitard, Self-assembly and characterization of layered double hydroxide/DNA hybrids. *Nano Letters*, 6 (2006), 199–204.
9. L. C. Du, B. J. Qu, and M. Zhang, Thermal properties and combustion characterization of nylon 6/MgAl-LDH nanocomposites via organic modification and melt intercalation. *Polymer Degradation and Stability*, 92 (2007), 497–502.
10. M. Lakraimi, A. Legrouri, A. Barroug, A. de Roy, and J. P. Besse, Removal of pesticides from water by anionic clays. *J. Chim. Phys. Phys. – Chim. Biol.*, 96 (1999), 470–78.
11. D. Yan, J. Lu, M. Wei, J. Ma, D. G. Evans, and X. Duan, A combined study based on experiment and molecular dynamics: Perylene tetracarboxylate intercalated in a

layered double hydroxide matrix. *Physical Chemistry Chemical Physics*, 11 (2009), 9200–9209.

12. Y. Tian, G. Wang, F. Li, and D. G. Evans, Synthesis and thermo-optical stability of *o*-methyl red-intercalated Ni–Fe layered double hydroxide material. *Materials Letters*, 61 (2007), 1662–6.

13. A. V. Lukashin, A. A. Vertegel, A. A. Eliseev, M. P. Nikiforov, P. Gornert, and Y. D. Tretyakov, Chemical design of magnetic nanocomposites based on layered double hydroxides. *Journal of Nanoparticle Research*, 5 (2003), 455–64.

14. D. Mohan and C. U. Pittman, Arsenic removal from water/wastewater using adsorbents – A critical review. *Journal of Hazardous Materials*, 142 (2007), 1–53.

15. Yoshiyuki Sugahar, Norimasa Yokoyama, Kazuyuki Kuroda, and Chuzo Kato, AlN formation from a hydrotalcite–polyacrylonitrile intercalation compound by carbothermal reduction. *Ceramics International*, 14 (1988), 163–7.

16. F. Leroux and J. P. Besse, Polymer interleaved layered double hydroxide: A new emerging class of nanocomposites. *Chemistry of Materials*, 13 (2001), 3507–15.

17. E. M. Moujahid, J. P. Besse, and F. Leroux, Synthesis and characterization of a polystyrene sulfonate layered double hydroxide nanocomposite. In-situ polymerization vs. polymer incorporation. *Journal of Materials Chemistry*, 12 (2002), 3324–30.

18. O. C. Wilson, T. Olorunyolemi, A. Jaworski, L. Borum, D. Young, A. Siriwat, E. Dickens, C. Oriakhi, and M. Lerner, Surface and interfacial properties of polymer-intercalated layered double hydroxide nanocomposites. *Applied Clay Science*, 15 (1999), 265–79.

19. S. Rey, J. Merida-Robles, K. S. Han, L. Guerlou-Demourgues, C. Delmas, and E. Duguet, Acrylate intercalation and in situ polymerization in iron substituted nickel hydroxides. *Polymer International*, 48 (1999), 277–82.

20. C. Vaysse, L. Guerlou-Demourgues, E. Duguet, and C. Delmas, Acrylate intercalation and in situ polymerization in iron-, cobalt-, or manganese-substituted nickel hydroxides. *Inorganic Chemistry*, 42 (2003), 4559–67.

21. M. Tanaka, I. Y. Park, K. Kuroda, and C. Kato, Formation of hydrotalcite–acrylate intercalation compounds and their heat-treated products. *Bulletin of Chemical Society of Japan*, 62 (1989), 3442–5.

22. C. O. Oriakhi, I. V. Farr, and M. M. Lerner, Incorporation of poly(acrylic acid), poly(vinylsulfonate) and poly(styrenesulfonate) within layered double hydroxides. *Journal of Materials Chemistry*, 6 (1996), 103–7.

23. P. B. Messersmith and S. I. Stupp, Synthesis of nanocomposites – Organoceramics. *Journal of Materials Research*, 7 (1992), 2599–611.

24. P. B. Messersmith and S. I. Stupp, High-temperature chemical and microstructural transformations of a nanocomposite organoceramic. *Chemistry of Materials*, 7 (1995), 454–60.

25. Q. Z. Yang, D. J. Sun, C. G. Zhang, X. J. Wang, and W. A. Zhao, Synthesis and characterization of polyoxyethylene sulfate intercalated Mg–Al-nitrate layered double hydroxide. *Langmuir*, 19 (2003), 5570–74.

26. N. T. Whilton, P. J. Vickers, and S. Mann, Bioinorganic clays: Synthesis and characterization of amino- and polyamino acid intercalated layered double hydroxides. *Journal of Materials Chemistry*, 7 (1997), 1623–9.

27. T. Challier and R. C. T. Slade, Nanocomposite materials – Polyaniline-intercalated layered double hydroxides. *Journal of Materials Chemistry*, 4 (1994), 367–71.

28. H. B. Hsueh and C. Y. Chen, Preparation and properties of LDHs/polyimide nanocomposites. *Polymer*, 44 (2003), 1151–61.

29. W. Chen and B. J. Qu, Structural characteristics and thermal properties of PE-g-MA/ MgAl-LDH exfoliation nanocomposites synthesized by solution intercalation. *Chemistry of Materials*, 15 (2003), 3208–13.
30. W. Chen and B. J. Qu, Synthesis and characterization of PE-g-MA/MgAl-LDH exfoliation nanocomposite via solution intercalation. *Chinese Journal of Chemistry*, 21 (2003), 998–1000.
31. W. Chen, B. J. Qu, and L. Feng, Synthesis of PMA/ZnAl-LDH intercalation nanocomposite by in situ polymerization and its morphology [in Chinese]. *Chemical Journal of Chinese Universities*, 24 (2003), 1920–22.
32. B. G. Li, Y. Hu, J. Liu, Z. Y. Chen, and W. C. Fan, Preparation of poly (methyl methacrylate)/LDH nanocomposite by exfoliation–adsorption process. *Colloid and Polymer Science*, 281 (2003), 998–1001.
33. W. Chen, L. Feng, and B. J. Qu, Preparation of nanocomposites by exfoliation of ZnAl layered double hydroxides in nonpolar LLDPE solution. *Chemistry of Materials*, 16 (2004), 368–70.
34. W. Chen and B. J. Qu, LLDPE/ZnAlLDH-exfoliated nanocomposites: Effects of nanolayers on thermal and mechanical properties. *Journal of Materials Chemistry*, 14 (2004), 1705–10.
35. L. Z. Qiu, W. Chen, and B. J. Qu, Morphology and thermal stabilization mechanism of LLDPE/MMT and LLDPE/LDH nanocomposites. *Polymer*, 47 (2006), 922–30.
36. F. R. Costa, M. Abdel-Goad, U. Wagenknecht, and G. Heinrich, Nanocomposites based on polyethylene and Mg–Al layered double hydroxide. I. Synthesis and characterization. *Polymer*, 46 (2005), 4447–53.
37. F. R. Costa, B. K. Satapathy, U. Wagenknecht, R. Weidisch, and G. Heinrich, Morphology and fracture behaviour of polyethylene/Mg–Al layered double hydroxide (LDH) nanocomposites. *European Polymer Journal*, 42 (2006), 2140–52.
38. F. R. Costa, U. Wagenknecht, and G. Heinrich, LDPE/Mg-Al layered double hydroxide nanocomposite: Thermal and flammability properties. *Polymer Degradation and Stability*, 92 (2007), 1813–23.
39. F. R. Costa, U. Wagenknecht, D. Jehnichen, M. A. Goad, and G. Heinrich, Nanocomposites based on polyethylene and Mg–Al layered double hydroxide. Part II. Rheological characterization. *Polymer*, 47 (2006), 1649–60.
40. F. R. Costa, U. Wagenknecht, D. Jehnichen, and G. Heinrich, Nanocomposites based on polyethylene and Mg–Al layered double hydroxide: Characterisation of modified clay, morphological and rheological analysis of nanocomposites. *Plastics Rubber and Composites*, 35 (2006), 139–48.
41. U. Costantino, A. Gallipoli, M. Nocchetti, G. Camino, F. Bellucci, and A. Frache, New nanocomposites constituted of polyethylene and organically modified ZnAl-hydrotalcites. *Polymer Degradation and Stability*, 90 (2005), 586–90.
42. A. Schonhals, H. Goering, F. R. Costa, U. Wagenknecht, and G. Heinrich, Dielectric properties of nanocomposites based on polyethylene and layered double hydroxide. *Macromolecules*, 42 (2009), 4165–74.
43. L. C. Du and B. J. Qu, Structural characterization and thermal oxidation properties of LLDPE/MgAl-LDH nanocomposites. *Journal of Materials Chemistry*, 16 (2006), 1549–54.
44. L. C. Du and B. J. Qu, Preparation of LLDPE/MgAl-LDH exfoliation nanocomposites with enhanced thermal properties by melt intercalation. *Chinese Journal of Chemistry*, 24 (2006), 1342–5.

45. P. Ding and B. J. Qu, Structure, thermal stability, and photocrosslinking characterization of HDPE/LDH nanocomposites synthesized by melt-intercalation. *Journal of Polymer Science, Part B: Polymer Physics*, 44 (2006), 3165–72.

46. C. Manzi-Nshuti, J. M. Hossenlopp, and C. A. Wilkie, Comparative study on the flammability of polyethylene modified with commercial fire retardants and a zinc aluminum oleate layered double hydroxide. *Polymer Degradation and Stability*, 94 (2009), 782–8.

47. C. Manzi-Nshuti, P. Songtipya, E. Manias, M. M. Jimenez-Gasco, J. M. Hossenlopp, and C. A. Wilkie, Polymer nanocomposites using zinc aluminum and magnesium aluminum oleate layered double hydroxides: Effects of LDH divalent metals on dispersion, thermal, mechanical and fire performance in various polymers. *Polymer*, 50 (2009), 3564–74.

48. C. Nyambo, D. Wang, and C. A. Wilkie, Will layered double hydroxides give nanocomposites with polar or non-polar polymers? *Polymers for Advanced Technologies*, 20 (2009), 332–40.

49. M. Saphiannikova, F. R. Costa, U. Wagenknecht, and G. Heinrich, Nonlinear behavior of polyethylene/layered double hydroxide nanocomposites under shear flow. *Polymer Science Series A*, 50 (2008), 573–82.

50. L. Z. Qiu, W. Chen, and B. J. Qu, Structural characterisation and thermal properties of exfoliated polystyrene/ZnAl layered double hydroxide nanocomposites prepared via solution intercalation. *Polymer Degradation and Stability*, 87 (2005), 433–40.

51. L. Z. Qiu, W. Chen, and B. J. Qu, Exfoliation of layered double hydroxide in polystyrene by in-situ atom transfer radical polymerization using initiator-modified precursor. *Colloid and Polymer Science*, 283 (2005), 1241–5.

52. L. Z. Qiu and B. J. Qu, Preparation and characterization of surfactant-free polystyrene/layered double hydroxide exfoliated nanocomposite via soap-free emulsion polymerization. *Journal of Colloid and Interface Science*, 301 (2006), 347–51.

53. P. Ding and B. J. Qu, Synthesis and characterization of exfoliated polystyrene/ZnAl layered double hydroxide nanocomposite via emulsion polymerization. *Journal of Colloid and Interface Science*, 291 (2005), 13–18.

54. P. Ding and B. J. Qu, Synthesis and characterization of polystyrene/layered double-hydroxide nanocomposites via in situ emulsion and suspension polymerization. *Journal of Applied Polymer Science*, 101 (2006), 3758–66.

55. P. Ding, M. Zhang, J. Gai, and B. J. Qu, Homogeneous dispersion and enhanced thermal properties of polystyrene layered double hydroxide nanocomposites prepared by in situ reversible addition–fragmentation chain transfer (RAFT) polymerization. *Journal of Materials Chemistry*, 17 (2007), 1117–22.

56. C. Manzi-Nshuti, D. Chen, S. P. Su, and C. A. Wilkie, Structure–property relationships of new polystyrene nanocomposites prepared from initiator-containing layered double hydroxides of zinc aluminum and magnesium aluminum. *Polymer Degradation and Stability*, 94 (2009), 1290–97.

57. L. J. Wang, S. P. Su, D. Chen, and C. A. Wilkie, Variation of anions in layered double hydroxides: Effects on dispersion and fire properties. *Polymer Degradation and Stability*, 94 (2009), 770–81.

58. C. Nyambo, E. Kandare, D. Y. Wang, and C. A. Wilkie, Flame-retarded polystyrene: Investigating chemical interactions between ammonium polyphosphate and MgAl layered double hydroxide. *Polymer Degradation and Stability*, 93 (2008), 1656–63.

59. L. C. Du, B. J. Qu, Y. Z. Meng, and Q. Zhu, Structural characterization and thermal and mechanical properties of poly(propylene carbonate)/MgAl-LDH exfoliation nanocomposite via solution intercalation. *Composites Science and Technology*, 66 (2006), 913–18.

60. T. M. Wu, eS. F. Hsu, Y. F. Shih, and C. S. Liao, Thermal degradation kinetics of biodegradable poly (3-hydroxybutyrate)/layered double hydroxide nanocomposites. *Journal of Polymer Science, Part B: Polymer Physics*, 46 (2008), 1207–13.

61. S. F. Hsu, T. M. Wu, and C. S. Liao, Isothermal crystallization kinetics of poly(3-hydroxybutyrate)/layered double hydroxide nanocomposites. *Journal of Polymer Science, Part B: Polymer Physics*, 44 (2006), 3337–47.

62. S. F. Hsu, T. M. Wu, and C. S. Liao, Nonisothermal crystallization behavior and crystalline structure of poly(3-hydroxybutyrate)/layered double hydroxide nanocomposites. *Journal of Polymer Science, Part B: Polymer Physics*, 45 (2007), 995–1002.

63. J. Liu, G. M. Chen, and J. P. Yang, Preparation and characterization of poly(vinyl chloride)/layered double hydroxide nanocomposites with enhanced thermal stability. *Polymer*, 49 (2008), 3923–7.

64. F. A. He, L. M. Zhang, F. Yang, L. S. Chen, and Q. Wu, New nanocomposites based on syndiotactic polystyrene and organo-modified ZnAl layered double hydroxide. *Journal of Polymer Research*, 13 (2006), 483–93.

65. M. Kotal, T. Kuila, S. K. Srivastava, and A. K. Bhowmick, Synthesis and characterization of polyurethane/Mg-Al layered double hydroxide nanocomposites. *Journal of Applied Polymer Science*, 114 (2009), 2691–9.

66. K. L. Dagnon, H. H. Chen, L. H. Innocenti-Mei, and N. A. D'Souza, Poly[(3-hydroxybutyrate)-co-(3-hydroxyvalerate)]/layered double hydroxide nanocomposites. *Polymer International*, 58 (2009), 133–41.

67. P. Ding and B. J. Qu, Synthesis of exfoliated PP/LDH nanocomposites via melt-intercalation: Structure, thermal properties, and photo-oxidative behavior in comparison with PP/MMT nanocomposites. *Polymer Engineering and Science*, 46 (2006), 1153–9.

68. M. Zhang, P. Ding, and B. J. Qu, Flammable, thermal, and mechanical properties of intumescent flame retardant PP/LDH nanocomposites with different divalent cations. *Polymer Composites*, 30 (2009), 1000–1006.

69. M. Zhang, P. Ding, B. J. Qu, and A. G. Guan, A new method to prepare flame retardant polymer composites. *Journal of Materials Processing Technology*, 208 (2008), 342–7.

70. C. Manzi-Nshuti, P. Songtipya, E. Manias, M. D. Jimenez-Gasco, J. M. Hossenlopp, and C. A. Wilkie, Polymer nanocomposites using zinc aluminum and magnesium aluminum oleate layered double hydroxides: Effects of the polymeric compatibilizer and of composition on the thermal and fire properties of PP/LDH nanocomposites. *Polymer Degradation and Stability*, 94 (2009), 2042–54.

71. M. Zammarano, S. Bellayer, J. W. Gilman, M. Franceschi, F. L. Beyer, R. H. Harris, and S. Meriani, Delamination of organo-modified layered double hydroxides in polyamide 6 by melt processing. *Polymer*, 47 (2006), 652–62.

72. Y. D. Zhu, G. C. Allen, J. M. Adams, D. Gittins, M. Herrero, P. Benito, and P. J. Heard, Dispersion characterization in layered double hydroxide/Nylon 66 nanocomposites using FIB imaging. *Journal of Applied Polymer Science*, 108 (2008), 4108–13.

73. L. Ye, P. Ding, M. Zhang, and B. J. Qu, Synergistic effects of exfoliated LDH with some halogen-free flame retardants in LDPE/EVA/HFMH/LDH nanocomposites. *Journal of Applied Polymer Science*, 107 (2008), 3694–701.

74. L. Ye and B. J. Qu, Flammability characteristics and flame retardant mechanism of phosphate-intercalated hydrotalcite in halogen-free flame retardant EVA blends. *Polymer Degradation and Stability*, 93 (2008), 918–24.

75. M. Zhang, P. Ding, L. C. Du, and B. J. Qu, Structural characterization and related properties of EVA/ZnAl-LDH nanocomposites prepared by melt and solution intercalation. *Materials Chemistry and Physics*, 109 (2008), 206–11.

76. C. Nyambo, E. Kandare, and C. A. Wilkie, Thermal stability and flammability characteristics of ethylene vinyl acetate (EVA) composites blended with a phenyl phosphonate-intercalated layered double hydroxide (LDH), melamine polyphosphate and/or boric acid. *Polymer Degradation and Stability*, 94 (2009), 513–20.

77. C. Nyambo and C. A. Wilkie, Layered double hydroxides intercalated with borate anions: Fire and thermal properties in ethylene vinyl acetate copolymer. *Polymer Degradation and Stability*, 94 (2009), 506–12.

78. P. J. Pan, B. Zhu, T. Dong, and Y. Inoue, Poly(L-lactide)/layered double hydroxides nanocomposites: Preparation and crystallization behavior. *Journal of Polymer Science, Part B: Polymer Physics*, 46 (2008), 2222–33.

79. W. D. Lee, S. S. Im, H. M. Lim, and K. J. Kim, Preparation and properties of layered double hydroxide/poly(ethylene terephthalate) nanocomposites by direct melt compounding. *Polymer*, 47 (2006), 1364–71.

80. S. Martinez-Gallegos, M. Herrero, C. Barriga, F. M. Labajos, and V. Rives, Dispersion of layered double hydroxides in poly(ethylene terephthalate) by in situ polymerization and mechanical grinding. *Applied Clay Science*, 45 (2009), 44–9.

81. A. Sorrentino, G. Gorrasi, M. Tortora, V. Vittoria, U. Constantino, F. Marmottini, and F. Padella, Incorporation of Mg–Al hydrotalcite into a biodegradable poly (epsilon, caprolactone) by high energy ball milling. *Polymer*, 46 (2005), 1601–8.

82. M. Zubitur, M. A. Gomez, and M. Cortazar, Structural characterization and thermal decomposition of layered double hydroxide/poly(*p*-dioxanone) nanocomposites. *Polymer Degradation and Stability*, 94 (2009), 804–9.

83. B. G. Li, Y. Hu, R. Zhang, Z. Y. Chen, and W. C. Fan, Preparation of the poly(vinyl alcohol)/layered double hydroxide nanocomposite. *Materials Research Bulletin*, 38 (2003), 1567–72.

84. G. A. Wang, C. C. Wang, and C. Y. Chen, Preparation and characterization of layered double hydroxides–PMMA nanocomposites by solution polymerization. *Journal of Inorganic and Organometallic Polymers and Materials*, 15 (2005), 239–51.

85. C. Manzi-Nshuti, J. M. Hossenlopp, and C. A. Wilkie, Fire retardancy of melamine and zinc aluminum layered double hydroxide in poly(methyl methacrylate). *Polymer Degradation and Stability*, 93 (2008), 1855–63.

86. C. Manzi-Nshuti, D. Y. Wang, J. M. Hossenlopp, and C. A. Wilkie, Aluminum-containing layered double hydroxides: The thermal, mechanical, and fire properties of (nano)composites of poly(methyl methacrylate). *Journal of Materials Chemistry*, 18 (2008), 3091–3102.

87. C. Manzi-Nshuti, D. Y. Wang, J. M. Hossenlopp, and C. A. Wilkie, The role of the trivalent metal in an LDH: Synthesis, characterization and fire properties of thermally stable PMMA/LDH systems. *Polymer Degradation and Stability*, 94 (2009), 705–11.

88. C. Nyambo, D. Chen, S. P. Su, and C. A. Wilkie, Variation of benzyl anions in MgAl-layered double hydroxides: Fire and thermal properties in PMMA. *Polymer Degradation and Stability*, 94 (2009), 496–505.

89. C. Nyambo, D. Chen, S. P. Su, and C. A. Wilkie, Does organic modification of layered double hydroxides improve the fire performance of PMMA? *Polymer Degradation and Stability*, 94 (2009), 1298–1306.
90. C. Nyambo, P. Songtipya, E. Manias, M. M. Jimenez-Gasco, and C. A. Wilkie, Effect of MgAl-layered double hydroxide exchanged with linear alkyl carboxylates on fire-retardancy of PMMA and PS. *Journal of Materials Chemistry*, 18 (2008), 4827–38.
91. G. A. Wang, C. C. Wang, and C. Y. Chen, The disorderly exfoliated LDHs/PMMA nanocomposite synthesized by in situ bulk polymerization. *Polymer*, 46 (2005), 5065–74.
92. L. J. Wang, S. P. Su, D. Chen, and C. A. Wilkie, Fire retardancy of bis[2-(methacryloyloxy)ethyl] phosphate modified poly(methyl methacrylate) nanocomposites containing layered double hydroxide and montmorillonite. *Polymer Degradation and Stability*, 94 (2009), 1110–18.
93. Y. Y. Ding, Z. Gui, J. X. Zhu, Y. Hu, and Z. Z. Wang, Exfoliated poly(methyl methacrylate)/MgFe-layered double hydroxide nanocomposites with small inorganic loading and enhanced properties. *Materials Research Bulletin*, 43 (2008), 3212–20.
94. S. O'Leary, D. O'Hare, and G. Seeley, Delamination of layered double hydroxides in polar monomers: New LDH-acrylate nanocomposites. *Chemical Communications* (2002), 1506–7.
95. Z. Matusinovic, M. Rogosic, and J. Sipusic, Synthesis and characterization of poly(styrene-co-methyl methacrylate)/layered double hydroxide nanocomposites via in situ polymerization. *Polymer Degradation and Stability*, 94 (2009), 95–101.
96. H. B. Hsueh and C. Y. Chen, Preparation and properties of LDHs/epoxy nanocomposites. *Polymer*, 44 (2003), 5275–83.
97. Y. N. Chan, T. Y. Juang, Y. L. Liao, S. A. Dai, and J. J. Lin, Preparation of clay/epoxy nanocomposites by layered-double-hydroxide initiated self-polymerization. *Polymer*, 49 (2008), 4796–801.
98. M. Zammarano, M. Franceschi, S. Bellayer, J. W. Gilman, and S. Meriani, Preparation and flame resistance properties of revolutionary self-extinguishing epoxy nanocomposites based on layered double hydroxides. *Polymer*, 46 (2005), 9314–28.
99. R. C. Xie, B. J. Qu, and K. L. Hu, Dynamic FTIR studies of thermo-oxidation of expandable graphite-based halogen-free flame retardant LLDPE blends. *Polymer Degradation and Stability*, 72 (2001), 313–21.
100. Q. Wu and B. J. Qu, Synergistic effects of silicotungistic acid on intumescent flame-retardant polypropylene. *Polymer Degradation and Stability*, 74 (2001), 255–61.
101. S. Miyata, T. Hirose, and N. Iizima, *Fire-Retarding Thermoplastic Resin Composition*, United States Patent 4,085,088 (1978).
102. J. W. Gilman, C. L. Jackson, A. B. Morgan, R. Harris, E. Manias, E. P. Giannelis, M. Wuthenow, D. Hilton, and S. H. Phillips, Flammability properties of polymer–layered-silicate nanocomposites. Polypropylene and polystyrene nanocomposites. *Chemistry of Materials*, 12 (2000), 1866–73.
103. X. X. Zheng and C. A. Wilkie, Nanocomposites based on poly(epsilon-caprolactone) (PCL)/clay hybrid: Polystyrene, high impact polystyrene, ABS, polypropylene and polyethylene. *Polymer Degradation and Stability*, 82 (2003), 441–50.
104. M. Zanetti, T. Kashiwagi, L. Falqui, and G. Camino, Cone calorimeter combustion and gasification studies of polymer layered silicate nanocomposites. *Chemistry of Materials*, 14 (2002), 881–7.

105. C. M. Jiao, Z. Z. Wang, X. L. Chen, and Y. A. Hu, Synthesis of a magnesium/ aluminum/iron layered double hydroxide and its flammability characteristics in halogen-free, flame-retardant ethylene/vinyl acetate copolymer composites. *Journal of Applied Polymer Science*, 107 (2008), 2626–31.
106. J. W. Gilman, Flammability and thermal stability studies of polymer layered-silicate (clay) nanocomposites. *Applied Clay Science*, 15 (1999), 31–49.
107. L. Z. Qiu, R. C. Xie, P. Ding, and B. J. Qu, Preparation and characterization of Mg(OH)(2) nanoparticles and flame-retardant property of its nanocomposites with EVA. *Composite Structures*, 62 (2003), 391–5.

14

Flame retardant SBS–clay nanocomposites

MAURO COMES-FRANCHINI,[a] MASSIMO MESSORI,[b] GUIDO ORI,[b]
AND CRISTINA SILIGARDI[b]

[a]Dipartimento di Chimica Organica "A. Mangini," Facoltà di Chimica Industriale
[b]Department of Materials and Environmental Engineering, University of Modena and Reggio Emilia

14.1 Layered silicates

The idea of flame retardant materials dates back to about 450 BC, when the Egyptians used alum to reduce the flammability of wood. The Romans (in about 200 BC) used a mixture of alum and vinegar to reduce the combustibility of wood. Today, there are more than 175 chemicals classified as flame retardants. The major groups are inorganic, halogenated, organic, organophosphorus, and nitrogen-based flame retardants, which account for 50%, 25%, 20%, and >5% of the annual production, respectively [1].

In many cases, existing flame retardant systems show considerable disadvantages. The application of aluminum trihydrate and magnesium hydroxide requires a very high portion of the filler to be deployed within the polymer matrix; filling levels of more than 60 wt% are necessary to achieve suitable flame retardancy, for example, in cables and wires. Clear disadvantages of these filling levels are the high density and the lack of flexibility of end products, the poor mechanical properties, and the problematic compounding and extrusion steps. In addition, intumescent systems are relatively expensive and electrical requirements can restrict the use of these products [2, 3]. Because of the toxicological and environmental problems caused by traditional halogen flame retardants, increasing attention has been focused on halogen-free flame retardants.

A new class of materials, called nanocomposites, can avoid the disadvantages of traditional flame retardant systems. Generally, the term "nanocomposite" describes a two-phase material with a suitable nanofiller (organoclay, nanoparticles, carbon nanotubes, etc.) dispersed in the polymer matrix at the nanometric scale [3, 4].

Among all the potential nanocomposite precursors, clay and layered silicates have been most widely investigated, probably because starting clay materials are easily available and because their intercalation chemistry has been studied for a long time [5, 6].

Layered silicates used in the synthesis of nanocomposites are natural or synthetic minerals, consisting of very thin layers that are usually bound together with counterions.

Among layered silicates, 2:1 phyllosilicates (smectites) have received the most attention in regard to nanocomposite applications. The main reasons are their potentially high aspect ratio and unique intercalation/exfoliation characteristics. Furthermore, smectites have been used extensively to prepare nanocomposites because of properties such as

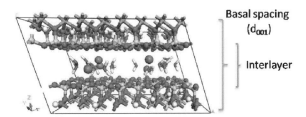

Figure 14.1 Ideal model of Na-Wyoming montmorillonite with water content of 0.049 g_{water}/g_{clay}. Reproduced from Ref. [8].

high cation exchange capacity, swelling behavior, absorption properties, and large surface area [5].

Smectite is a three-layer clay mineral. This three-layer clay has two continuous tetrahedral silica sheets joined to a central octahedral sheet. Ideally, each tetrahedron consists of a cation (silicon) coordinated to four oxygen atoms and linked to adjacent tetrahedra by sharing three corners (basal oxygen atoms) to form an infinite two-dimensional hexagonal mesh pattern along the a, b crystallographic directions. In octahedral sheets connections of each octahedron to neighboring octahedra are made by sharing edges [7]. Edge-shared octahedra form sheets of hexagonal or pseudo-hexagonal symmetry. Octahedra show different topologies related to (OH) position, that is, *cis* and *trans* orientation [7]. The free corners (the tetrahedral apical oxygen atoms) of all tetrahedra point to the same side of the sheet and connect the tetrahedral and octahedral sheets to form a common plane with octahedral anionic position O_{oct} (OH). O_{oct} anions lie near the center of each tetrahedral sixfold ring, but are not shared with tetrahedra [7, 8]. The octahedral sheets may be dominantly occupied by either trivalent cations (dioctahedral smectites) or divalent cations (trioctahedral smectites). Cation substitutions mainly take place in the octahedral sheets and may induce enormous changes in the physicochemical properties of clay minerals [7–12].

Swelling and hydration of smectite clay minerals impact a wide variety of environmental and engineering processes (see Figure 14.1). The use of modified clays with traditional clays as low-cost, effective catalysts for the removal of metal ions from industrial effluents and wastewaters through ion exchange or surface complexation has been of great relevance to both applications and our understanding of natural processes [13–19]. Clays are considered suitable materials for the geological storage of toxic and radioactive waste. Such practical issues have motivated extensive experimental and theoretical investigations. Furthermore, surface modifications of clay minerals have received attention because they allow the creation of new materials and new applications [5, 7, 8].

Most of the technological uses of smectite are related to reactions that take place in the interlayer space. Na^+, K^+, Ca^{2+}, and Mg^{2+}, which balance the negative 2:1 layer charge, are commonly hydrated and exchangeable. Smectites contain water in several forms.

The most important end members of dioctahedral smectites have the following general compositions:

Montmorillonite	$(M_x^+ \cdot nH_2O)(Al^3{}_{2-x}Mg_x{}^{2+})Si_4{}^{4+}O_{10}(OH)_2$
Beidellite	$(M_x^+ \cdot nH_2O)Al^{3+}{}_2(Si_{4-x}{}^{2+}Al_x{}^{3+})O_{10}(OH)_2$
Nontronite	$(M_x^+ \cdot nH_2O)Fe^{3+}{}_2(Si_{4-x}{}^{2+}Al_x{}^{3+})O_{10}(OH)_2$
Volkonskoite	$(M_x^+ \cdot nH_2O)Cr^{3+}{}_2(Si_{4-x}{}^{2+}Al_x{}^{3+})O_{10}(OH)_2$

The most important species of trioctahedral smectites are the following:

Hectorite	$(M_x^+ \cdot nH_2O)(Mg_{3-x}{}^{2+}Li_x{}^+)Si_4{}^{4+}O_{10}(OH)_2$
Saponite	$(M_x^+ \cdot nH_2O)Mg_3{}^{2+}(Si_{4-x}{}^{2+}Al_x{}^{3+})O_{10}(OH)_2$
Sauconite	$(M_x^+ \cdot nH_2O)Zn^{2+}{}_3(Si_{4-x}{}^{2+}Al_x{}^{3+})O_{10}(OH)_2$

The rock in which montmorillonite and similar minerals are dominant is bentonite. Bentonites that are used industrially are predominantly composed of either sodium montmorillonite or calcium montmorillonite and to a much lesser extent of hectorite. Smectite minerals have significantly different physical and chemical properties, which dictate their industrial utilization to a large degree.

There is considerable substitution of Fe and Mg for Al in the octahedral sheet, which creates a charge deficiency in the layer (octahedrally charged). Also, there is some substitution of aluminum for silicon in the tetrahedral sheets, which again creates a charge balance (tetrahedrally charged). This net positive charge deficiency is balanced by exchangeable cations adsorbed between the unit layers and around edges. The layer thickness is around 1 nm and the lateral dimensions may vary from 300 Å to several micrometers, or even larger, depending on the particulate silicate, the source of the clay, and the method of preparation (e.g., clays prepared by milling typically have lateral platelet dimensions of approximately 0.1–1.0 μm). Therefore, the aspect ratio of these layers (length/thickness ratios) can be particularly high, with values greater than 1,000. As mentioned before, the potential aspect ratios of silicates are one of the crucial factors that make them important candidate for nanocomposites. Moreover, a particular feature of the structure of smectites is that the layers (Tetra–Octa–Tetra) are held together by relatively weak forces, so that water and other polar molecules can enter between the unit layers, causing the lattice to expand [5, 7, 8].

The cation exchange capacity and the distribution of the isomorphous substitution (mainly tetrahedral or octahedral) of these silicates, and its correlation to the negative charge of the layers, are other important factors that can influence smectite behavior during cation exchange reactions. These characteristics are the aspects of cation-exchanged smectites that mainly affect the type of product resulting. Furthermore, the nature (in terms mainly of hydration characteristics) of the pristine inorganic cations present in the interlayer (or adsorbed on the external surface) can also affect the exchange reaction [7, 20]. These different points have to be taken into consideration in relation to exchange reactions both with inorganic or organic cations and with nonpolar molecules [5, 8, 21]. However, to fully understand the complex structure of these materials and the critical mechanisms of intercalation, it has become necessary to use modern molecular simulation methods, especially because many layered materials are restricted to nanosized morphologies and are less suitable for conventional experimental analysis [22].

To obtain more detailed insight into these processes, the extended molecular modeling work performed so far must also be taken into account to have a more specific vision of the process from the atomic/molecular point of view. Classical methods in computational chemistry usually take the form of molecular dynamics (MD) or Monte Carlo (MC) simulation, with the aim of sampling the phase space of the system once it has reached thermodynamic equilibrium. For simulations of layered clay minerals, a preliminary issue is whether to treat the mineral lattice as a rigid framework, or to allow flexibility of bonds, angles, and dihedra within the clay [22]. The rigid framework approach has the advantage of lower computational cost and is easier to parameterize, because only nonbonded interactions between the clay and interlayer species are included. However, such simplified models are inherently limited by the immobility of the lattice atoms, which precludes complete exchange of energy and momentum between the interacting atoms of the mineral substrate and the interlayer or surface molecules [22].

In addition to interlayer structure and dynamics, an underlying motivation for simulations of hydrated clays is to better understand the phenomenon of intracrystalline swelling [22]. The basal *d*-spacing of natural clays such as Na-montmorillonite increases stepwise with increasing relative humidity, corresponding to an integral number of water layers (up to three) between adjacent clay layers. *d*-spacing values from simulation have been compared with the corresponding experimental X-ray diffraction values under different conditions [21–25]. Inexpensive computing power has led to a second generation of force fields for simulating clays, so that flexibility within the clay lattice can now be considered. The structure and dynamics of interlayer species is more accurately simulated when momentum transfer between those molecules and the clay lattice is included. Additionally, model clay systems may be expanded to include hydroxylated surfaces, which cannot be accurately modeled using a rigid lattice.

In nanocomposite applications all this information has to be considered to hypothesize/design better silicates for specific polymer matrices, considering the possible silicate–interlayer cation (also inorganic–organic)–solvent (water)–polymer matrix interactions [26, 27]. Concerning the specific case of SBS-based nanocomposites, to our knowledge, only smectitic clays (bentonite or montmorillonite) have been used as layered silicate starting materials. Systematic work to evaluate possible improvement in the correspondent clay-modified system and/or nanocomposite final properties has been done in other cases, where the potential crucial role of the selection of the starting layered silicate has been emphasized.

14.2 Organic modification

The dispersion of nanofillers in polymers is generally rather poor, because of their incompatibility with polymers and large surface-to-volume ratio. Therefore, the addition of organic

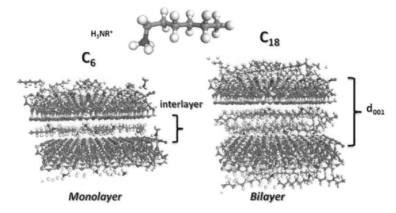

Figure 14.2 Snapshots of an ideal model of Wyoming-type montmorillonite with intercalated C_6- and C_{18}-alkylammonium ions (NH_3^+ as head group). Reproduced from Ref. [8].

molecules is needed to improve the dispersion. A well-known example is layered silicate surfaces (clays), which are hydrophilic and for which organic modifiers are needed; the obtained product is known as "organoclay." Organoclays can be more or less easily delaminated into nanoscale platelets by polymer molecules, leading to the formation of polymer–layered silicate nanocomposites (PLSNs) [28, 29]. Compared with pure polymers, PLSNs can show improved properties; the amount of nanofiller needed within the polymer matrix is often in the range of 2–10 wt%. PLSNs have attracted much attention from scientists and engineers in recent years because of their excellent properties and, among these, enhanced fire retardancy is one of the most important [30, 31]. Very often organoclays are used after exchange of the original inorganic small cations within the galleries of the layers for more bulky organic cations such as quaternary ammonium cations. This exchange of cations increases the gap between the layers and also makes the interlayer organophilic, which allows polymers to penetrate into the galleries (Figure 14.2).

Unfortunately, not all organoclays can be nanodispersed in all polymers. A significant amount of research material has been published indicating that a polar polymer matrix is very helpful and that the length and the chemical structure of the quaternary organic ammonium compound also play an important role in the successful production of a nanocomposite. Polymers such as ethylene vinyl acetate (EVA) and polyamide (PA) easily form nanocomposites, whereas this is considerably more difficult for nonpolar polymers such as polyethylene (PE) and polypropylene (PP). In this context, theoretical models are increasingly being applied, with some success, to processes involving organic–mineral interfaces, such as exchange reactions and biomineralization [32–34].

To better understand the properties of the interfacial region between layered silicates and organic molecules (such as surfactants and polymers), the separation of the clay mineral layers has been simulated using all-atom models, and the interaction between these different systems has been simulated using coarse-grain models. Chemically specific conclusions, however, are mostly drawn from all-atom MD simulations [22, 35].

As an example, the role and the effect of alkylammonium ions in the montmorillonite interlayer in relation to the interaction (but also cleavage) energy between the components were investigated by MD simulation. It was confirmed that the organic surface modification of montmorillonites increases the likelihood of exfoliation of high–aspect ratio aluminosilicate layers in polymer matrices. The molecular interpretation on the basis of the simulation is supported by various experimental observations, such as cleavage energies and surface tensions [26, 27, 36].

However, there are also problems to be solved with nanocomposites. For example, in certain cases, the thermal stability of the organoclays must be improved. This is due to rapid thermal decomposition of the quaternary ammonium compounds with long alkylic chains, because it is known that they present the drawback of poor thermal stability at $>200\,^\circ$C [3, 37–40].

It has been proposed that the thermal degradation of organoclays, based on organic ammonium salts, includes desorption of organic ions and/or fragmentation of the organic moiety itself, proceeding by Hofmann elimination, SN_2 nucleophilic substitution, and other processes. The type of degradation mechanism of ammonium-based organoclay depends on many variables, including the chemical and physical characteristics both of the clay and of the organic molecules [41, 42].

On the other hand, it is well known that nitrogen–containing heterocycles show improved thermal stability [43–47]. Bottino *et al.* and Awad and co-workers reported the preparation of organoclays using more thermally stable (than ammonium salts), reactive and not, surfactant imidazolium salts [37, 48–53].

Furthermore, it is nearly established that melamine organic- and inorganic-based derivatives present enhanced thermal stability and also inherent fire retardancy, decomposing at high temperature (also with endothermic processes) in compounds that act as obstacles for fire processes with different mechanisms [54].

From the viewpoint of fire retardancy, the most significant advantage offered by PLSNs is reduction in the *peak heat release rate* (PHRR, from cone calorimetry). However, in most cases the nanocomposite ignites faster than the virgin polymer. The most likely hypothesized cause is the excess of organic surfactant (alkylammonium) used to disperse the clay, which increases the probability of early ignition. Moreover, a major limitation of nanocomposites is that they only work in the condensed phase and do nothing to inhibit the flame in the gas phase [55]. Indeed, it has become necessary to develop novel synergistic flame retardant systems in order to achieve high fire performance levels, based on the optimization of formulations of different flame retardant agents.

In this effort, the role of organic molecules seems to be crucial. Therefore, a thorough understanding of changes in the structure and properties under the processing conditions imposed in conjunction with novel flame retardants is essential to tailor nanocomposites for specific applications.

14.3 Styrenic block copolymers

Thermoplastic elastomers (TPEs), also named thermoplastic rubbers, represent a commercially relevant class of polymers. TPEs can be defined as biphasic materials possessing the combined properties of glassy or semicrystalline rigid thermoplastics and soft elastomers, which enable rubbery materials to be processed as thermoplastics. This behavior arises from the molecular structure, in particular diblock, triblock, or segmented block copolymer architectures, in which thermoplastic segments able to form rigid domains at a nanometric scale are covalently linked to soft segments providing a rubbery matrix in which the rigid domains reside. The rigid domains form a three-dimensional network of physical cross-link sites, thanks to covalent bonding between two thermodynamically incompatible segments. Many mechanical properties of TPEs are comparable to those of vulcanized rubbers (with irreversible chemical cross links), with the peculiar characteristic that the network is thermally reversible, permitting their use for high-throughput thermoplastic processes such as extrusion and injection molding.

Different TPEs grades are usually characterized by their resistance to abrasion, cutting, scratching, and local strain and wear, but one of their major peculiar properties is their hardness, which covers a wide range (from 30 Shore A to 60 Shore D hardness), bridging the range from thermoset rubbers to hard thermoplastics.

Among the different generic classes of commercially available TPEs, styrenic block copolymers are most widely used. The molecular architecture of styrenic block copolymers is characterized by hard/rigid polystyrene segments (about 30–40 wt%) bonded to soft elastomeric segments. Styrenic TPEs are usually poly(styrene-*b*-butadiene-*b*-styrene) (SBS), poly(styrene-*b*-(ethylene-*ran*-butylene)-*b*-styrene) (SEBS), or poly(styrene-*b*-isoprene-*b*-styrene) (SIS). The lateral polystyrene segments form domains that act as thermally (physically) reversible cross links, allowing thermoplasticity during melt-processing. After cooling, the polystyrene domains harden, reforming the cross-linking points of the rubber matrix.

The main applications of TPEs are in the footwear industry, for molded shoe soles and other uses, in the field of film/sheet extrusion and wire coverings, and as pressure-sensitive adhesives (PSA) and hot-melt adhesives. Other uses are as viscosity index–improver additives in lubricating oils, resins, and asphalt modifiers. TPEs are also used as grips, kitchen utensils, clear medical products, and personal care products. Styrenic TPEs are useful in particular in adhesive compositions in web coatings.

Some examples of commercially available SBS are Asaflex (AKelastomers), Asaprene (AKelastomers), Badaflex (Bada AG), Dryflex (VTC TPE Group), Europrene (Polimeri Europa), Evoprene (AlphaGary), Kraton (Kraton Polymers LLC), Sofprene T (SO.F.TER. SPA), Tufprene (AKelastomers), and Vector (Dexco Polymers LP).

The main drawbacks of TPEs compared with conventional rubber are the relatively high cost, the general inability to load TPEs with low-cost fillers (for example, carbon black, preventing their use in automobile tires), low chemical and heat resistance, high compression set, and poor thermal stability.

14.4 Flame retardant SBS–clay nanocomposites

Despite the enormous number of scientific papers and patents published recently concerning the preparation and characterization of polymer matrix nanocomposites reinforced and modified with intercalated or exfoliated clays, systems based on SBS as polymeric matrix are relatively poorly investigated.

Nanocomposites of SBS containing exfoliated organophilic silicates were prepared using techniques such as melt-mixing, solution-casting, and a process in which a highly filled masterbatch is mixed into the melt with the pure polymer [56]. To achieve the highest dispersion of the nanoclay, it was shown that it is necessary to use solution-casting as one of the stages in nanocomposite preparation. The masterbatch process was shown to be an effective method for dispersing the clay because it combines the advantages of the solution-casting technique, which is good for the dispersion of coarse particle aggregates, and the shear stress provided by melt-mixing. The same authors recently reported a further study in which the rheological behavior of the materials was investigated [57].

Exfoliation in SBS–clay composites was also reported after modification of montmorillonites by in situ emulsion polymerization of organic monomers [58]. Upon the incorporation of the modified clay, the tensile strength, elongation at break, and tear strength of the nanocomposites increased from 22.6 to 31.1 MPa, from 608% to 948%, and from 45.32 to 55.27 N/mm, respectively. The low glass transition temperature of the products was about $-77\ ^\circ$C, almost constant, but the high glass transition temperature increased from 97 to $106\ ^\circ$C. Furthermore, the nanocomposites of SBS and modified montmorillonites showed good resistance to thermal degradation and aging.

An exfoliated SBS/montmorillonite nanocomposite was prepared by anionic polymerization [59]. The introduction of organophilic montmorillonite (OMMT) resulted in a small high–molecular weight fraction of SBS in the composites, leading to a slight increase in the average molecular weight as well as the polydispersity index. ^{1}H NMR analysis showed that the introduction of OMMT almost did not affect the microstructure of the copolymer when the OMMT content was lower than 4 wt%. The exfoliated nanocomposite exhibited higher thermal stability, glass transition temperature, elongation at break, and storage modulus than pure SBS.

Intercalated composites consisting of a commercial SBS and a commercial organophilic clay containing a dioctadecyldimethylammonium salt have been prepared by Laus *et al.* by melt-mixing [60]. No interactions of unfunctionalized clay with the polymer matrix have been observed. It is interesting to observe that the storage modulus, in the plateau region between the glass transition processes of the polybutadiene and the polystyrene blocks, and the glass transition temperature of the polystyrene block domain increase as both the organophilic clay content and the annealing time increase. In contrast, the glass transition process of the polybutadiene block domain is practically unaffected by the filler content and the annealing treatments.

The preparation of nanocomposites intercalatedwith star-shaped and linear block thermoplastic SBS/organophilic montmorillonite clays prepared by solution-mixing [61] and using a twin-screw extruder [62] was also reported. The mechanical properties of nanocomposites with star-shaped SBS/OMMT were significantly enhanced. The incorporation of OMMT also provided an increase of the elongation, the dynamic storage modulus, the dynamic loss modulus, and the thermal stability of nanocomposites. The increased elongation of nanocomposites indicated that SBS has retained good elasticity.

The use of a commercial Cloisite 20A organoclay to prepare SBS-based nanocomposites by melt processing was recently reported [63]. In this case, the nanocomposite morphology was characterized by a combination of intercalated and partly exfoliated clay platelets, with occasional clay aggregates present at higher clay content. For this particular thermoplastic elastomer nanocomposite system, well-dispersed nanoclays lead to enhanced stiffness and ductility, suggesting promising improvements in nanocomposite creep performance. The use of stearic acid as a surface modifier of montmorillonite clay to effectively improve the clay dispersion in the SBS matrix and the mechanical properties of the SBS–clay nanocomposites was reported [64].

Apart from the usually reported morphological analysis (electron microscopy and rheological analysis) and evaluation of the mechanical properties of the prepared nanocomposites, some authors have published more specific papers on the use of SBS–montmorillonite nanocomposites as toughening additives for polypropylene [65], as rheological modifiers for asphalts [66], and as irradiation (UV–light, γ-rays, and electron beams) stabilizers [67].

The flame resistance behavior of SBS–clay nanocomposites is only partially investigated and reported in the patent literature, together with that of several other polymeric matrices.

The preparation of soft flame retardant thermoplastic compounds (preferably halogen-free) including a styrenic block copolymer was recently claimed by Lee [68]. The reported formulations exhibited high char formation upon burning and also desirable vertical burn characteristics. Lee and Worley [69] also claimed the preparation of flame-retardant polymeric blends based on the synergistic combination of flame retardants, including a halogenated flame retardant, and nanoparticulates, such as an organically modified clay or a natural nanoclay.

Flame retarded polypropylene–SBS blends have been claimed after incorporation of organically modified clays and metal hydroxides [70]. The blends showed good heat resistance and self-extinguishing properties according to the UL 94 standard.

In our recent paper, we reasoned that the combination of the melamine skeleton containing a long alkylic chain with the ammonium salt would lead to an interesting synergistic effect in fire retardancy for SBS copolymers [71]. This paper describes the preparation and characterization of organophilic bentonites with a synthetically modified melamine containing a long alkylic chain with amine salts and with octadecyltrimethylammonium bromide (OTAB) as a comparative model. The preparation of SBS/organobentonite nanocomposites obtained through melt-blending and their characterization are also reported, with particular attention to the resulting fire retardancy properties.

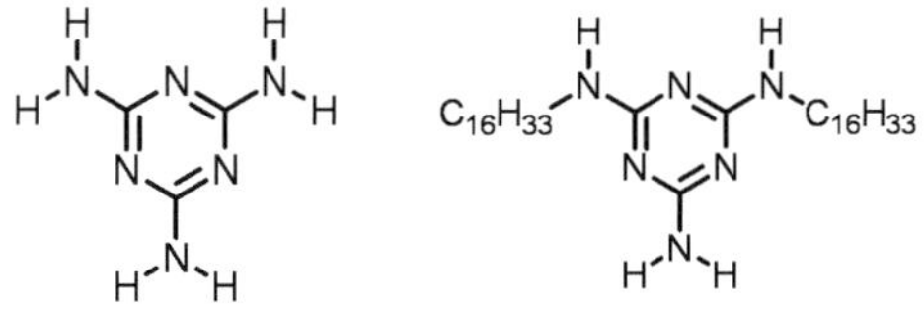

Figure 14.3 Molecular structures of melamine and *N*2,*N*4-dihexadecyl-1,3,5-triazine-2,4,6-triamine.

14.5 Case study: Bentonite-based organoclays as innovative flame retardant agents for SBS copolymer

14.5.1 Materials and characterization methods

Pristine bentonite (BentoNAT) was provided by Rostich spa (Neuquen, Patagonia, Republica Argentina). Copoly(styrene–butadiene–styrene) linear block copolymer (SBS, Calprene 500, Dynasol), with a melt flow index of 5 g/10 min (5 kg, 190 °C), containing 30 wt% polystyrene and 70 wt% polybutadiene, was supplied by Softer spa (Forli, Italy). Octadecyltrimethylammonium bromide (OTAB), melamine (Figure 14.3), hexadecylamine, and solvents were purchased from Sigma-Aldrich. Morphological studies have been carried out on BentoNAT, organoclays, and composites prepared by X-ray powder diffraction (XRPD) and Transmission Electron Microscopy (TEM, Philips EM400, The Netherlands). XRPD experiments were performed at room temperature on a Philips XPert Pro diffractometer (40 kV, 30 mA) equipped with an X'ccelerator detector system using Cu$K\alpha$ radiation ($\lambda = 1.54$ Å) in the range of 2°–40°. For TEM analysis, specimens (BentoNAT and organoclays) were prepared by dispersing the as-obtained powder in an epoxy resin (Conchem LC 202, Italy) at high temperature by strong mixing, cutting it with an ultramicrotome, and then placing it on a copper grid coated with a carbon film.

14.5.2 Layered silicates

The elemental composition of BentoNAT was determined by X-ray fluorescence (Philips PW 1480, using a Philips PW 1510 auto sampler). Wt%: SiO_2, 56.36; Al_2O_3, 17.98; Na_2O, 1.91; K_2O, 0.308; CaO, 1.60; MgO, 2.70; Fe_2O_3, 4.14; TiO_2, 0.203; loss on ignition (LOI, 1000 °C), 14.90; major volatiles: H_2O and CO_2. The cation exchange capacity (CEC) was determined by exchanging interlayer cations with triethylenetetramine copper (II) complex (Cu(trien)$^{2+}$) and measuring the decrease in Cu(trien)$^{2+}$ solution concentration photometrically (Cary 1E Varian UV–vis Spectrophotometer, $\lambda_{max} = 576$ nm). The CEC obtained is 87 ± 3 mmol/100 g of clay. The size distribution of BentoNAT was determined by laser granulometry (Malvern Instrument, model Hydro 2000S): 10 vol% <1.31 μm, 50 vol% < 4.08 μm, and 90 vol% <17.9 μm.

BentoNAT presents the typical elemental composition of a bentonite useful for the synthesis of PLSNs. In particular, there is an interlayer rich in Na, Ca, and Mg cations, which are easily exchangeable for organic cations. The basal spacing of the BentoNAT

Table 14.1 *Basal spacing for BentoNAT and organoclays from XRD and TEM analysis*

	XRD d_{001} (nm)	TEM d_{001} (nm)	Organic wt%[a]
BentoNAT	1.25	1.30	–
BentOTAB	1.90	2.60	25.33[b]
BentoDEDMEL	2.60	2.93	30.58[b]

[a]Calculated from C wt% content.

[b]Determined from the mass of the protonated form of the organic modifier.

is 1.25 nm, which is calculated from the d_{001} peak position using the Bragg equation (Table 14.1).

14.5.3 Organoclays

14.5.3.1 Preparation

Two different organic modifiers were used to render the clay organophilic: octadecyltrimethylammonium bromide (OTAB) for comparison experiments and N2,N4-dihexadecyl-1,3,5-triazine-2,4,6-triamine (DEDMEL) synthesized by a slight modification of the Zerweck and Keller procedure [71, 72]. For DEDMEL synthesis and characterization see Ref. [71]. The bentonite modified with OTAB (BentOTAB) was prepared by a cation exchange reaction between the pristine bentonite and OTAB in water/EtOH solution. A quantity of 20 g of pristine bentonite was dispersed in 400 mL of water/EtOH (2:1 by volume) by mechanical stirring. A quantity of 11.78 g of OTAB (30.01 mmol, ~1.7 × CEC) was solubilized in 150 mL water/EtOH (1:1 by volume) and added to the bentonite solution. The reaction mixture was stirred at 65 °C for 22 h. The exchanged bentonite was filtered and washed with a hot solution of water/EtOH (1:1 by volume) until no bromine ions were detected by titration with 0.1 M $AgNO_3$.The product was dried in the oven at 60 °C under vacuum for 1 h and finely powdered with a mortar. For the bentonite modified with DEDMEL (BentoDEDMEL), the same procedure was followed. A quantity of 10 g of pristine bentonite was dispersed in 200 mL of water/EtOH (1:1 by volume) by mechanical stirring. A quantity of 7.54 g of DEDMEL (13.11 mmol, ~1.5 × CEC) was solubilized in 300 mL of water by the slow addition of concentrated HCl (37 vol%) until dissolution was observed, and NaOH (10 vol%) was used to maintain the pH at approximately 7. This solution was then added to the bentonite suspension. Then the procedure used for BentOTAB was followed, giving a finely powdered gray product. The amount of organic modifier in the organoclays was determined with an NC elemental analyzer (Flash 1112 Series EA) by analyzing for organic C and N content. BentOTAB: C 20.46; N 1.14. BentoDEDMEL:

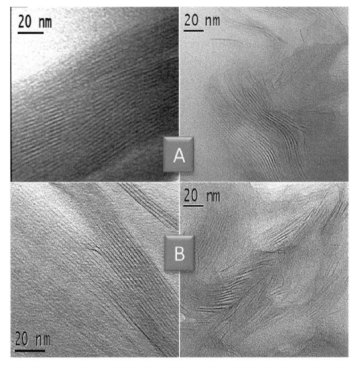

Figure 14.4 TEM images of BentOTAB (A) and BentoDEDMEL (B). Reproduced from Ref. [71] with permission of American Scientific Publishers.

C 22.24; N 4.46. Before composites were prepared, the organoclays were finely powdered with a spherical balls mill (using balls of agate) to obtained a size distribution like that of pristine bentonite (10 vol% <1.00 μm, 50 vol% <3.20 μm, and 90 vol% < 15.0 μm).

14.5.3.2 Characterization and discussion

After modification, the BentOTAB sample shows a broad diffraction peak that is shifted to 1.90 nm, whereas BentoDEDMEL shows a peak corresponding to 2.60 nm (Table 14.1) [71]. This confirms that the BentoNAT basal spacing is increased in both the two synthesized organoclays, and in the case of BentoDEDMEL, a greater expansion and a much more organophilic structure are obtained. Furthermore, the intensity of the peak (corresponding to d_{001}) becomes higher, implying that the layers of clay become much more well-ordered. The greater organic fraction of BentoDEDMEL than BentOTAB (Table 14.1), obtained from CN element analysis, confirm the much more organophilic interlayer of BentoDEDMEL. Some TEM images of organoclays are shown in Figure 14.4. The images show the typical structure of layered silicates and the elaboration (Digital Micrograph 3.1 Software) of a sampling of many images gives an evaluation of the basal spacing and thus an estimate of d_{001}, reported in Table 14.1.

For BentoNAT, a d_{001} value comparable to that obtained by XRD measurements is obtained, whereas a greater value is obtained for BentOTAB and BentoDEDMEL. We believe that the resin used in the preparation of samples for TEM analysis, considering the stressful conditions used, intercalates in the layered structure of the organoclays. For this reason, only the XRD d_{001} value was taking into account for the organoclays. BentOTAB present a d_{001} that is in good agreement with other experimental and modeling studies of organoclays modified with similar quaternary alkylammoniums, which correspond to a local bilayer arrangement of the long alkylic chain of OTAB inside the montmorillonite interlayer, in particular with the alkylic chain axes parallel to the silicate layers, whereas BentoDEDMEL shows a basal spacing comparable with organoclays containing an organic modifier that presents two long alkylic chains, corresponding to a local pseudo-trimolecular arrangement of the alkylic chains and to a much more organophilic interlayer [73–75].

14.5.4 SBS/organoclay micro–nanocomposites

14.5.4.1 Preparation

In this study, the composite materials were prepared by melt-mixing of linear SBS [Calprene 500, Dynasol, with a melt flow index of 5 g/10 min (5 kg, 190 °C), containing 30 wt% polystyrene and 70 wt% polybutadiene] with different amount of organoclays at 120 °C in an internal mixer (Haake Polylab Rheomix) with mixer chamber capacity approximately 45 cm^3 and equipped with roller rotors (600p). In the first part, samples were prepared by melt-mixing of SBS with 1, 3, and 5 wt% BentOTAB under two different process conditions: rotor speed 30 rpm for 30 min (A) or rotor speed 60 rpm for 10 min (B). A second set of samples was obtained by melt-compounding of SBS with 3 wt% of BentOTAB or BentoDEDMEL at rotor speed 60 rpm for 10 min. In all cases, the organoclay was added to the molten polymer after torque stabilization (approximately 2 min). All the obtained bulk materials were compression-molded at 120 °C (using a Carver heated press) into suitable test specimens. Samples of pure copolymer and composites with 3 wt% BentoNAT and 3 wt% melamine were processed in the same way and used for comparison for the flammability study.

14.5.4.2 Characterization and discussion

In the first part we have found the optimal process conditions for the preparation of SBS-BentOTAB nanocomposite. This optimization has been used to obtain SBS-BentoDEDMEL nanocomposite in the second part. Two different conditions were used for the preparation of SBS/BentOTAB composite, as suggested by previous works of Laus, Francescangeli, and Sandrolini [60].

Figure 14.5 illustrates XRD patterns of the SBS–BentOTAB composites obtained by method B. Different micro/nanostructures were formed. For the first series (A: 30 rpm, 30 min), the diffraction peaks have a position much lower than that of BentOTAB

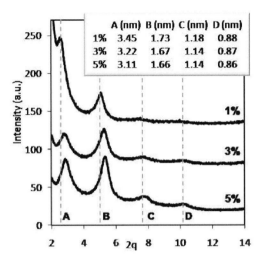

Figure 14.5 XRD patterns of SBS-BentOTAB prepared for 10 min at 60 rpm. Reproduced from Ref. [71] with permission of American Scientific Publishers.

(~1.73 nm) [71], which indicates that SBS did not intercalate into the layers of BentOTAB, and SBS-BentOTAB microcomposites were obtained.

Nevertheless, in the second series (B: 60 rpm, 10 min), a significant increase in the basal reflection $d001$ was observed (from 1.90 to ~3.30 nm, average value), which could be indicative of intercalated nanocomposites formation. A deeper investigation of the differences in the results (torque and XRD) obtained by the two methods used can be found in Ref. [71]. To further support this assumption, we recognize the presence of a series of peaks at approximately integral multiples of the lowest value (for example, 1 wt% series: $00l = 3.4$; $00(l + 1) = 3.4/2 = 1.7$; $00(l + 2) = 3.4/3 = 1.13$; $00(l + 3) = 3.4/4 = 0.85$). These values demonstrate the presence of reflections due to diffraction by different planes: inorganic planes of the clay (including the d-spacing), or different organic planes due to the layering of the polymeric chains. These results are in good agreement with the work of Laus *et al.* [60], Lim and Park [76], and Zhang *et al.* [67], which showed that it was possible to intercalate into the organoclay's structure the styrenic block of the copolymer and not the butadienic block. Accordingly, we could assume that method B allowed obtaining an intercalated nanocomposite rather than an exfoliated one, and furthermore that only the 30 wt% of the copolymer is constituted of polystyrene. Presumably, this might be because of chemical affinity between the polymer and the surface of the organoclay and the clay. Laus *et al.* [60] and Zhang *et al.* [67] demonstrated this hypothesis mainly based on XRD and thermal analysis, showing that the T_g of the polystyrene block domain increased with organoclay content, whereas the polybutadiene block domain T_g remained nearly constant in SBS nanocomposites [60]. Furthermore, Lim and Park proved that using an epoxidized SBS copolymer, rather than a pure SBS copolymer, a better exfoliate nanocomposite system is achieved [76]. A more detailed investigation of the morphology of the nanocomposites

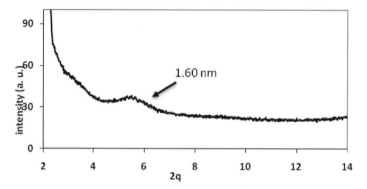

Figure 14.6 XRD patterns of SBS-BentoDEDMEL. Reproduced from Ref. [71] with permission of American Scientific Publishers.

is still ongoing, because of the difficulty encountered in the preparation of the sample for TEM analysis because of the nature of the polymeric matrix [71]. Accordingly, we used method (B) for the preparation of SBS-BentoDEDMEL nanocomposite with 3 wt% of BentoDEDMEL. The XRD pattern of the SBS/BentoDEDMEL composite (3 wt%, Figure 14.6) shows only a broad lower-intensity peak at 1.60 nm, which corresponds to a much lower $d001$ than the respective organoclay.

In this case we believe that during the melting process, a minimum part of the organo-clay structure collapsed and deintercalated part of the organic modifier DEDMEL. It is reasonable to emphasize that the peak at 1.60 nm shows an intensity much lower than that of the pristine organoclay. We are aware that more information is necessary but some authors, in analogous cases, explain the possible intercalated nanocomposite formation with XRD data alone [60, 67, 76–79].

14.5.5 Fire retardancy

14.5.5.1 Method

The evaluation of the flammability properties was carried out by cone calorimetry. Cone calorimetry is an increasingly important technique for assessing the fire behavior of materials. Typically, the subject material is irradiated at a heat intensity similar to that experienced in a fire situation (25–75 kW/m^2). In the present case, cone calorimetry was carried out at a heat flux of 35 kW/m^2, using a cone-shaped heater and specimens of dimensions $47 \times 47 \times 8$ mm. The heat release rate (HRR) was calculated from the oxygen consumption. Other parameters were also calculated: time to ignition (TTI), time of flame out (TOF), average heat release rate (HRR$_m$) as a function of time, peak heat release rate (PHRR), total heat release (THR), time to PHRR (TTP), and total smoke released (TSR). The average heat release rate is correlated with the heat released in a room where the flammable materials are not all ignited at the same time. Two other parameters were calculated separately: the fire performance index (FPI), because it is related to the time available for flashover, and the fire growth rate index (FIGRA). These two parameters have been accepted as representative of polymer combustion behavior in a real fire.

14.5.5.2 Results and discussion

All of the cone calorimetry data for the obtained SBS/clay composites are shown in Table 14.2, in comparison to pure SBS.

Figure 14.7 shows the influences on the HRR values of 3 wt% of BentoNAT and 3 wt% of BentOTAB in SBS composites. An interesting decrease in PHRR (~20%) is observable for both SBS-BentoNAT and SBS-BentOTAB composites, characterized by a constant slope of HRR between 130 and 220 s. This behavior might derive from the formation of a char layer over the surface of the material during the combustion process, which inhibits gaseous-fuel transport from the polymer to the flame front and therefore reduces the HRR of the burning surface according to the proposed combustion mechanism for clay-based nanocomposites during cone calorimetry experiments. Considering all the composites, an anticipation of the TTI for SBS-BentoNAT is evident that is probably because of a catalytic effect caused by the presence of transition metals and/or acid sites (both Lewis and Brønsted) in the pristine bentonite that are able to promote the degradation of copolymer matrix, considering that the temperature on the surface of the sample, before ignition, reaches about 400 °C.

However, it is well known that, for PLSNs, an acceleration of the decomposition of the polymer matrix is due to a catalytic effect of acidic sites originating from the Hofmann elimination of the organic modifier [46, 80, 81].

In the second set of experiments, we investigated the effect of the melamine skeleton by analyzing BentoDEDMEL and a microcomposite obtained from pure melamine with the block copolymer (SBS-MEL). The effects of 3 wt% of pure melamine and 3 wt% BentoDEDMEL in SBS micro-/nanocomposites on the HRR values are shown in Figure 14.7.

Different behavior was found for SBS-MEL microcomposite, which had a TTI similar to that of pure SBS (85 s), but after 100–110 s a decrease in the HRR was observed, resulting in a delay in the PHRR (TTP = 239 than 204 s). Moreover, SBSMEL shows a PHRR similar to that for pure SBS. The mechanism of the fire retardant action of melamine derivatives is under investigation by many academic and industrial research groups; however, it is well known that when exposed to heat and flames, melamine and derivatives decompose, absorbing heat and causing a cooling effect [46, 47].

Melamine also eliminates nitrogen during decomposition, which inhibits the spread of flames and the generation of toxic smoke; it is accepted that the formation of inert gases dilutes the fuel gases [46, 47]. The behavior of SBS-BentoDEDMEL is comparable to that of SBS-BentOTAB material in terms of both TTI and PHRR.

Considering FPI and FIGRA values, indices that have been introduced to assess the hazard of developing fires, it is evident that the loading of pure MEL in SBS matrices has s substantial effect on the FIGRA index, mainly due to its lowering the HRR [47].

Table 14.2 *Results obtained from cone calorimeter experiments*

Sample	TTI (s)	PHRR (kW/m²)[a]	TOF (s)	TTP (s)	HRR$_m$ (kW/m²)	THR (kW/m²)	TSR (m²/m²)	FPI[b]	FIGRA[b]
SBS[c]	94	1,929	274	206	1,028	232	8,151	21	7.1
SBS–BentoNAT-3B[d]	57	1,560	278	218	896	238	8,328	27	7.2
SBS–MEL–3B[c]	85	1,893	287	241	1,006	248	8,034	22	5.9
SBS–BentOTAB-3B[d]	73	1,501	283	202	922	239	8,926	21	7.4
SBS–BentoDEDMEL-3B[d]	79	1,539	294	294	961	249	8,814	19	7.3

Note. Reproduced from Ref. [71] with permission from American Scientific Publishers.

[a]%diff = 100 − [PHRR(composite)/PHRR(pure SBS) × 100].

[b]FPI = PHRR/TTI (kW/m² × s); FIGRA = pHRR/TTP (kW/m² × s).

[c]The cone calorimetry data reported are the averages of four replicate experiments.

[d]The cone calorimeter data reported are the averages of three replicate experiments.

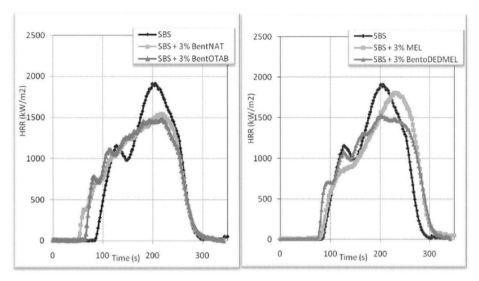

Figure 14.7 HRR plots at a heat flux of 35 kW/m^2. Reproduced from Ref. [71] with permission of American Scientific Publishers.

SBS-BentoNAT presents the worst FPI index, because of the negative effect of bentonite on the performance of the material, mainly caused by the shortening of the TTI. Clay/organoclay–SBS composites present similar FIGRA indices, and this behavior is essentially due to a balance between a lower TTP and a lower PHRR. SBS-BentoDEDMEL shows a decrease in PHRR similar to that for SBS-BentoNAT and SBS-BentOTAB, with a satisfactory TTI compared with SBSBentoNAT.

Another interesting result is the similar values of TOF obtained for SBS-BentoDEDMEL and SBS-MEL. This improvement might be due to a synergistic flame retardant effect of the organoclay in the complex and the melamine heart of the organic modifier used in the SBS nanocomposite, evident from observing the plots of HRR versus time for SBS-BentOTAB, SBS-MEL, and SBS-BentoDEDMEL. Also, the thinner organophilic protective layer that covered the montmorillonite stacks in the case of BentOTAB, in comparison with that derived from BentoDEDMEL (d_{001} for BentoDEDMEL of 2.60 nm) and hypothesizable from the structure of DEDMEL, presents one more long alkylic chain than OTAB.

14.6 Conclusions

The preparation of organic polymer/inorganic clay mineral nanocomposites offers a route to new materials with predefined structure and performance, such as SBS copolymers. The organic modification of the clay mineral is a necessary step for the compatibilization of the inorganic clay into the polymer matrix. Moreover, it also gives the opportunity to tailor the properties of the corresponding organoclay toward innovative functional applications in a deeper way. Thus, organic synthesis will play a key role in designing novel organic-based scaffolds. Furthermore, considering the results and information that can be obtained by computational simulations, it will be necessary in the future to employ a multidisciplinary approach to preparing advanced materials based on these systems. Economic and environmental aspects are among the reasons for the strong interest in clays. Clays are not expensive materials and are widespread in all countries. Even after the required treatments, their price is still acceptable in view of the vast improvement in their properties; therefore, they will be materials of choice for improving fire retardancy in SBS copolymers.

References

1. M. Alaeea, P. Ariasb, A. Sjodinc, and A. Bergman, An overview of commercially used brominated flame retardants, their applications, their use patterns in different countries/regions and possible modes of release. *Environment International*, 29 (2003), 683–9.
2. G. Beyer, Nanocomposites: A new class of flame retardants for polymers. *Plastics Additives and Compounding*, 4 (2002), 22–8.
3. G. Beyer, Nanocomposites offer new way forward for flame retardants. *Plastics Additives and Compounding*, 7 (2005) 32–5.
4. J. Murphy, Flame retardants: Trends and new developments. *Plastics Additives and Compounding*, 3 (2001), 16–20.
5. F. Bergaya and B. K. G. Theng. In *Handbook of Clay Science*, ed. F. Bergaya, B. K. G. Theng, and G. Lagaly (Amsterdam: Elsevier, 2006), pp. 1–19.
6. S. Pavlidou and C. D. Papaspyrides, A review on polymer–layered silicate nanocomposites. *Progress in Polymer Science*, 33 (2008), 1119–98.
7. M. F. Brigatti, E. Galan, B. K. G. Theng, G. Lagaly, M. Ogawa, and I. Dékany. In *Handbook of Clay Science*, ed. F. Bergaya, B. K. G. Theng, and G. Lagaly (Amsterdam: Elsevier, 2006), pp. 19–87.
8. G. Ori, Experimental and computational approaches towards functional glass- and clay based materials. Ph.D. thesis, University of Modena and Reggio Emilia, 2009.
9. M. Stadler and P. W. Schindler, Modeling of H+ and Cu+ adsorption on calcium-montmorillonite. *Clays Clay Minerals*, 41 (1993), 288–96.
10. J. Comets and L. Kevan, Adsorption of ammonia and pyridine on copper(II)-doped magnesium-exchanged smectite clays studied by electron spin resonance. *Journal of Physical Chemistry*, 97 (1993), 466–9.
11. R. Van Bladel, H. Halen, and P. Cloos, Calcium–zinc and calcium–cadmium exchange in suspensions of various types of clays. *Clay Minerals*, 28 (1993), 33–8.
12. T. Matsuda, K. Yogo, C. Pantawong, and E. Kikuchi, Catalytic properties of copper-exchanged clays for the dehydrogenation of methanol to methyl formate. *Applied Catalysis A: General*, 126 (1995), 177–86.
13. J. Wagner, H. Chen, B. J. Brownawell, and J. C. Westall, Use of cationic surfactants to modify soil surfaces to promote sorption and retard migration of hydrophobic organic compounds. *Environmental Science and Technology*, 28 (1994), 231–7.
14. M. F. Brigatti, F. Corradini, G. C. Franchini, S. Mazzoni, L. Medici, and L. Poppi, Interaction between montmorillonite and pollutants from industrial waste-waters: Exchange of Zn^{2+} and Pb^{2+} from aqueous solutions. *Applied Clay Science*, 9 (1995), 383–95.
15. M. Auboiroux, P. Bailif, J. C. Touray, and F. Bergaya, Fixation of Zn^{2+} and Pb^{2+} by Ca-montmorillonite in brines and dilute solutions: Preliminary results. *Applied Clay Science*, 11 (1996), 117–26.

16. G. Sheng, S. Xu, and S. A. Boyd, Mechanisms controlling sorption of neutral organic contaminants by surfactant-derived and natural organic matter. *Environmental Science and Technology*, 30 (1996), 1553–7.

17. E. Á. Ayuso and A. G. Sánchez, Removal of heavy metals from waste waters by natural and Na-exchanged bentonites. *Clays Clay Minerals*, 51 (2003), 475–80.

18. O. Abollino, A. Giacomino, M. Malandrino, and E. Mentasti, Interactions of metal ions with montmorillonite and vermiculite. *Applied Clay Science*, 38 (2008), 227–36.

19. K. G. Bhattacharyya and S. S. Gupta, Adsorption of a few heavy metals on natural and modified kaolinite and montmorillonite: A review. *Advances in Colloid Interface Science*, 140 (2008), 114–31.

20. P. M. Dove and C. J. Nix, The influence of the alkaline earth cations, magnesium, calcium, and barium on the dissolution kinetics of quartz. *Geochimica et Cosmochimica Acta*, 61 (1997), 3329–40.

21. M. F. Brigatti, T. Manfredini, M. Montorsi, G. Ori, and C. Siligardi, Cation-exchanged smectites: Preparation, characterization and modeling. Submitted for publication.

22. R. T. Cygan, J. A. Greathouse, H. Heinz, and A. G. Kalinichev, Molecular models and simulations of layered materials. *Journal of Materials Chemistry*, 19 (2009), 2470–81.

23. T. J. Tambach, P. G. Bolhuis, and B. Smit, A molecular mechanism of hysteresis in clay swelling. *Angewandte Chemie International Edition*, 43 (2004), 2650–52.

24. X. D. Liu and X. C. Lu, A thermodynamic understanding of clay-swelling inhibition by potassium ions. *Angewandte Chemie International Edition*, 45 (2006), 6300–6303.

25. D. E. Smith, Y. Wang, and H. D. Whitley, Molecular simulations of hydration and swelling in clay minerals. *Fluid Phase Equilibria*, 222–3 (2004), 189–94.

26. H. Heinz, R. A. Vaia, R. Krishnamoorti, and B. L. Farmer, Self-assembly of alkylammonium chains on montmorillonite: Effect of chain length, head group structure, and cation exchange capacity. *Chemistry of Materials*, 19 (2007), 59–68.

27. Y.-T. Fu and H. Heinz, Cleavage energy of alkylammonium-modified montmorillonite and relation to exfoliation in nanocomposites: Influence of cation density, head group structure, and chain length. *Chemistry of Materials*, 22 (2010), 1595–1605.

28. M. Mainil, M. Alexandre, F. Monteverde, and P. Dubois, polyethylene organo-clay nanocomposites: The role of the interface chemistry on the extent of clay intercalation/exfoliation. *Journal of Nanoscience Nanotechnology*, 6 (2006), 337–44.

29. M. A. J. Rajan, T. Mathavan, A. Ramasubbu, A. Thaddeus, V. F. Latha, T. S. Vivekanandam, and S. Umapathy, Thermal properties of PMMA/montmorillonite clay nanocomposites. *Journal of Nanoscience Nanotechnology*, 6 (2006), 3993–6.

30. J. G. Zhang, D. D. Jiang, and C. A. Wilkie, Fire properties of styrenic polymer–clay nanocomposites based on an oligomerically-modified clay. *Polymer Degradation and Stability*, 91 (2006), 358–66.

31. J. G. Zhang, D. D. Jiang, and C. A. Wilkie, Thermal and flame properties of polyethylene and polypropylene nanocomposites based on an oligomerically-modified clay. *Polymer Degradation and Stability*, 91 (2006), 298–304.

32. C. I. Sainz-Diaz, A. H.-Laguna, and M. T. Dove, Modeling of dioctahedral 2:1 phyllosilicates by means of transferable empirical potentials. *Physics and Chemistry of Minerals*, 28 (2001), 130–41.

33. C. L. Freeman, J. H. Harding, D. J. Cooke, J. A. Elliott, J. S. Lardge, and D. M. Duffy, New forcefields for modeling biomineralization processes. *Journal of Physical Chemistry C*, 111 (2007), 11943–51.

34. J. H. Harding, D. M. Duffy, M. L. Sushko, P. Mark Rodger, D. Quigley, and J. A. Elliott, Computational techniques at the organic–inorganic interface in biomineralization. *Chemical Reviews*, 108 (2008), 4823–54.

35. T. Manfredini, M. Montorsi, G. Ori, and C. Siligardi, Alkylammonium– and alkylimidazolium–montmorillonite interactions: A MD study. Submitted for publication.

36. H. Heinz, R. A. Vaia, and B. L. Farmer, Interaction energy and surface reconstruction between sheets of layered silicates. *Journal of Chemical Physics*, 124 (2006), 224713.

37. F. A. Bottino, E. Fabbri, I. L. Fragalà, G. Malandrino, A. Orestano, F. Pilati, and A. Pollicino, Polystyrene–clay nanocomposites prepared with polymerizable imidazolium surfactants. *Macromolecular Rapid Communications*, 24 (2003), 1079–84.

38. N. H. Kim, S. V. Malhotra, and M. Xanthos, Modifications of cationic nanoclays with ionic liquids. *Microporous and Mesoporous Materials*, 96 (2006), 29–35.

39. Y. C. Chua, S. Wu, and X. Lu, Polye(thylene naphthalene)/clay nanocomposites based on thermally stable trialylimidalium-treated montmorillonite: Thermal and dynamic mechanical properties. *Journal of Nanoscience Nanotechnology*, 6 (2006), 3985–8.

40. P. S. G. Krishnan, M. Joshi, P. Bhargava, S. Valiyaveettil, and C. He, Effect of heterocyclic based organoclays on the properties of polyimide–clay nanocomposites. *Journal of Nanoscience Nanotechnology*, 5 (2005), 1148–57.

41. F. Bellucci, G. Camino, A. Frache, and A. Sarra, Catalytic charring–volatilization competition in organoclay nanocomposites. *Polymer Degradation and Stability*, 92 (2007), 425–36.

42. J. M. Cervantes-Uc, J. V. Cauich-Rodriguez, H. Vazquez-Torres, L. F. Garfias-Mesias, and D. R. Paul, Thermal degradation of commercially available organoclays studied by TGA-FTIR. *Thermochimica Acta*, 457 (2007), 92–102.

43. G. E. Zaikov and S. M. Lomakin, Ecological issue of polymer flame retardancy. *Journal of Applied Polymer Science*, 86 (2002), 2449–62.

44. S. V. Levchik, G. F. Levchink, A. I. Balabanovich, G. Camino, and L. Costa, Mechanistic study of combustion performance and thermal decomposition behaviour of nylon 6 with added halogen-free retardants. *Polymer Degradation and Stability*, 54 (1996), 217–22.

45. H. Horacek and R. Grabner, Advantages of flame retardants based on nitrogen compounds. *Polymer Degradation and Stability*, 54 (1996), 205–15.

46. E. D. Weil and V. Choudhary, Flame-retarding plastics and elastomers with melamine. *Journal of Fire Sciences*, 13 (1995), 104–26.

47. A. B. Morgan and C. A. Wilkie, *Flame Retardant Polymer Nanocomposites*, 1st ed. (Hoboken, NJ/Chichester: Wiley–VCH, 2007).

48. W. H. Awad, J. W. Gilman, M. Nyden, R. H. Harris, J. T. Sutto, J. Callahan, P. C. Trulove, H. C. De Long, and D. M. Fox, Thermal degradation studies of alkyl-imidazolium salts and their application in nanocomposites. *Thermochimica Acta*, 409 (2004), 3–11.

49. D. M. Fox, W. H. Awad, J. W. Gilman, P. H. Maupin, H. C. De Long, and P. C. Trulove, Flammability, thermal stability, and phase change characteristics of several trialkylimidazolium salts. *Green Chemistry*, 5 (2003), 724–7.

50. D. M. Fox, J. W. Gilman, H. C. De Long, and P. C. Trulove, TGA decomposition kinetics of 1-butyl-2,3-dimethylimidazolium tetrafluoroborate and the thermal effects of contaminants. *Journal of Chemical Thermodynamics*, 37 (2005), 900–905.

51. J. Langat, S. Bellayer, P. Hudrlik, A. Hudrlik, P. H. Maupin, J. W. Gilman, and D. Raghavan, Synthesis of imidazolium salts and their application in epoxy montmorillonite nanocomposites. *Polymer*, 47 (2006), 6698–709.

52. L. Cui, J. E. Bara, Y. Brun, Y. Yoo, P. J. Yoon, and D. R. Paul, Polyamide- and polycarbonate-based nanocomposites prepared from thermally stable imidazolium organoclay. *Polymer*, 50 (2009), 2492–502.

53. D. M. Fox, J. W. Gilman, A. B. Morgan, J. R. Shields, P. H. Maupin, R. E. Lyon, H. C. De Long, and P. C. Trulove, Flammability and thermal analysis characterization of imidazolium-based ionic liquids. *Industrial Engineering Chemistry Research*, 47 (2008), 6327–32.

54. F. Laoutid, L. Bonnaud, M. Alexandre, J.-M. Lopez-Cuesta, and Ph. Dubois, New prospects in flame retardant polymer materials: From fundamentals to nanocomposites. *Materials Science and Engineering: Reports*, 63 (2009), 100–125.

55. M. Si, V. Zaitsev, M. Goldman, A. Frenkel, D. G. Peiffer, E. Well, J. C. Sokolov, and M. H. Rafailovich, Self-estinguishing polymer/organoclay nanocomposites. *Polymer Degradation and Stability*, 92 (2007), 86–93.

56. D. J. Carastan and N. R. Demarquette, Microstructure of nanocomposites of styrenic polymers. *Macromolecular Symposia*, 233 (2006), 152–60.

57. D. J. Carastan, N. R. Demarquette, A. Vermogen, and K. M.-Varlot, Linear viscoelascticity of styrenic block copolymers–clay nanocomposites. *Rheologica Acta*, 47 (2008), 521–36.

58. Z. Chen and R. Feng, Preparation and characterization of poly(styrene-b-butadiene-b-styrene)/montmorillonite nanocomposites. *Polymer Composites*, 30 (2009), 281–7.

59. Z. Zhang, L. Zhang, Y. Li, and H. Xu, Styrene-butadiene/montomorillonite nanocomposites synthesized by anionic polymerization. *Journal of Applied Polymer Science*, 99 (2006), 2273–8.

60. M. Laus, O. Francescangeli, and F. Sandrolini, New hybrid nanocomposites based on organophilic clay and poly(styrene-b-butadiene) copolymers. *Journal of Materials Research*, 12 (1997), 3134–9.

61. M. Liao, J. Zhu, H. Xu, Y. Li, and W. Shan, Preparation and structure and mechanical properties of poly(styrene-b-butadiene)/clay nanocomposites. *Journal of Applied Polymer Science*, 92 (2004), 3430–34.

62. H. Xu, Y. Li, and D. Yu, Studies on the poly(styrene-b-butadiene-b-styrene)/clay nanocomposites prepared by melt intercalation, *Journal of Applied Polymer Science*, 98 (2005), 146–52.

63. S. Lietz, J.-L. Yang, E. Bosch, J. K. W. Sandler, Z. Zhang, and V. Altstadt, Improvement of the mechanical properties and creep resistance of SBS block copolymers by nanoclay fillers. *Macromolecular Materials and Engineering*, 292 (2007), 23–32.

64. T. Yamaguchi and E. Yamada, Preparation and mechanical properties of clay/polystyrene-block-polybutadiene-block-polystyrene triblock copolymer (SBS) intercalated nanocomposites using organoclay containing stearic acid. *Polymer International*, 55 (2006), 662–7.

65. P. A. da Silva, M. M. Jacobi, L. K. Schneider, R. V. Barbosa, P. A. Countinho, R. V. B. Oliverira, and R. S. Mauler, SBS nanocomposites as toughening agent for polypropylene. *Polymer Bulletin*, 64 (2010), 245–57.

66. G. Polacco, P. Kriz, S. Filippi, J. Stastna, D. Biondi, and L. Zanzotto, Rheological properties of asphalt/SBS/clay blends. *European Polymer Journal*, 44 (2008), 3512–21.

67. W. Zhang, J. Zeng, L. L. Fang, and Y. Fang, A novel property of styrene–butadiene–styrene/clay nanocomposites: Radiation resistance. *Journal of Material Chemistry*, 14 (2004), 209–13.

68. B.-L. Lee, *Soft Zero Halogen Flame Retardant Thermoplastic Elastomers*, United States Patent Application 0124743 A1 (2009).
69. B.-L. Lee and D. C. Worley, II, *Flame-Retardant Thermoplastic Elastomer Compositions* (assigned to Teknor Apex Company), United States Patent Application 0048382 A1 (2009).
70. M. Watanabe, H. Hashimoto, and S. Tokuda, *Flame-Retardant Resin Compositions and Insulated Electric Wires Coated Therewith* (assigned to the Furukawa Electric Co.), Japanese Patent Application 2004075993 (2004).
71. M. Comes Franchini, P. Fabbri, A. Frache, G. Ori, M. Messori, C. Siligardi, and A. Ricci, Bentonite-based organoclays as innovative flame retardants agents for SBS copolymer. *Journal of Nanoscience and Nanotechnology*, 8 (2008), 6316–24.
72. W. Zerweck and K. Keller, United States Patent 2,228,161 (1941).
73. G. Lagaly, M. Ogawa, and I. Dékany. In *Handbook of Clay Science*, ed. F. Bergaya, B. K. G. Theng, and G. Lagaly (Amsterdam: Elsevier, 2006), pp. 309–79.
74. M. A. Osman, M. Ploetze, and U. W. J. Suter, Surface treatment of clay minerals – thermal stability, basal-plane spacing and surface coverage. *Journal of Material Chemistry*, 13 (2003), 2359–66.
75. Q. H. Zeng, A. B. Yu, G. Q. Lu, and R. K. Standish, Molecular dynamics simulation of the structural and dynamic properties of dioctadecyldimethyl ammoniums in organoclays. *Journal of Physical Chemistry B*, 108 (2004), 10025.
76. Y. T. Lim and O. O. Park, Microstructure and rheological behavior of block copolymer/clay nanocomposites. *Korean Journal of Chemical Engineering*, 18 (2001), 21–5.
77. M. Alexandre and P. Dubois, Polymer-layered silicate nanocomposites: Preparation, properties and uses of a new class of materials. *Materials Science and Engineering: Reports*, 28 (2000), 1–63.
78. R. K. Shah and D. R. Paul, Nylon 6 nanocomposites prepared by a melt mixing masterbatch process. *Polymer*, 45 (2004), 2991–3000.
79. J. T. Yoon, W. H. Jo, M. S. Lee, and M. B. Ko, Effects of co-monomers and shear on the melt intercalation of styrenics/clay nanocomposites. *Polymer*, 42 (2001), 329–36.
80. A. Leszczynska, J. Njuguna, K. Pielichowski, and J. R. Banerjee, Polymer/montmorillonite nanocomposites with improved thermal properties. Part I. Factors influencing thermal stability and mechanisms of thermal stability improvement. *Thermochimica Acta*, 453 (2007), 75–96.
81. W. Xie, Z. Gao, W.-P. Pan, D. Hunter, A. Singh, and R. Vaia, Thermal degradation chemistry of alkyl quaternary ammonium montmorillonite. *Chemistry Materials*, 13 (2001), 2979–90.

Index